Particulate Composites

Randall M. German

Particulate Composites

Fundamentals and Applications

 Springer

Randall M. German
San Diego State University
San Diego, CA, USA

ISBN 978-3-319-29915-0   ISBN 978-3-319-29917-4   (eBook)
DOI 10.1007/978-3-319-29917-4

Library of Congress Control Number: 2016933672

© Springer International Publishing Switzerland 2016
This work is subject to copyright. All rights are reserved by the Publisher, whether the whole or part of the material is concerned, specifically the rights of translation, reprinting, reuse of illustrations, recitation, broadcasting, reproduction on microfilms or in any other physical way, and transmission or information storage and retrieval, electronic adaptation, computer software, or by similar or dissimilar methodology now known or hereafter developed.
The use of general descriptive names, registered names, trademarks, service marks, etc. in this publication does not imply, even in the absence of a specific statement, that such names are exempt from the relevant protective laws and regulations and therefore free for general use.
The publisher, the authors and the editors are safe to assume that the advice and information in this book are believed to be true and accurate at the date of publication. Neither the publisher nor the authors or the editors give a warranty, express or implied, with respect to the material contained herein or for any errors or omissions that may have been made.

Printed on acid-free paper

This Springer imprint is published by Springer Nature
The registered company is Springer International Publishing AG Switzerland

Dedicated to John Lee Johnson

Preface

Engineering materials continue to encounter challenges for new property combinations, and these challenges are leading to many hybrid materials known as composites. Particulate composites are formed using powders to create property combinations not available in traditional engineering compositions. This book provides engineers with information relevant to meeting the challenges on synthesis, selection, fabrication, and design associated with applying composites to demanding applications. Growth in the field is accelerating as we learn more about how to deliver novel property combinations. Further, the shaping flexibility associated with particles provides compelling economic benefit.

Growth in the particulate composites is evident in patents, journal publications, and press releases. For example, a new composite of gold and titanium nitride is employed in the newest smart watches. The 18-karat watch has exceptional wear resistance beyond that of traditional jewelry. The 75 wt% Au (46 vol.%) gives elegance, while the hard, gold-colored TiN provides 400 HV hardness, tenfold higher than pure gold. This is just one of many ideas. Millions of formulations are possible, with tremendous flexibility in composition, microstructure, properties, and performance. Although the field is challenging to master, at the same time the performance gains are most attractive. Only now are we appreciating the full spectrum of opportunities.

Although particulate composites are everywhere, the field is poorly organized. Predictive calculations are ignored or missing. Consequently, product development efforts rely on the empirical approach to see what works. In this book, the underlying principles are emphasized with attention to the critical factors. Important aspects are detailed with regard to the relations between phases, composition, powder characteristics, microstructure, fabrication, and properties. The resulting composites are isotropic, unlike long fiber graphite and fiberglass composites. Accordingly, a wide array of products are treated as part of this book. The intent is to introduce particulate composites in a manner relevant to training the next generation of innovators.

San Diego, CA, USA   Randall M. German

Acknowledgements

I am grateful to several individuals for support on this project. The idea for the book arose in discussions with a number of individuals associated with the Center for Innovative Sintered Products at Penn State University. But it took several years to realize the end product. I am especially appreciative of the help from the following experts: Sundar Atre, Animesh Bose, Louis Campbell, Robert Dowding, Zak Fang, Arun Gokhale, Mark Greenfield, Anthony Griffo, Ozkan Gulsoy, Anita Hancox, John Johnson, John Keane, Young Sam Kwon, Todd Leonhardt, Jason Liu, Yixiong Liu, Kathy Lu, Hideshi Muira, Neal Myers, Seong Jin Park, Leo Praksh, Joe Sery, Ivi Smid, Jose Torralba, Anish Upadhyaya, Ridvan Yamanoglu, and Rudy Zauner.

A draft copy of this book was used by students at SDSU who provided useful comments, especially helpful were detailed comments from Hamood Almoshawah, Carolos-Jose Andallo, Sonny Birtwistle, Jonathan Briend, Vanessa Bundy, Justin Camba, Jeremiah Cox, Sam Delaney, Michael Diomartich, Abhi Dhanani, Natalia Ermolaeva, Britney Finley, Ian Gardner, Rebecca Gloria, Deanna Key, Akash Khatawate, Adam Lane, Leo Malmeryd, Falon Mannus, Robert McClellan, Cesar Ocampo, Nicholas Schwall, Nika Sedghi, Lauren Parrett, Nathan Reed, Matthew Revie, Marc Schumacher, Amritpal Singh, Arturo Sotomayorgayosso, Meena Wasimi. Oleg Usherenko, and Wes Yeagley.

Contents

Author Biography

Professor German's research and teaching deal with the net-shape fabrication of engineering materials for applications ranging from wire drawing dies to jet engine bearings. He is Professor of Mechanical Engineering at San Diego State University, having previously served as Associate Dean, Chaired Professor, and Center Director at Rensselaer Polytechnic Institute, Pennsylvania State University, and Mississippi State University. In addition he was Director of Research for two companies and staff member at Sandia National Laboratories.

Rand obtained his PhD from the University of California at Davis, MS from The Ohio State University, and an honors BS degree from San Jose State University. He completed management development programs at Hartford Graduate Center and Harvard.

He has an honorary doctorate from the Universidad Carlos III de Madrid and is a Fellow of three professional organizations. His awards include the Tesla Medal, Nanyang Professorship, Japan Institute for Materials Research Lectureship, Penn State Engineering Society Premiere Research Award, University of California at Davis Distinguished Engineering Alumni Award, The Ohio State University Distinguished Engineering Alumnus Award, San Jose State University Award of Distinction, and Honorary Member of Alpha Sigma Mu. He served on three National Academy review boards and helped form a dozen start-up companies.

Rand supervised more than 100 theses and 200 postdoctoral fellows. These efforts resulted in 1025 published articles, 25 patents, and 18 books, including *Sintering Theory and Practice*. He edited 19 books and co-chaired 30 conferences. His publications have been cited more than 22,000 times.

Chapter 1
Introduction

Particulate composites deliver property and cost combinations not available from tradi- tional single-phase materials. These composites involve many combinations where at least one phase starts as a powder. Basic concepts are introduced in this chapter and first details are given on nomenclature, property models, and some compositions in use. The benefits from starting with coated particles are introduced.

Context

Composites use two or more phases to attain property combinations not possible from either phase alone. Chocolate is an everyday example of a composite consisting of sugar, milk solids, cocoa, and cocoa butter. Different ratios of the ingredients lead to semi-sweet, milk chocolate, or dark chocolate.

Particulate composites are around us everywhere. Concrete with steel reinforce- ment bars is a widely used construction material. Rebar strengthened concrete consists of rock and sand mixed with hydrated calcium silicate. If we performed random spot chemical analysis, at some points the composition is low carbon steel (Fe-0.4C), other places it is silica (SiO_2), and other places it is hydrated calcium silicate (a mixture of calcia (CaO), silica (SiO_2), and water). Engineers ignore this "granularity" and treat concrete as a homogeneous material.

Some composites involve fibers, while others rely on particles, including elon- gated particles (whiskers) and flat particles (flakes) [1–10]. Especially important are particulate composites with metal or ceramic phases enmeshed in a metal, ceramic, or polymer. The added phase is selected to improve performance of the continuous matrix phase.

The number of possible phase combinations is enormous. This book treats the subject using a wide range of compositions. Some attention is given to polymer matrix composites; however, interfacial bonding to the matrix requires an active interface that is often absent from polymer composites. Thus, more attention is

© Springer International Publishing Switzerland 2016   1
R.M. German, *Particulate Composites*, DOI 10.1007/978-3-319-29917-4_1

Fig. 1.1 Samples of
various kitchen counter tops
with different colors and
swirl patterns. The
composites are a mixture of
alumina trihydrate and
acrylic polymer

directed toward systems where either a chemical or thermal treatment induces a strong bond between phases. The resulting applications are diverse, and include electrical switches, metal cutting tools, electric bicycles, golf clubs, computer servers, automobile engines, practice bullets, concrete cutting tools, and kitchen counter tops. Often these very common applications are forgotten, yet they are frequently used every day. For example, Fig. 1.1 is a photograph of different styles and colors of manmade composites used in kitchens. Durability is quite important to the commercial success of this composite.

Composites

A composite is a mixture of two or more phases. Each phase remains distinct in the structure as evident using microscopy. Properly formulated composites deliver attractive property combinations. As an example, iron particles coated with a polymer are consolidated to form magnets used in electric bicycles. Iron provides magnetic behavior while the polymer matrix makes the composite nonconductive. This composite avoids eddy current energy losses in high frequency electric motors. Eddy currents arise when a moving magnetic field induces electrical current flow in the structure. Thus, the composite has desirable magnetic response without the inefficiency from eddy currents. Likewise, composites used for cutting concrete, marble, or granite rely on hard diamond particles dispersed in a tough metallic phase such as cobalt. The combination delivers exceptional cutting performance for

Table 1.1 Examples of particulate composites

System	Application	Ingredients
Paint	Spreadable surface coating	Mixture of solvent, opaque particles, and polymer such as an acrylic emulsion
Ink	Printing on paper	Small graphite particles in a mixture of solvent and polymer
Porcelain	Dishes, dental crowns, electrical insulators	Mixture of oxide ceramic crystals and glass phases
Electrical contacts	Make-break circuit switches	Arc resistant refractory phase (W, WC, Mo) and high electrical conductivity phase (Ag, Cu)
Heat sinks	Redistribution of heat in computers, rocket engines, high intensity lighting	High conductivity phase (Cu, Ag) with low thermal expansion phase (W, Mo, WC)
Brake pads	Transformation of kinetic energy into heat to stop mechanical systems	Mixtures of graphite, polymers, metals, and ceramics
Electromagnetic shields	Absorption of radio wave interference in devices such as computers	Polypropylene or other polymers with electrically conductive dispersed conductors of nickel and graphite
Permanent magnets	Flexible magnets for use in head-phones, stereo speakers, electric motors	Polymer mixed with high capacity magnetic compound
Correction fluid	Opaque cover up for typographical errors or drawing mistakes on paper	White titania (TiO_2) particles dispersed in a solvent-softened polymer
Cemented carbide	Provide hard surfaces for drawing, machining, drilling, shearing, extrusion of metals	Interlocked network of hard carbide (WC) particles in a tough metal matrix (Co)
Wear resistant aluminum	Air conditioner rotors, endurance horseshoes, sporting equipment	Mixture of hard silicon carbide (SiC) particles in aluminum alloy matrix
Inertial weights	Selective mass to balance gyroscopes, aircraft wings, helicopter rotors, vibrators, fishing, and golf club weights	Composite consisting of mostly tungsten (W) mixed with transition metals, such as Cu, Fe, Ni, Mn, Co
Low toughness projectiles	Lead-free frangible ammunition where the bullet has sufficient strength for firing but disintegrates on target impact	Variants include tungsten (W) bonded with nylon or copper (Cu) bonded with tin (Sn)
Foamed ceramic	Insulation for high temperature heating pipes with low thermal conductivity up to 1000 °C	High porosity foamed hydrous calcium silicate with a density near 0.2 g/cm^3

hard structures. Neither the diamond or cobalt alone would survive the harsh conditions, but the composite proves exceptionally durable. Particulate composites arise in many fields as illustrated in Table 1.1 for various compositions and applications.

Fig. 1.2 A cross-section microscope image of a two phase composite formed from iron (Fe) and a complex oxide compound known as cordierite ($2MgO \cdot 2Al_2O_3 \cdot 5SiO_2$), where both phases form interlaced networks [courtesy L. Shaw]

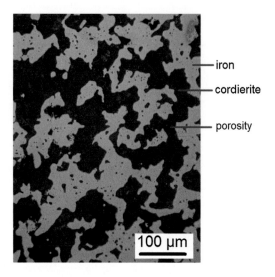

iron

cordierite

porosity

100 μm

Within a composite, each phase is distinct with regard to composition and atomic structure. The microscopic difference between phases is readily evident. Figure 1.2 is an image of an iron-cordierite composite. In this case there is some residual porosity. This cross-section image contrasts the phases based on a difference in reflectivity. The phases form interlaced networks. In terms of properties, the iron is soft, conductive, magnetic, and ductile while the cordierite consists is an oxide compound ($2MgO-2Al_2O_3-5SiO_2$) that is hard, nonconductive, nonmagnetic, brittle, and stiff, with a lower thermal expansion coefficient. The two phases are insoluble in one another, so the composite is magnetic, but low in thermal expansion. This combination of properties is desirable for use in the automated assembly of electronic diodes.

Some composites form naturally, such as bones, seashells, bamboo, and wood. These biological composites have a broad array of properties, ranging from soft skin to hard dental enamel [11, 12]. Biological composites provide evidence of the importance to the phase arrangement, what is known as the morphology. For example, abalone shell consists of calcium carbonate and rubbery biopolymer. Neither phase is noteworthy, but the composite provides significant toughness. Intentional control of phase morphology is part of composite design.

Microstructure refers to the image in a light or electron microscope. Each phase has easily identifiable attributes. Microstructure can dominate properties. For dilute compositions, one phase is dispersed in the other phase. At higher concentrations, both phases are connected, resulting in an interlaced three-dimensional structure. The level of phase connectivity is critical to properties. A composite microstructure is captured in Fig. 1.3 using a polished cross-section. The cross-section shows low thermal expansion Invar (Fe-36Ni) as the darker phase and high thermal conductivity silver (Ag) as the lighter phase. This composite is known as *Silvar*. The two phases are distinct, forming three-dimensional, intertwined networks. At 40 wt%

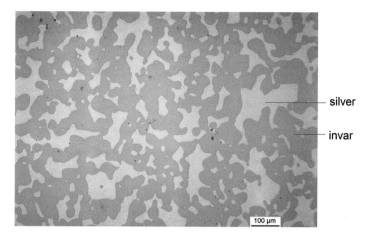

Fig. 1.3 Microstructure of a two phase particulate composite consisting of low thermal expansion Invar (Fe-36Ni) as the dark phase and high thermal conductivity silver (Ag) as the light phase. Both phases are interconnected in three dimensions [courtesy B. Lograsso]

Table 1.2 Thermal properties of silver, Invar, and *Silvar* composite (*Silvar* = Invar-40Ag)

Property	Silver (Ag)	Invar Fe-36Ni)	Silvar (Invar-40Ag)
Density, g/cm^3	10.5	8.2	8.9
Tensile strength, MPa	180	500	200
Thermal conductivity, W/(m °C)	420	14	160
Thermal expansion, 10^{-6} 1/K	20	1	8

silver (34 vol% silver) a desirable combination of properties arises, as summarized in Table 1.2. These properties are useful in heat dissipation devices that cool microelectronic chips. The high thermal conductivity reduces heating while the low thermal expansion coefficient minimizes thermal fatigue. The on-off cycles in a computer expand and contract the semiconductors, leading to fatigue failure after repeated cycles. To avoid thermal fatigue failure, the heat spreader must match the semiconductor expansion and contraction strains.

Interest in particulate composites derives from novel property and cost combinations. In general there are four options:

1. properties are dominated by one phase,
2. properties are intermediate between the two phases,
3. properties are synergistically advanced over that attainable with either phase,
4. properties are degraded below that of either phase.

Most often composites are formulated to deliver improved properties. Using hardness, Table 1.3 compiles examples for each of the four situations listed above. In the first case of WC-8Co, the composite is nearly the same as harder phase; tungsten carbide dominates the composite hardness. In the second case of

Table 1.3 Hardness (HV) for particulate composites (composition in wt%)

Property level	Composite	Hard phase	Soft phase	Composite
Dominated by one phase	WC-8Co	WC 1850	Co 180	1800
Intermediate between phases	Polyethylene-20 mica	Mica 133	Polyethylene 3	16
Advanced over both phases	Al_2O_3-20Si_3N_4	Al_2O_3 1800	Si_3N_4 1800	2400
Degraded below both phases	Al_2O_3-10ZrO_2	Al_2O_3 1800	ZrO_2 1300	700

polyethylene-mica, the composite hardness is intermediate between that of the two components. In this system hardness is a linear function of the mica content.

In the third case listed in Table 1.3, corresponding to alumina (Al_2O_3) with 20 % silicon nitride (Si_3N_4), the composite hardness is higher than either of the constituents. In the last case, corresponding to alumina-zirconia, the composite properties are inferior to either of the constituents. This sometimes happens in systems optimized for one property while a separate property is sacrificed. In the case of the alumina-zirconia composite, the facture toughness is optimized with a sacrifice of the hardness.

There are applications where the fourth option of inferior properties is desired. Practice ammunition is an example. The idea is to replace copper jacketed lead with a bullet similar in mass and aerodynamics. Practice bullets desirably shatter on striking the target to avoid ricochets. A copper-tin composite is one variant designed to shatter on target impact. Shattering occurs because a brittle intermetallic is intentionally formed to cause disintegration on striking a target.

Applications for composites require a balance of properties and cost. One idea for automobile bodies was to reinforce steel with titanium boride (TiB) to lower density and increase stiffness and toughness. Unfortunately, the composite was costly so commercial efforts were discontinued. In contrast, polypropylene reinforced with clay is a favorite for automotive exteriors. The composite delivers improved toughness and strength using low-cost clay particles. In addition to engineering properties, cost is a persistent design requirement.

Significant property gains are possible through microstructure adjustment. Consider the example of alumina (Al_2O_3) processed to either a 1 μm grain size or 0.1 μm (100 nm) grain size. Figure 1.4 compares the fracture strengths, illustrating how pure alumina doubles with the nanoscale grain size. Reinforcing the micrometer grain size alumina with 25 vol% SiC delivers more strength, reaching 900 MPa. However, just 5 vol% SiC in the nanoscale alumina results in 1130 MPa strength.

Nanoscale composites are an area of intense research. Very small grain sizes deliver extraordinary strength and hardness. Two fabrication option are in use for the nanoscale composites. The first involves fast consolidation to avoid microstructure coarsening during consolidation. Fast consolidation involves rapid heating and

Fig. 1.4 Fracture strength for alumina (Al$_2$O$_3$) at two grain sizes, without and with silicon carbide (SiC). The highest strength is associated with a nanoscale grain size and just 5 vol% added silicon carbide

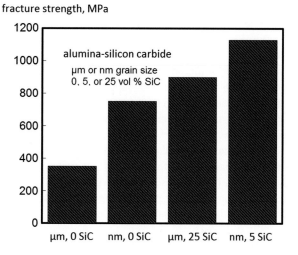

high applied pressures while keeping process temperatures low and hold times short. An alternative relies on seeded microstructures and traditional consolidation cycles. Here a core particle is coated with nanoscale second phase. The core particle hinders coarsening during consolidation under traditional conditions, retaining a small grain size in the final structure. In one implementation, 2 μm diamond particles are coated with nanoscale WC-10Co. When consolidated using traditional sintering at 1300 °C for 1 h, the structure is preserved. The product gives a sixfold longer life as a machine tool.

Laminates, Fibers, Whiskers, Particles

Composites include laminates, continuous fibers, shorter discontinuous whiskers, and particles. Sheet materials are laminated to form layered composites. Some applications are in armor (metal-ceramic-metal), automotive windshields (glass-polymer-glass), kitchen cookware (stainless-copper-stainless), and electronic packaging (copper-molybdenum-copper). These are flat or curved shapes. Because of the lamination, the properties are anisotropic, implying the properties change with orientation. That is useful in kitchen cookware. The two outer stainless layers provide corrosion resistance while the intermediate high conductivity copper layer delivers uniform heating.

Composites consisting of long fibers are also anisotropic. Since the fibers are aligned or woven into specific patterns, usually one direction has the highest strength. Wood is a fiber composite where the ease of cutting varies with orientation. Fiber alignment during manufacturing slows fabrication and restricts design features. Much attention is given to long fiber graphite-epoxy composites, especially in aircraft, bicycles, and high performance automobiles.

Whiskers are short, discontinuous fibers. Composites based on whiskers are about 20 % lower in strength versus continuous fiber composites. Casting and molding are common fabrication routes where molten matrix and whiskers are combined into a slurry for shaping at low cost. Whisker composites such as Al-SiC are used in automotive components due to the good strength at a low cost. Another application relies on carbon whiskers added to plastics to provide electrical conductivity.

Unfortunately, whiskers represent health hazards. The problem is especially acute for diameters below about 1 μm that are most susceptible to inhalation. A large health hazard arose from worker exposure to asbestos whiskers in the fabrication of cement water pipes, automotive brakes, insulation, flooring and roofing. Because extra caution is required in handling whiskers, many whiskers are intentionally manufactured to larger sizes to prevent inhalation.

Particulate composites offer tremendous composition and component design flexibility. A wide range of processing options enable low-cost fabrication of complex shapes. Fibers are recognized for good load transfer between the phases, but interpenetrating phase particle composites offer similar load transfer. Additionally, the particulate composite is isotropic in structure.

A summary comparison of attributes for laminate, fiber, whisker, and particle composites is presented in Table 1.4. For many cases cost and fabrication flexibility are dominant considerations. Due to the anisotropic character of aligned fiber composites, engineering design requires more testing when compared to isotropic particulate composites. Thus, development costs are lower for the particulate composite.

History

Early particulate composites predate written history [1, 13]. Archeological treasures identify some of the early successes. Roman concrete is a good example, evident by a trip to Rome even in modern times. Durable concrete was used to construct large buildings. The Roman formulation contained brick fragments, stones, sand, volcanic ash, and limestone. Cost was reduced using both large and small pieces. These reinforcement phases were mixed with calcia (CaO), silica (SiO$_2$), and water to form cement. After casting the hydrated calcium-silicate cured to form 2 to 30 μm crystals to bond the ingredients. Concrete based on Portland cement is a modern version of this idea. Current global cement production exceeds 20 billion tons per year.

Porcelain dinnerware was another early composite. First developments date from about 3500 years ago. The minerals were mined and mixed with water to form a paste for shaping into dishes and decorative objects. One key to high quality porcelain was in the kiln design to reach high firing temperature, this was first

Table 1.4 Contrast and comparison of attributes for major composite classifications

Type	Laminate	Continuous fiber	Discontinuous whisker	Particle
Character	Phase as layers	Aligned fibers in matrix phase	Short fibers in matrix phase	Intermixed phases
Interpenetrating	No	Matrix only	Sometimes	Yes, possible
Isotropic	No	No	Somewhat	Yes
Example matrix material	Stainless Glass Copper	Polyethylene epoxy polyester	Polypropylene aluminum calcium silicate	Tungsten alumina cobalt
Example reinforcing material	Copper Polyester Molybdenum	Aramid glass graphite	Graphite (CNT) silicon carbide asbestos	Copper zirconia diamond
Example application	Glass-polymer-glass	Graphite-epoxy	Aluminum-silicon carbide	Diamond-cobalt
	Windshields	Golf club shaft	Brake caliper	Stone cutting
Fabrication cost	Low	High	Intermediate	Low
Shape range	2D, flat, curved	2D, flat, round	3D, simple	3D, complex
Load transfer	Excellent	Excellent	Intermediate	Variable
Flaw sensitivity	Low	High	Intermediate	Low
Global market	$1 billion	$7 billion	$1 billion	$25 billion

mastered in China, then Korea and Japan, and finally reached Europe by 1300 AD. Dental porcelain for false teeth arose in the late 1700s [14].

Porcelain is formed from oxide minerals—nominally quartz (SiO_2), feldspar ($KAlSi_3O_8$-$NaAlSi_3O_8$-$CaAl_2Si_2O_8$), and kaolinite ($Al_2Si_2O_5(OH)_4$). After firing at 1400 °C (1673 K) the structure consists of quartz (SiO_2) and mullite ($Al_6Si_2O_{13}$) bonded by glass. Modern porcelain consists of about 60 % SiO_2, 32 % Al_2O_3, 4 % K_2O, 2 % Na_2O, with traces of oxides of iron, titanium, calcium, and magnesium. Early charcoal kilns failed to reach sufficiently high temperatures to form strong porcelain. The required high firing temperatures required innovations in kiln design, fuel, and supports inside the furnace [15].

Metallic composites emerged in South America about 300 BC in the form of gold-platinum products [15–17]. Two compositions were popular, high gold with 12 wt% platinum or high platinum with 15–40 wt% gold. The Incas fused gold and platinum powder mixtures using temperatures near 1100 °C (1373 K), sufficient for melting gold to bond the higher melting temperature solid platinum. The composite was formed into needles, spoons, fish hooks, jewelry, and safety pins using forging and annealing steps. After the early European exploratory voyages to the Americas, this gold-platinum composite idea spread to Spain and Spanish jewelry.

Engineered particulate composites grew as manmade powders emerged in the early 1900s. Mixed powder composites were patented for use in bearings, lamp filaments, and abrasives. Indeed, the important cemented carbide composite, consisting of hard tungsten carbide with a metallic bond, was commercialized in the late 1920s [18]. Many of those early formulations are still in use.

Polymeric matrix composites emerged in the 1950s, soon after thermoplastics were accepted in engineering structures. Although thermosetting polymers saw earlier commercialization, little benefit arose from the addition of a second phase particle, so most of the composites relied on thermoplastics.

Although particulate composites have ancient origins [4–6, 10, 19–25], the past 100 years produced rapid growth [2, 3, 8, 26]. Although an infinite variety of combinations are possible, the more impressive successes come from complimentary combinations, such as hard-tough, flexible-wear resistant, strong-stiff, conductive-erosion resistant, magnetic-nonconductive, and conductive-high melting temperature. Now functional gradients used to further customize properties. Continuous property changes occur across a component based on composition changes with position. One example might be a metal on one side with each layer containing a higher ceramic content. The opposite side could be pure ceramic. Functional gradients allow for performance customization as desired in engine components, biomedical structures, metal cutting tools, electronic and optical devices, and metal-to-ceramic or metal-to-glass bonds.

Shorthand Notations

Shorthand notations provide a means to quickly convey information. This section gives a few early concepts.

Composition is given in weight percent, the common means used to formulate a composite. Conceptual learning is enhanced by using volume percent, especially understanding phase morphology. The major constituent is listed first, usually without a numerical value since it constitutes the "balance". For example, WC-10Co corresponds to 90 wt% tungsten carbide and 10 wt% cobalt. Often ingredients under 1 wt% are not listed. No periods or spaces are used in the abbreviation.

In the WC-10Co example, Co indicates pure cobalt. Stoichiometric compounds are common in particulate composites. For example, WC indicates equal atoms of tungsten and carbon; some other examples are Al_2O_3, SiC, and Ni_3Al corresponding to 2:3, 1:1, and 3:1 atomic ratios.

If one phase is an alloy or solution, then it is stated that way. For example 316L-5Y_2O_3, indicates 95 wt% 316L stainless steel (consisting of Fe-18Cr-12Ni-2Mo-2Mn with low levels of carbon, phosphorus, sulfur, and silicon) and 5 wt% yttria (a stoichiometric compound of yttrium and oxygen). The abbreviations "wt%" and "vol%" correspond to weight and volume percentages; the atomic percent is identified as "at.%".

In some instances a common name is used to designate a phase with a complex composition. For example borosilicate glass is a homogeneous mixture of silica (SiO_2), boric acid (B_2O_3), sodium oxide (Na_2O), and alumina (Al_2O_3) in a wt% ratio of 80:13:4:3. When no significant information is lost, these phase is identified by its common name.

Particulate composites are mostly isotropic, meaning the properties are the same in each direction. Anisotropic behavior arises when the composite is formed with phase orientation. This occurs mostly with whiskers or when orientation is intentionally induced during forming. Sometimes symbols are used to designate the grain shape, P = particle, F = fiber, W = whisker (length about 20 times the diameter), and D = disk. For example, Al-15SiC(W) designates 15 wt% silicon carbide whiskers dispersed in aluminum. The P, F, W, D nomenclature is not needed in this book since most systems are isotropic.

The transform from weight fraction to volume fraction, or volume percent, relies on the theoretical density of each phase. Fabrication processes rely on weight, but property models rely on volume fraction. The volume fraction of a phase V is calculated from the constituent weight W and densities ρ. Assuming two phases, the calculation of the volume fraction of phase 2, denoted V_2, is as follows:

$$V_2 = \frac{\left(W_2 / \rho_2 \right)}{\left[W_2 / \rho_2 + W_1 / \rho_1 \right]} \qquad (1.1)$$

For example, consider of WC-10Co. The tungsten carbide theoretical density is 15.7 g/cm^3 and the cobalt theoretical density is 8.96 g/cm^3. Using WC and Co subscripts to identify tungsten carbide and cobalt gives the specific equation as,

$$V_{Co} = \frac{\left(W_{Co} / \rho_{Co} \right)}{\left[W_{Co} / \rho_{Co} + W_{WC} / \rho_{WC} \right]} \qquad (1.2)$$

The weight of each phase is symbolized by the subscripted W and the density of each phase is represented by the subscripted ρ;

W_{Co} = 10 g (wt% on 100 g basis)
W_{WC} = 90 g (wt% on 100 g basis)
ρ_{Co} = 8.96 (g/cm^3)
ρ_{WC} = 15.7 (g/cm^3).

Substitution of these values into Eq. (1.2) gives the volume fraction of cobalt as 0.163. Often this is converted into a percentage, giving 16.3 vol% cobalt and by difference 83.7 vol% tungsten carbide, because $V_{WC} = 100 - V_{Co}$. This assumes zero porosity.

The theoretical density of the composite ρ_C is calculated from the sum of the constituent volume fractions and densities, as follows:

$$\rho_C = [V_{Co}\, \rho_{Co} + V_{WC}\, \rho_{WC}] \qquad (1.3)$$

Equation (1.3) says the relative contributions are linearly additive. It is also known as the rule of mixtures. In some composites solubility allows for one phase to partly dissolve into the other phase. When this happens, Eq (1.3) is slightly less accurate due to a change in the theoretical density.

A few examples summarize nomenclature and illustrate calculations of density and volume fraction for the phases:

- Mg-15SiC has 15 wt% silicon carbide in magnesium, corresponding to 8.9 vol% silicon carbide and based on the constituents at 1.74 g/cm^3 for Mg and 3.15 g/cm^3 for SiC, the calculated composite density is 1.85 g/cm^3,
- W-7Ni-3Fe has 90 wt% W with 7 wt% Ni and 3 wt% Fe, with a theoretical density of 16.97 g/cm^3 based on 79.2 vol% W, 15.1 vol% Ni, and 5.7 vol% Fe,
- polyimide with 15 wt% graphite filler has a density of 1.5 g/cm^3 and consists of 10 vol% graphite and 90 vol% polyimide,
- 6061Al-20SiC has 80 wt% aluminum 6061 alloy (Al with about 0.5 wt% each Cr, Cu, and Si) and 20 wt% SiC, with a theoretical density of 2.79 g/cm^3 based on 82.6 vol% alloy and 17.4 vol% ceramic,
- diamond dispersed in cobalt corresponding to Co-15D or 15 wt% diamond has a theoretical density of 7.23 g/cm^3 and consists of 30.9 vol% diamond in 69.1 vol% cobalt matrix,
- iron coated with 2 wt% cellulose corresponds to a theoretical density of 7.02 g/cm^3 and 12.4 vol% cellulose separating 87.6 vol% iron particles,
- W-40Cu has 60 wt% W and 40 wt% Cu with a theoretical density of 13.16 g/cm^3 corresponding to 41 vol% W and 59 vol% Cu.

Phase Connectivity

Figure 1.5 sketches the microstructure variation with composition for a random mixture of two phases. Generally composites are associated with at least 5 vol% of each phase, ignoring structures with low concentrations of second phase. Within the 5–95 vol% composition range, a second subdivision arises depending on the three-dimensional phase arrangement. At low concentrations one phase is isolated in a continuous matrix of the other phase. This is shown by the sketches on the left and right. Below about 20 vol% the minor phase grains are nominally separated. At intermediate concentrations, roughly from 20 to 80 vol%, both phases are intertwined, as illustrated by the central sketch. These composition ranges correspond to randomly mixed phases. Important differences in properties occur when a phase is connected, termed percolated, compared to when it is isolated.

Coated powders are a potent means to modify connectivity [27, 28]. For example, brittle phases are separated to avoid forming weak fracture paths in the composite. On the other hand, if the application requires strength or stiffness, then usually both phases are connected. To generalize,

- composites generally range from 5 to 95 vol% of each phase,
- phase connectivity usually occurs with more than 20 vol% of a phase,
- coated grains avoid forming connected structures even at high concentrations,

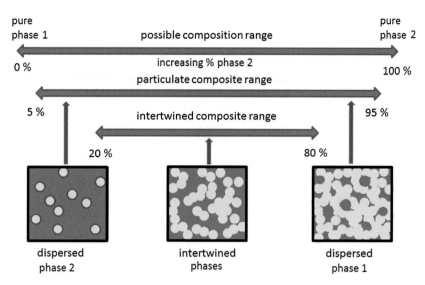

Fig. 1.5 Composition ranges and the relative behavior of particulate composites using mixed powders. For any two phases the composition possibilities range from 0 to 100 % of either phase. Particulate composites nominally correspond to a composition range from 5 to 95 vol%. For mixed powders, phase connectivity giving intertwined phases generally between about 20 and 80 vol% for each phase. The lower concentration phase is dispersed and not connected at concentrations less than 20 vol%. Coated powders allow phase separation over the entire composite range

• elongated grains allow connections at less than 20 vol%.

Idealized connectivity variants are illustrated in Fig. 1.6, showing combinations corresponding to no connectivity for one phase (0), one-dimensional connectivity (1), two-dimensional connectivity (2), and three-dimensional connectivity (3). Each variant is possible for either the white or the black phase, leading to ten combinations designated 0–0 to 3–3 [29]. Connections in three-dimensions are difficult to visualize since most materials are opaque. Interfaces control many properties, so understanding phase connectivity and the relative proportion of each interface is important to predicting composite properties.

Novel Properties

Many engineering composites are introduced in this book. Some systems are ubiquitous, such as the white filling in sandwich cookies that is based on added TiO_2 particles, and other systems are valuable, such as diamond-carbide composites used for oil well drilling.

Property models rely on input parameters that include composition and constituent attributes. Good models exist for static properties, such as density, hardness,

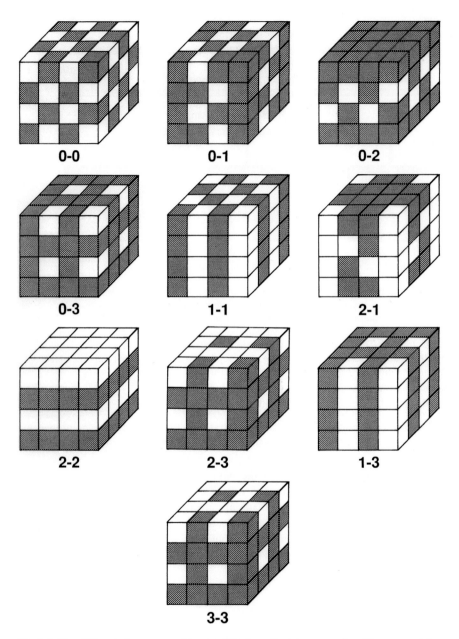

Fig. 1.6 Ten different phase connections possible in particulate composites. The nomenclature corresponds to *0* = no connectivity, *1* = one-dimensional or rod-like connectivity, *2* = two-dimensional or planar connectivity, and *3* = three-dimensional connectivity

Fig. 1.7 Elastic modulus for alumina (Al$_2$O$_3$)–zirconia (ZrO$_2$) composites; the *square* symbols are experimental determinations. The *upper curve* represents model predictions assuming isostrain where both phases expand the same, and the *lower curve* represents model predictions assuming isostress where both phases carry the same stress. An average of those two models provides the best experimental behavior representation

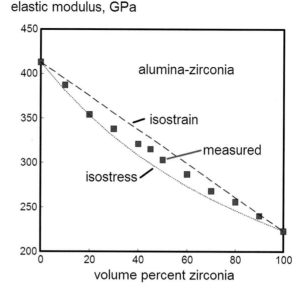

and elastic modulus. Consider the elastic modulus data plotted in Fig. 1.7 for alumina (Al$_2$O$_3$) and zirconia (ZrO$_2$) across the full composition range. The elastic modulus relates stress to strain prior to deformation. Alumina is stiff but zirconia adds toughness to the composite. Included on the plot are two property models:

1. isostrain, assuming the strain is uniform in both phases,
2. isostress, assuming the stress is uniform in both phases.

Neither is quite correct as the measured behavior is intermediate between the two models. Indeed, the average of the two models results in almost perfect fit to the experimental results, with an error of 1 GPa. The experimental data are about this accurate, so better agreement is not reasonable.

Besides density, hardness, and elastic modulus, properties related to deformation, such as toughness, are dominated by the interface strength. Weak or porous interfaces allow easy failure. A strong interface results in high fracture resistance. Thus, for a single composition small changes to the interface might have profound impact. Hence, for properties related to deformation and fracture, information is required on the interface attributes. Likewise, high temperature properties are difficult to predict without information on any deleterious reactions that occur between the phases. Thus, as we find later in this book some properties are predicted with reasonable precision, but other properties largely require experimental tests.

Because particulate composites are relatively new to engineering, more time and effort is required to qualify for an application. This increases the expense due to the prolonged testing versus well-established materials. Even so, much opportunity is recognized in the field. A biomedical example illustrates the opportunities. The composite under evaluation consists of porous titanium and hydroxyapatite

($Ca_{10}\cdot6PO_4\cdot2OH$). It is designed to be an implant replacement for a skeletal components such as a knee or hip. Three phases make up the implant—titanium, hydroxyapatite, and pores. The hydroxyapatite is chemically similar to bone, but weak with strength near 40 MPa. Human bone is variable in properties, but typically has a tensile strength over 130 MPa, elastic modulus around 20 GPa, and fracture toughness of 2–12 MPa$\sqrt{m}$ [11]. The properties vary between individuals, so these are only representative properties. Titanium is strong (300–900 MPa depending on alloying), corrosion resistant, and stiff (elastic modulus around 110 GPa). Indeed titanium is undesirably strong and stiff, to the point where it damages surrounding bone. However, 40 % dense titanium has compatible strength and elastic modulus, close to human bone. Adding hydroxyapatite makes the structure biologically similar to bone. Thus, the combination of porous titanium with hydroxyapatite enables customized replacement bones compatible with humans. When this composite is formed using additive manufacturing techniques, a customized implant is made to match the specific needs for each patient. New efforts are making the structure absorbable over time using magnesium to avoid any need for surgical removal.

Porous titanium scaffolds with hydroxyapatite are an example of how particulate composites operate in a multiple-factor design space, as conceptually outlined in Fig. 1.8. This octahedron links the key factors. Performance largely results from the properties, which in turn depend on the composition (phases and relative amounts), structure (grain size, spacing, connectivity, and interface), and processing (time, temperature, pressure, and heating and cooling rates). This book attends to the intersection of these factors. Each apex of the polyhedron provides a means to customize the composite to meet long-felt needs.

Particulate composites span most known materials, both manmade and naturally occurring. A few ingredients and combinations dominate the field. However, many fabrication approaches are in use, and a few are popular due to low cost [1, 9, 19–21, 30]. Several of the fabrication routes are detailed later in this book. Consider some of the options as follows:

- mixed particles are consolidated using techniques such as hot isostatic pressing, hot pressing, or other approaches that combine temperature and pressure,
- particles are dispersed in a molten matrix and the mixture is formed by casting or molding, such as the replica of Rodin's *Thinker* shown in Fig. 1.9,
- particles are formed into a porous preform that is subsequently infiltrated with a second phase, for example to form the automotive valve seats shown in Fig. 1.10,
- coated particles are consolidated to surround core particles with a matrix phase, consolidated by spark sintering to give fracture resistant structures.

Coated particles are a new variant; a second phase envelopes the core particle prior to consolidation. The approach allows for specific placement of each phase to customize properties to the application [27–37]. Usually a hard phase is dispersed in a ductile phase. When consolidated, the core hard phase is dispersed, even at high concentrations. Such a structure reduces the fracture susceptibility associated with

Fig. 1.8 A particulate composite relies on several factors in the typical design space. Performance in an application reflects the properties, and in turn properties depend on the composition (phases, amount of each), structure (interfaces, microstructure), and processing (thermo-mechanical fabrication cycle)

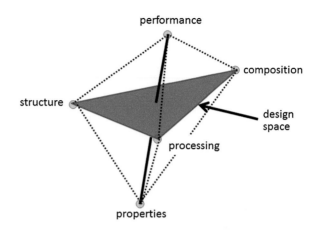

Fig. 1.9 A cast replica of Rodin's Thinker, fabricated using bronze powder and epoxy, where the slurry is cast into a mold. The final artwork has the look and feel of bronze, but is fabricated at room temperature. The original statuary from 1902 is nearly 1.9 m in height, cast from high temperature bronze

the hard phase, resulting in improved fracture resistance. Figure 1.11 is a scanning electron micrograph of a coated particle composite that exhibits a high wear resistance from the hard phase with a high fracture resistance due to the tough phase separating the hard grains.

Particulate composite fabrication occurs using several manufacturing options. Most frequently fabrication relies on molding or pressing to form the shape, followed by heating to bond the constituents. Some productive and low-cost forming approaches rely on manufacturing concepts adopted from plastic processing, such as extrusion, injection molding, and compression molding. The production of final component shapes with minimal machining is critical to keeping manufacturing costs low.

Fig. 1.10 Automotive valve seats fabricated using tool steel powder preform infiltrated with molten copper alloy. The tool steel provides excellent wear resistance without deterioration from the fuel or combustion gases and the infiltrant fills pores to prevent pressure loss during operation

Fig. 1.11 An industrial wear composite consisting of block alumina (Al_2O_3) grains with intermediate small grains of tungsten carbide with cobalt (WC-10Co). This fracture surface shows the alumina dispersed in the matrix based on consolidation of coated powders. Alumina provides high wear resistance while the matrix provides high toughness. The composite delivers significant gains in tool life when operated under compressive loading conditions

10 µm

Summary

Particulate composites start with a mixture of phases. Each phase has a distinct composition and atomic arrangement. One phase might be an element such as copper, compound such as B_4C, or solution such as bronze. The composite is heterogeneous at the particle level. Pores are an additional option, since they significantly modify properties. Many variants exist, with ample opportunity to customize performance. Thus, particulate composites are characterized by the following attributes:

1. they consist of a mixture of two or more phases,

2. each phase is different in composition or crystal structure, with differing properties, including pores,
3. each phase is resolvable using microscopy,
4. the composite has at least 5 vol% of each phase,
5. they are fabricated form powders, including powder-liquid mixtures,
6. the microstructure is random, generally with equiaxed grains, giving isotropic properties,
7. one phase has unique properties, such as high hardness
8. the other phase has complimentary features not available with the first, such as high toughness,
9. the composite delivers novel property combinations.

Accordingly, many systems and applications are evident, including the following examples:

- polymer-based systems where a filler provides improved stiffness at a low cost, as in glass-filled nylon for automotive components,
- ceramic-glass systems, used in porcelain dinnerware, dental restorations, and electronic substrates,
- metal-metal systems as encountered in electrical contacts and computer heat sinks, such as tungsten-copper
- ceramic metal systems, such as tungsten carbide—cobalt intended for wear applications,
- ceramic-ceramic systems such as silicon nitride reinforced with silicon carbide for metal cutting tools.

Mixed phases provide a means to synthesize new property combinations using established manufacturing tools. One novel approach is spark sintering, where mixed powders are rapidly consolidated [38]. An intense electrical discharge induces heating while external stress ensures densification with little reaction between the phases.

To summarize, composites arise in many forms. Much opportunity exists to tailor the composite to specific application needs. This book organizes around the belief that a new science base is arising in particulate composites. Emerging from the field are cost-effective engineering solutions to broad problems in important fields.

Study Questions

1.1. What are the ingredients in white household paint? What is the role of each phase? Is this an example of a particulate composite?
1.2. Composites form the thermal protection tiles on vehicles returning from space. How are these tiles engineered to survive the heat and erosion associated with the intense heating?

1.3. Calculate the volume fraction of silicon carbide in a composite of Mg-15SiC, assuming a density of 1.74 g/cm^3 for the magnesium and 3.15 g/cm^3 for silicon carbide.

1.4. For the Mg-15SiC composite, what is the expected theoretical density? What is the measured density in g/cm^3 with 10 % porosity?

1.5. The elastic modulus for magnesium is about 45 GPa and for silicon carbide the elastic modulus is about 414 GPa. Assume a linear relation between elastic modulus and volume fraction, then what weight percent of silicon carbide is required to reach a composite elastic modulus of 100 GPa, assuming full density?

1.6. For the 100 GPa elastic modulus Mg-SiC composite in the last problem, what is the specific elastic modulus (elastic modulus divided by density).

1.7. Figure 1.3 is an image of the Invar-40Ag composite known as *Silvar* and Table 1.2 provides constituent and composite properties. Reflect on the strength and any basis you might have for estimating the strength. What factors are influencing the composite strength?

1.8. Various studies have consolidated mixtures of conductive and nonconductive powders to full density, seeking the minimum concentration of conductor giving electrical conduction to the composite. The findings show a change from nonconductive to conductive composite generally occurs between 10 and 25 vol% conductor powder. What factors beyond concentration need to be considered to explain the variation in critical concentration for conduction?

1.9. Cemented carbide compositions rely on tungsten carbide (WC) and cobalt (Co). Plot an estimate of the heat capacity, thermal expansion coefficient, and elastic modulus versus composition for cemented carbides ranging from 0 to 15 vol% cobalt.

1.10. For a metal cutting application it is proposed to use a composite of alumina (density 3.96 g/cm^3) with 30 wt% titanium carbide (density 4.91 g/cm^3). The composite density is reported at 4.52 g/cm^3. Is this a realistic density?

1.11. For the composite of alumina and titanium carbide described in the last question, the alumina hardness is 1800 HV and the titanium carbide hardness is 3200 HV. What is a first estimate for the composite hardness?

1.12. Fire retardants are added as particles to polymers. Identify one of the inorganic retardants added as a particle and describe how it works to prevent burning.

References

1. M. Balasubramanian, *Composite Materials and Processing* (CRC Press, Boca Raton, 2014)
2. S.K. Bhattacharya, *Metal Filled Polymers* (Marcel Dekker, New York, 1986)
3. F. Carmona, Conducting filled polymers. Phys. A Stat. Mech. Appl. **157**, 461–469 (1989)
4. K.K. Chawla, *Composite Materials Science and Engineering*, 3rd edn. (Springer, New York, 2012)

5. D.D.L. Chung, *Composite Materials: Functional Materials for Modern Technologies* (Springer, New York, 2003)
6. W.F. Gale, T.C. Totemeier (eds.), *Smithells Metals Reference Book*, 3rd edn. (Elsevier Butterworth-Heinemann, Oxford, 2004)
7. F. Matthews, R.D. Rawlings, *Composite Materials: Engineering and Science* (Woodhead, Cambridge, 1999)
8. J.V. Milewski, H.S. Katz (eds.), *Handbook of Reinforcements for Plastics* (Van Nostrand Reinhold, New York, 1987)
9. D.B. Miracle, S.L. Donaldson, in *Introduction to Composites*, ASM Handbook of Composite Materials, vol 21 (ASM International, Materials Park, 2001)
10. R. Riedel (ed.), *Handbook of Ceramic Hard Materials* (Wiley-VCH, Weinheim, 2000)
11. J.W.C. Dunlop, P. Fratzl, Biological composites. Annu. Rev. Mater. Res. **40**, 1–24 (2010)
12. Q. Chen, G.A. Thouas, Metallic implant biomaterials. Mater. Sci. Eng. **R87**, 1–57 (2015)
13. W.D. Kingery, in *Sintering From Prehistoric Times to the Present*, eds. by A.C.D. Chaklader, J.A. Lund. Sintering'91 (Trans Tech Publ., Brookfield, 1992), pp. 1–10
14. J.R. Kelly, I. Nishimura, S.D. Campbell, Ceramics in dentistry: historical roots and current perspectives. J. Prosthet. Dent. **75**, 18–32 (1996)
15. R.M. German, *Sintering: From Empirical Observations to Scientific Principles* (Elsevier, Oxford, 2014)
16. J.A.P. Elorz, J.I. Verdja-Gonzalez, J.P. Sancho-Martinez, N. Vilela, Melting and sintering platinum in the 18th Century: the secret of the Spanish. J. Met. **51**(10), 9–12 (1999). 41
17. M. Noguez, R. Garcia, G. Salas, T. Robert, J. Ramirez, About the Pre-Hispanic Au-Pt 'Sintering' technique. Int. J. Powder Metall. **43**(1), 27–33 (2007)
18. K.J.A. Brookes, Half a century of hardmetals. Met. Powder Rep. **50**(12), 22–28 (1995)
19. A. Bose, *Advances in Particulate Materials* (Butterworth-Heinemann, Boston, 1995)
20. S.J. Schneider, in *Ceramics and Glasses*, Engineered Materials Handbook, vol 4 (ASM International, Materials Park, 1991)
21. P. Samal, J. Newkirk (eds.), *Powder Metallurgy*, ASM Handbook, vol 7 (ASM International, Materials Park, 2015)
22. R. Morrell, *Handbook of Properties of Technical and Engineering Ceramics* (Her Majesty's Stationery Office, London, 1987)
23. P. Schwarzkopf, R. Kieffer, W. Leszynski, F. Benesovsky, *Refractory Hard Metals: Borides, Carbides, Nitrides, and Silicides,* (MacMillan, New York, 1953)
24. D.J. Green, *Introduction to the Mechanical Properties of Ceramics* (Cambridge University Press, Cambridge, 1998)
25. J.B. Watchman, *Mechanical Properties of Ceramics* (Wiley, New York, 1996)
26. A.B. Strong, *Plastics—Materials and Processing*, 3rd edn. (Prentice Hall, Upper Saddle River, 2006)
27. N. Barbat, K. Zangeneh-Madar, Preparation of Ti-Ni binary powder via electroless nickel plating of titanium powder. Powder Metall. **57**, 97–102 (2014)
28. C.L. Hu, M.N. Rahaman, Factors controlling the sintering of ceramic particulate composites: II, coated inclusion particles. J. Am. Ceram. Soc. **75**, 2066–2070 (1992)
29. D.S. McLachlan, M. Blaszkiewicz, R.E. Newnham, Electrical resistivity of composites. J. Am. Ceram. Soc. **73**, 2187–2203 (1990)
30. R.M. German, *Powder Metallurgy and Particulate Materials Processing* (Metal Powder Industries Federation, Princeton, 2005)
31. A. Elsayed, W. Li, O.A. El Kady, W.M. Daoush, E.A. Olevsky, R.M. German, Experimental investigation on the synthesis of W-Cu nanocomposite through spark plasma sintering. J. Alloys Compd. **639**, 373–380 (2015)
32. T. Schafter, J. Burghaus, W. Pieper, F. Petzoldt, M. Busse, New concept of Si-Fe based sintered soft magnetic composite. Powder Metall. **58**, 106–111 (2015)

33. W.M. Daoush, H.S. Park, S.H. Hong, Fabrication of TiN/cBN and TiC/diamond coated
    particles by titanium deposition process. Trans. Nonferrous Met. Soc. China **24**, 3562–3570
    (2014)
34. J.W. Kim, Y.D. Kim, Sintering of Nd-Fe-B magnets from Dy coated powder. J. Korean
    Powder Metall. Inst. **20**, 169–173 (2013)
35. E.A. Anumol, B. Viswanath, P.G. Ganesan, Y. Shi, G. Ramanath, N. Ravishankar, Surface
    diffusion driven nanoshell formation by controlled sintering of mesoporous nanoparticle
    aggregates. Nanoscale **2**, 1423–1425 (2010)
36. B. Ozkal, A. Upadhyaya, M.L. Ovecoglu, R.M. German, Comparative properties of
    85 W-15Cu powder prepared using mixing, milling, and coating techniques. Powder Metall.
    53, 236–243 (2010)
37. X. Liang, C. Jia, K. Chu, H. Chen, J. Nie, W. Gao, Thermal conductivity and microstructure of
    Al/diamond composites with Ti coated diamond particles consolidated by spark plasma
    sintering. J. Comp. Mater. **46**, 1127–1136 (2011)
38. Z.A. Munir, D.V. Quach, M. Ohyanagi, Electric current activation of sintering: A review of the
    pulsed electric current sintering process. J. Am. Ceram. Soc. **94**, 1–19 (2011)

Chapter 2
Background Definitions

A standardized language helps specify the attributes desired from a composite. This brief chapter provides starting definitions for the more common terms used in the field.

Key Terms Used in Composites

Chapter 1 introduced several concepts to lay in place definitions for key terms. The definitions given here establish core concepts used in this book based on details given in several references [1–20]. The goal is to ensure the terms are understood in subsequent chapters that give more detail to the engineering principles.

Presented below are the key phrases where each phrase is coupled with a brief explanation.

Composite: A composite is a structure consisting of two or more phases, where each phase remains different in crystallography or composition and is resolvable using tools such as X-ray diffraction, microscopy, or chemical analysis. An example composite is imaged in Fig. 2.1, showing polycrystalline Ni_3Al as the matrix or continuous phase, reinforced with a dispersed SiC phase. In this case the silicon carbide is coated with a layer of alumina to avoid reaction with the Ni_3Al during consolidation.

Particulate Composite: The particulate composites are formed with at least one phase starting as particles. In some cases the second phase is a fluid such as molten polymer or liquid metal, while even occasionally a vapor is employed to fill the voids between particles. For the majority of cases the starting structure consists of two or more mixed solid powders.

Phase: A distinct region in the microstructure is termed a phase. Each phase is characterized by specifics such as composition, amount, crystal structure, and grain size. Some phases might be amorphous. Typical is Fig. 2.2, a two phase composite imaged based on differences in the microstructure evident using microscopy.

© Springer International Publishing Switzerland 2016
R.M. German, *Particulate Composites*, DOI 10.1007/978-3-319-29917-4_2

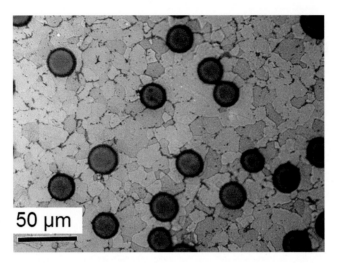

Fig. 2.1 A high magnification cross-section image of a nickel aluminide intermetallic (Ni$_3$Al) matrix composite reinforced with silicon carbide (SiC) to improve high temperature strength. The silicon carbide is coated to avoid reaction with the matrix during high temperature use [courtesy D. Alman]

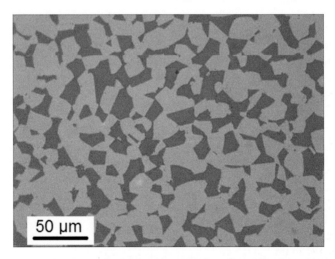

Fig. 2.2 Microstructure of a two phase composite consisting of equal fractions of both phases, one phase is iron–rich and the other contains chromium-boride. Differences in the properties between the phases provide a means to generate contrast for microscopic analysis [courtesy C. Toennes]

Grain: A grain represents a single crystal region within a polycrystalline body. Particles might be composed of several grains held together in the form of agglomerates with weak polymer bonds or aggregates with strong sinter bonds. Once grains are consolidated the structure is a solid consisting of multiple grains. When a grain structure is etched, boundaries between grains are attacked preferentially and

Fig. 2.3 This scanning electron microscope image shows zinc oxide grains after removal from a ZnO-13BaO composite. The BaO phase was dissolved to leave these ZnO grains. These grains have faces where they made contact with one another in the composite, but rounded corners where they were in contact with the BaO [courtesy S. C. Yang]

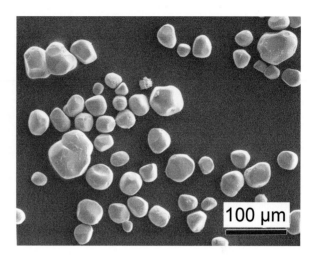

enable visualization in microscopy. Figure 2.3 is a scanning electron microscope picture of zinc oxide grains after dissolution of the surrounding BaO matrix; the composite consisted of ZnO-13BaO. The single crystal ZnO grains exhibit flat faces where they bonded to neighboring grains.

Particle: A particle is a discrete solid with a maximum dimension smaller than 1 mm or 1000 μm [4, 5, 14]. Particles come in many forms, ranging from sizes smaller than blood cells to larger than grains of sand. Engineering particles are measured using the micrometer (μm) scale, which is 10^{-6} m. Most engineering particles are from 0.1 to 200 μm in size, with ceramic particles tending toward the smaller sizes and plastic particles tending toward the larger sizes. For reference, human hair nominally has a diameter of 100 μm and white paint pigment is in the 0.3 μm range.

Powder: A large collection of particles is called a powder as long as the particles are not bonded together. A powder might consist of multiple components, reflecting how powders are mixed to give a desired composition. Powders are able to flow and exhibit properties that are similar to both solids and liquids.

Grain Boundary: The interface zone between two crystal grains is the grain boundary. It represents the narrow zone where atomic bonding is disrupted by misalignment of the crystalline grains as sketched in two dimensions in Fig. 2.4. In some cases the grain boundary zone might be amorphous if a glassy film forms. Usually the disrupted bonding at the grain boundary is about five to ten atoms across. Grain boundaries are sites for impurity segregation and are active paths for atomic motion at high temperature.

Green: The term green comes from ceramics where it is used to describe a formed shape that has not been fired [4]. This same idea applies to any powder structure after shaping but prior to heating. The green density is the unfired mass divided by volume, reported as g/cm^3, but might also be given as a fraction or percentage of theoretical density. For example, iron has a theoretical density of 7.86 g/cm^3 but when iron powder is compacted to 6.8 g/cm^3 the fractional green

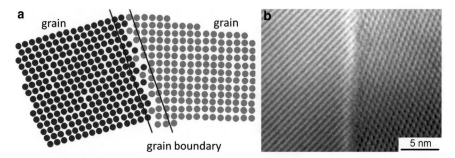

Fig. 2.4 The grain boundary represents the thin region of disrupted atomic bonding where impurity atoms segregate and where rapid atomic motion occurs; (**a**) is a two-dimensional schematic of the disrupted atomic crystal structure at the interface between two grains of the same phase, with different crystal alignments, and (**b**) is a high magnification transmission electron microscopy image of the grain boundary in composite of Si_3N_4-$6Y_2O_3$ [courtesy Q. Wei]

density is 0.865 or 86.5 % of theoretical. Green strength is the handling strength prior to firing.

Consolidation: Particulate composites are formed into green bodies from powders. Consolidation might rely on liquids, temperature, or pressure to bond the particles. Most consolidation approaches deliver work to the powder to increase density and strength. Consolidation approaches include sintering, extrusion, hot pressing, die compaction, forging, hot isostatic pressing, and spark sintering [4].

Sintering: Sintering is a thermal treatment that caused neighboring particles to bond together to improve strength. Sinter bonding occurs by high temperature atomic diffusion, often performed close to the melting temperature of the matrix phase. Interparticle bonds form by atomic motion and over several minutes the bonds become evident via microscopy. Figure 2.5 illustrates the sinter bonding of spherical particles. Sintering typically strengthens a powder compact by about 200-fold over the green strength. Early sintering called the process induration to reflect the substantial hardening after exposure to a high temperature heating cycle. Pressure-assisted sintering involves the simultaneous application of high pressure during heating.

Connectivity: Properties change with the degree of bonding between similar phases in the composite microstructure [15]. Connectivity is the integer number of bonds with grains of the same phase evident in two-dimensional cross-sections. A grain totally surrounded by second phase has zero connectivity. It is measured using two-dimensional microscopy by selecting one grain and counting the number of contacts it has with similar grains. In Fig. 2.6 the grain in the center has four contacting neighbor grains of the same composition. For this grain the connectivity is 4. Repeated measure over many grains leads to an average connectivity for the composite.

Contiguity: Contiguity is related to connectivity in terms of measuring grain contacts with similar grains; however, contiguity reports the relative fraction of grain perimeter in contact with similar phase grains [5]. It is also measured on

Fig. 2.5 A scanning
electron microscopy picture
of bronze spheres bonded
together after sintering. The
particles were initially loose
and not bonded [courtesy
G. A. Shoales]

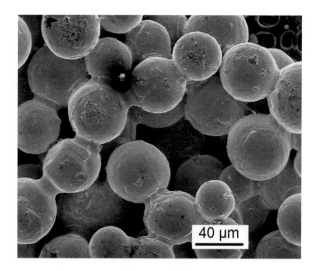

Fig. 2.6 Connectivity
depicted in two-dimensions
where the central grain has
contacts with four
neighboring grains in this
random cross-section

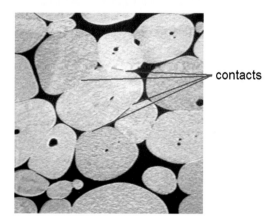

contacts

two-dimensional cross-sections. Quantitatively it is the fraction of grain perimeter
that is occupied by the same phase. For example in Fig. 2.6, the contiguity for the
central grain is about 0.2, reflecting 20 % of the perimeter is occupied by grains of
the same phase as the central grain.

Percolation: The idea of a connected phase is treated by percolation theory
[15]. Percolation is illustrated by considering a mixture of conductor and nonconductor particles. Percolation occurs at the conductor composition where the whole
mixture is conductive. At lower concentrations of conductor the mixture is
nonconductive. The percolation limit corresponds to the threshold concentration
of conductor particles needed to make the mixture conductivity. Besides concentration, the percolation limit depends on factors such as the particle size ratio,
particle shape, and porosity. In mixed spherical particles of the same size for the

conductor and nonconductor a percolated structure requires about 20 vol% of conductor phase.

Matrix: Composites have at least one phase connected throughout the structure and this phase is called the matrix [6–8, 13, 16]. For a brick wall, the mortar forms a continuous network, separating the bricks. Because the mortar is continuous, it is the matrix phase. As in the case of bricks, the matrix is not necessarily the majority phase. It is an error to call the matrix a binder, since binders are sacrificial phases used in shaping and lack the permanency of a matrix.

Filler: Low cost minerals added to polymers, largely to lower the cost, are termed fillers. Filler phases might not benefit properties. As such the connectivity of the filler is intentionally not developed. For polymers, metallic or graphite fillers are added in some cases to provide conductivity and have minor impact on mechanical properties [11].

Reinforcement: The reinforcement phase is added to the matrix to improve properties, usually mechanical properties [1, 3, 10, 16]. Reinforcements come in a variety of forms, including particles, hollow spheres, disks, flakes, whiskers, and fibers. For whiskers and fibers it is possible to orient the structure to produce anisotropic properties. In doing so, the composite properties are anisotropic, implying they change with testing orientation.

Hard Material (Hard Metal): Cemented carbides are also known as hard metals or hard materials. The term is broadly applied to hard composites consisting of carbides, borides, oxides, or nitrides; the field is dominated by tungsten carbide composites. Commonly, components fabricated from hard particles use a cementing metallic phase; for example TiC is bonded by Ni. This composite is also termed a cermet. Such hard materials are targeted at applications requiring wear resistance. Cermets, cemented carbides, hard metals, and hard materials are similar in that they are all composites involving hard phases.

Metal Matrix Composite: Metal matrix composites have a continuous metal phase that makes them electrically conductive [8, 13, 16]. A ceramic phase is dispersed in the metal to add stiffness or wear resistance. Aluminum with added silicon carbide is a case where the matrix is the aluminum and the reinforcement is silicon carbide. In many cases metal matrix composites have less than 50 vol% ceramic. A special class of metal matrix composites focuses on intermetallic matrix phases such as TiNi-SiC or TiAl-Al_2O_3 [7].

Ceramic Matrix Composite: Composites with a continuous ceramic matrix phase with dispersed second phase grains are termed ceramic matrix composites [17, 18]. Alumina with dispersed zirconia is an example ceramic matrix composite. Composite toughness, measured by the resistance to crack propagation, is a key focus for ceramic matrix composites.

Polymer Matrix Composite: The polymer matrix identifies the continuous phase as a polymer, with a filler that might be mineral, glass, ceramic, or metal [11, 12]. Generally the polymer is more than 50 vol% of the composition. A few polymer-polymer composites are known, but the most significant hardness and strength gains are seen using nanoscale ceramic or mineral particles [6, 19].

Fig. 2.7 A
tetrakaidecahedron is a
14-sided polyhedron that is
used to represent the shape
of a single grain in a fully
dense structure

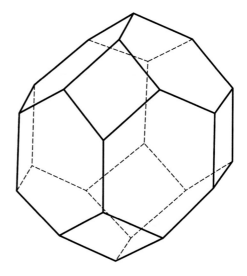

Particle Coordination Number: Coordination refers to the number of touching neighbors. For particles, this is the three dimensional average number of contacting neighbors [5]. Particles form random packings, resulting in a distribution to the coordination number. Larger particles have a higher number of contacts and smaller particles have fewer contacts. Further, for particles the coordination number increases with packing density. For many particles that pack to about 0.6 fractional density, the average coordination number is near seven. For nanoscale particles the coordination number drops to an average of two contacts per particle with a corresponding low density, often near 0.04. When particles are consolidated to full density, the coordination number approaches 12–14. The resulting polygonal grain shape is depicted in Fig. 2.7, showing in this case a 14-sided polyhedron representation of a single grain at full density. This grain shape is known as a tetrakaidecahedron.

Particle Hardness: The hardness of individual particles is the particle hardness. It is measured by scratch, indentation, or microhardness tests, giving a particle characteristic that best links to wear resistance. The composite hardness might be much lower depending on the composition and porosity. Pores have no hardness, so residual porosity results in a lower hardness versus the particle hardness.

Particle Size: Powder dimensions are represented by a characteristic size known as the particle size. Particle size is calculated from measurements on the particles, such as length, cross-sectional area, volume, mass, or surface area. The analysis is performed using screens, laser light scattering, sedimentation analysis, or other techniques. Powders are distributed in size, so a common measure of particle size is the median size. This corresponds to half the particles being larger and half being smaller, occurring at the 50 % point on a cumulative particle size distribution [5]. This size is identified as D50. Other sizes corresponding to 10 and 90 % on the cumulative particle size distribution are designated D10 and D90; D10 is

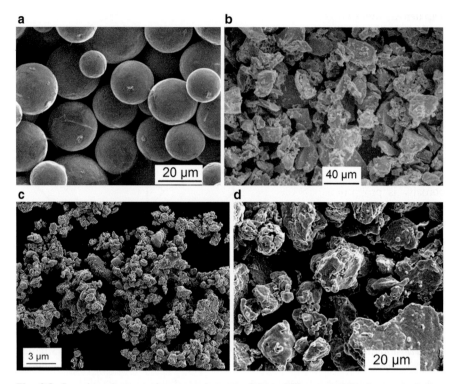

Fig. 2.8 Scanning electron microscopy images of four different particle shapes, including (**a**) spherical, (**b**) angular, (**c**) agglomerated polygonal, and (**d**) rounded

smaller than D50 and D90 is larger than D50. In some cases the size is given based on the maximum size (negative sign) or minimum size (positive sign) for the powder. For example the designation -325 mesh implies all of the particles passed through a 325 mesh screen so they are smaller than 45 µm. The designation $-200/$ $+325$ indicates the powder is all smaller than 75 µm but larger than 45 µm.

Particle Shape: Both descriptive parameters and quantitative measures are used to convey a sense of the particle shape. The descriptive names include spherical, dendritic, irregular, porous, polygonal, or ligamental. The images in Fig. 2.8 show scanning electron microscopy images for a few variants, including spherical, angular, polygonal, and rounded particles. Particle shape is quantified using the aspect ratio, defined as the longest dimension divided by smallest width.

Particle Packing: Tests that measure how a powder fills a container provide information on particle packing characteristics. When loosely filled into a container without vibration or pressure, the powder mass divided by the container volume gives the apparent density. Random packing structures typically produce 60 % of theoretical density for spheres without agitation. Irregular particles pack to just 30 % of theoretical. Low apparent densities are associated with small, irregular powders. In the case of nanoscale particles, packing densities are as low as 4–6 %.

Tap density relies on vibration to induce closer packing, reaching near 64 % density for larger spherical particles. Packing structures are influenced by the powder size distribution, particle shape, and surface chemistry.

Porosity: The relative amount of void space in a powder compact is the porosity. Pores have no mass, but dilate the volume. Under microscopic examination pores appear as holes in the microstructure. Porosity is determined by the measured density and how much it falls below the theoretical density. Thus, 10 % porosity implies the measured density is just 90 % of theoretical.

Pore Former: When intentional pores are desired in a consolidated powder, a sacrificial substance is added prior to powder consolidation [4]. The pore former is removed during processing, leaving behind pores of the desire size. The residual porosity is proportional to the amount of pore former, while the pore size is proportional to the size of the sacrificial addition. Examples of sacrificial particles include wax, salt, cellulose, ice, stearates, fatty acids, potassium chloride, and ammonium carbonate.

Microstructure: The phase arrangement observed in a microscope is termed the microstructure. Both optical and electron microscopes are used to image the phases in particulate composites. Microstructure evaluation includes observations on the sizes of the grains, as well as grain shape, phase connectivity, and homogeneity of the phases. In addition, microstructures show gradients and evidence of pores, including pore size distribution.

Nanoscale: The word nano means 10^{-9} and in particle systems this corresponds to 1/1000 micrometers or 10 Angstrom [19]. By convention, nanoscale particle structures are associated with sizes smaller than 100 nm. Nanoscale particles of titania (TiO_2), silver, and carbon black are widely used in products such as paint, antibacterial ointments, and rubber tires. Extraordinary properties arise when the nanoscale structure is preserved in the composite, especially hardness and strength [12, 19]. However, properties such as conductivity tend to degrade at the nanoscale due to the high concentration of interfaces.

Volume Fraction: Composites are usually reported on a weight basis, but it is helpful to understand the relative volume of each phase. Accordingly, volume fraction is the portion of component volume taken up by one phase. It is easier to reconcile behavior and property models to the volume fraction since it ignores the effects when the two phases are very different in density.

Homogeneity: Homogeneity is a description of phase uniformity. A homogeneous structure appears similar when examined at different locations [5]. As the magnification increases, a particulate composite eventually appears inhomogeneous. Thus, homogeneity as a measure of uniformity cannot look too closely. Most models for the properties of particulate composites assume homogeneous structures. The properties are assumed to be the same in all directions; homogeneous composites are isotropic.

Interface: Phases contacts create interfaces that represent changes in composition with disrupted atomic bonding. Grain boundaries are interfaces between grains of the same phase while interfaces are boundaries between dissimilar phases. Composite properties are sensitive to the interface structure and chemistry, leading

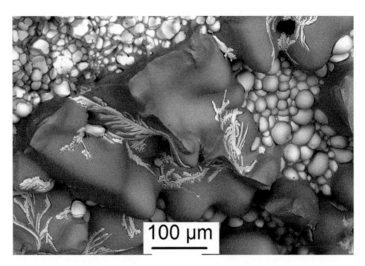

Fig. 2.9 This scanning electron micrograph shows a particulate composite fracture surface. This picture gives evidence of the underlying heterogeneous structure and weak interfaces associated with low strength

to large potential variations if not properly controlled. Figure 2.9 is a picture generated using scanning electron microscopy. It shows a fracture surface resulting from contaminated, poor quality interfaces that provided an easy fracture path, resulting in low strength and ductility.

Micromechanics: The mechanical properties at a small size scale corresponding to microstructure elements are treated by micromechanical models [1, 8, 20]. Primary focus is on mechanical load transfer between phases. Unlike bulk constitutive behavior, where a single stiffness is assumed, the micromechanical approach combines phase properties and microstructure to estimate the interaction. The main concerns are models for elastic, plastic, and thermal properties.

Study Questions

2.1. Where did the term "induration" come from?
2.2. The term "cermet" is often used to identify what type of composite? How did this term arise?
2.3. High impact polystyrene (HIPS) is a polymer-polymer composite. What is the other ingredient besides polystyrene?
2.4. What is a typical volume fraction of polystyrene in HIPS?
2.5. For a particulate composite of Al-15SiC, make a sketch of the microstructure and on that sketch indicate an interface and grain boundary—how are they different?

2.6. Many foods contain relatively inert particles, such as reduced iron (cereal), titania (white frosting), or calcium carbonate (hot chocolate mix). Look at the ingredient list on some packaged food items and identify inorganic particles added to adjust properties. For example, aluminosilicate is added to some foods to prevent clogging. Identify something similar?

2.7. A friend says that milk chocolate is a particulate composite. Take a position on this statement and defend that position.

2.8. Is ice cream a particulate composite (before it melts)?

2.9. The white filling in sandwich cookies contains titania (TiO_2) particles of about 0.25 μm diameter. Is this a particulate composite based on the definitions given here?

2.10. An early composite was formed using a mixture of broken bricks, stones, limestone, sand, and volcanic ash. Any idea where it was used and if it was successful?

2.11. Concrete is a low cost and widely used composite. It is not very good under tensile loading, so what provision is used in many designs?

2.12. Sunscreen relies on dispersed particles to disrupt sunlight and polymers to convert the sunlight into heat. What particle ingredients are common to sunscreen? One product says it contains titanium. Is this accurate?

References

1. K.K. Chawla, *Composite Materials Science and Engineering* (Springer, New York, 2012)
2. D.D.L. Chung, *Composite Materials Science and Applications* (Springer, London, 2010)
3. S. Ahmed, F.R. Jones, A review of particulate reinforcement theories for polymer composites. J. Mater. Sci. **25**, 4933–4942 (1990)
4. R.M. German, *A-Z of Powder Metallurgy* (Elsevier Scientific, Oxford, 2005)
5. R.M. German, S.J. Park, *Mathematical Relations in Particulate Materials Processing* (Wiley, Hoboken, 2008)
6. F. Hussain, M. Hojjati, M. Okamoto, R.E. Gorga, Polymer-matrix nanocomposites, processing, manufacturing, and application: an overview. J. Comp. Mater. **40**, 1511–1575 (2006)
7. C.M. Ward-Close, R. Minor, P.J. Doorbar, Intermetallic matrix composites—A review. Intermetallics **4**, 217–229 (1996)
8. T.W. Clyne, P.J. Withers, *An Introduction to Metal Matrix Composites* (Cambridge University Press, Cambridge, 1993)
9. F.L. Matthews, R.D. Rawlings, *Composites Materials: Engineering and Science* (CRC Press, Boca Raton, 1999)
10. D.B. Miracle, S.L. Donaldson (eds.), *Composites*, ASM Handbook vol 21, (ASM International, Materials Park, 2001)
11. D.M. Bigg, in *Electrical Properties of Metal-Filled Polymer Composites*, ed. by S.K. Bhattacharya, Metal-Filled Polymers, (Marcel Dekker, New York, 1986), pp. 165–226
12. F. Hussain, M. Hojjati, M. Okamoto, R.E. Gorga, Polymer-matrix nanocomposites, processing, manufacturing, and applications: an overview. J. Compos. Mater. **40**, 1512–1575 (2006)
13. A. Mortensen, J. Llorca, Metal matrix composites. Annu. Rev. Mater. Res. **40**, 243–270 (2010)

14. P. Samal, J. Newkirk (eds.), *Powder Metallurgy*, ASM Handbook vol 7 (ASM International, Materials Park, 2015)
15. D.R. Clarke, Interpenetrating phase composites. J. Am. Ceram. Soc. **75**, 739–759 (1992)
16. S.C. Tjong, Z.Y. Ma, Microstructural and mechanical characteristics of in situ metal matrix composites. Mater. Sci. Eng. **R29**, 49–113 (2000)
17. K. Xia, T.G. Langdon, The toughening and strengthening of ceramic materials through discontinuous reinforcement. J. Mater. Sci. **29**, 5219–5231 (1994)
18. M. Ruhle, A.G. Evans, High toughness ceramics and ceramic composites. Prog. Mater. Sci. **33**, 85–167 (1989)
19. K. Lu, *Nanoparticulate Materials: Synthesis, Characterization, and Processing* (Wiley, Hoboken, 2013)
20. S.Y. Fu, X.Q. Feng, B. Lauke, Y.W. Mai, Effects of particle size, particle/matrix interface adhesion and particle loading on mechanical properties of particulate-polymer composites. Compos. Part B **39**, 933–961 (2008)

Chapter 3
Analysis Techniques

The analysis of composite components follows traditional lines of thinking, realizing certain tests are favored because they are easier to perform or better relate to performance. This chapter reviews the approaches used to quantify properties, with an emphasis on standardized procedures relevant to performance goals. Comments are made on accuracy and cross-test comparisons.

Introduction

This chapter addresses the means to measure properties in particulate composites. These properties range from geometric aspects to behavior aspects. As a first principle, the testing applied to particulate composites is not different from that encountered elsewhere in engineering. However, some variants are easier to perform, so those tests are favored [1]. The selection of one measurement approach over another occurs because it is easier, less expensive, or more relevant to the application [2]. For example, metallic strength is traditionally measured in the uniaxial tensile test using machined samples. However, for particulate composites strength tests are commonly performed in bending using the transverse rupture strength. Due to differences in the stress condition, the transverse strength is about 60 % higher than the tensile strength. Thus, caution is required when comparing tensile strength to transverse rupture strength.

This chapter introduces the measurement approaches used to quantify properties. The first topic is dimensional aspects and means for quantification of component size variation. Attention is directed to distributions in measured properties. This is followed by introducing the tests for composition, density, porosity, hardness, and other attributes. Only the more common situations are included, in line with the Pareto Principle or 80–20 rule; 20 % of the test procedures account for 80 % of the testing.

© Springer International Publishing Switzerland 2016
R.M. German, *Particulate Composites*, DOI 10.1007/978-3-319-29917-4_3

Dimensional Testing

Dimensional measurements are the most common check employed in component production. An engineering design specifies the geometric parameters, including upper and lower size ranges for each dimension. Other common specifications might include surface roughness or feature location. Dimensional tests rely on simple fit fixtures, handheld calipers, micrometers, plug gauges, and might include measurements by laser imaging or touch probe coordinate measuring machines. Surface roughness is measured using surface profile meters. Spatial relations and angles are extracted using coordinate measuring machines, video systems, or optical comparative devices. These are the same devices applied to components fabricated by traditional manufacturing approaches, so the particulate composite provides no challenge in that regard.

Not all dimensions and not all components are necessarily measured. Often a go/no-go test gauge is sufficient to determine proper fit. For example on a hole, a plug gauge at the smaller end of the allowed size range is inserted to ensure the hole is at least larger than the stated minimum. Next a plug gauge at the larger end of the allowed size range is used to ensure the hole is smaller than this size. The approach is simple and easily automated.

Formal inspection protocols guide the frequency of testing, usually with a declining sampling rate as quality is established. In one situation involving an automotive turbocharger component, the first 3 months of production required 100 % inspection. As confidence in the quality increased, inspection slowly declined to one out of 25,000. However, should a nonconforming component be detected, the test frequency immediately increased.

Dimensional control is inherently an issue with particulate composites, especially if the fabrication process involves several steps. Production might involve mixing, molding, sintering, and finishing. Conceptually the combination of steps, each with inherent variability, leads to a stack-up in dimensional variation versus single step fabrication routes, such as machining.

Statistical analysis is employed to understand the scatter between components, production machines, and operators. Additional scatter arises with new raw material lots. Experiments during production help quantify each factor and its relative impact based on the average and the variability around the average. Statistical analysis can isolated the causes of variability, including factors such as the furnace location, room temperature changes, or humidity swings. A useful parameter is the coefficient of variation C_V, defined as the standard deviation σ divided by average or mean size U_A, ($C_V = \sigma/U_A$). It is dimensionless so it is often cited as a percentage.

Reduced dimensional variation often demands final machining or grinding steps to remove scatter on critical features. Efforts to isolate sources of dimensional variation eventually conclude everything is a possible factor. For example in study of a powder uniaxial pressing facility, an audit showed a capability to hold dimensions to ± 0.140 mm (one standard deviation) in the vertical direction and ± 0.025 mm (one standard deviation) in the perpendicular direction. Detailed analysis concluded the largest cause of variation traced to changes in the incoming powder lots. Another factor arose from frictional heating of the compaction tools. After a shutdown, the tooling slowly heated during repeated strokes of the press. The heating caused slight thermal expansion to the tooling, resulting in drifting component size over a production shift.

Tolerances specify the permissible dimensional deviation, usually given by the mean size and range. The dimensional control capability in production is captured by the dimensionless process capability, Cp. It is a measure of the ability of an operation to stay within a specified tolerance range. It measures the ratio of the process spread, based on three standard deviations, and compares this spread with the allowed tolerance range,

$$Cp = \left| \frac{U_M - U_A}{3\,\sigma} \right| \tag{3.1}$$

where U_M is the maximum or minimum control limit, U_A is the mean, and σ is the standard deviation. In simpler terms, if you are throwing darts at a target, a high Cp indicates the shots are closely clustered, but not necessarily on the bullseye. Most desirable are closely clustered dimensions centered on the target size.

If the process mean is centered between the upper and lower bounds of the tolerance range, then the process is allowed to have the largest variation. However, typically a process skews toward one end or the other of the allowed dimensional range, necessitating a narrower scatter range. The distance from the mean to the closest tolerance provides a smaller, tighter one-sided parameter designated Cpk. The value of Cpk cannot exceed Cp. In other words, for throwing darts, a high Cpk implies the closely clustered shots are on the bullseye. High values of the process control capability Cpk are desirable and values of 1.33 are impressive and expected in sophisticated manufacturing operations.

The concept of a tolerance budget is useful in realizing the tradeoff between precision and cost. For a production process there is inherent variation, reflecting factors such as tool wear, frictional heating, process temperature variations, and humidity fluctuations. Some particle consolidation approaches inherently have large variations. For example, hot isostatic pressing typical holds wall thickness to $\pm 2\,\%$. With attention the variation reduces to $\pm 1\,\%$. When focus is given to a single dimension, variability reduces to $\pm 0.2\,\%$, a tenfold reduction. Since this requires much attention, often the decision is to oversize the component and rely on machining to reach final dimensions; this adds $40\,\%$ to the manufacturing cost.

Reduced dimensional variation requires attention to the root cause factors. In turn, cost increases with those efforts, offset by less material loss or lower machining costs. Consider the brute force approach where components are produced, $100\,\%$ inspected, and those with acceptable dimensions are shipped. Those outside the allowed tolerance range are scrapped. The narrower the tolerance band, the higher the scrap rate, so those shipped carry a high cost burden. The tolerance budget concept teaches tighter tolerances cause cost to increase, usually in a nonlinear manner. Accordingly, it is appropriate to relax some tolerances to compensate for any closely specified dimension. The tolerance budget B is the summation of all specified tolerances, captured by the allowed coefficient of variation C_V for each dimension,

$$B = \sum \log\left(\frac{1}{C_V}\right) \tag{3.2}$$

To hold to the same budget, any move toward a tighter tolerance is offset by a corresponding relaxation in other tolerances. In a related philosophy, cost increases with complexity. Thus, an increase in the number of features or the precision of those features rapidly increases cost.

Property Distributions

Dimensional scatter follows a normal distribution, described by two parameters—the mean or average U_A and the standard deviation σ. The normal distribution is the bell curve associated with grade distributions in school. It is attributed to the astronomer Gauss who quantified telescope errors in locating stars.

In any measurement there is an inherent variation characterized by scatter about the mean value. The scatter is described by the standard deviation. For a normal distribution, the mean, mode (most common), and median (half are smaller and half are larger) are equal. The normal distribution is a good model for variations in component mass and dimensions. It represents the impact of many random and independent fluctuations. For example in measuring the size of a component using 10 operators, random but subtle variations in alignment, instrument reading, and ambient temperature add to measurement scatter. The normal distribution is given by probability density function $P(X)$ corresponding to the occurrence of each measure X according to the following distribution:

$$P(X) = \frac{1}{\sigma\sqrt{2\pi}} \, exp\left[-\frac{(X - U_A)^2}{2\sigma^2}\right] \tag{3.3}$$

As mentioned U_A is the mean or expected value, $\sigma > 0$ is the standard deviation, and X is the measured value. This gives the bell curve when the probability is plotted versus the measured size, symmetric about the mean. Alternatively, the cumulative distribution $F(X)$ gives the probability of encountering a size X, or smaller:

$$F(X) = \frac{1}{2}\left[1 + erf\left(\frac{X - U_A}{\sqrt{2}\,\sigma}\right)\right] \tag{3.4}$$

where *erf* is the error function. The cumulative fraction F ranges from 0 to 1 as the size parameter X varies from several standard deviations below U_A to several standard deviations above U_A, recognizing the ratio of σ/U_A is the coefficient of variation. This function proves useful in determining the fraction smaller than a given size.

While the normal distribution is good for fitting component size data, the Weibull distribution is used to represent property variations [3]. A good example is in measuring strength. A normal distribution assumes any variation in repeated tests is simply due to random operator, sample, or machine errors. However, in the case of strength a larger variation arises from material defects, leading to a Weibull distribution [4]. This distribution assumes a characteristic strength for the material with variations between samples arising due to a population of flaws. That distribution depends on S_O, termed the characteristic strength and the Weibull modulus M. In this model, the cumulative probability of failure F at an applied stress S is given by,

$$F(S) = 1 - exp\left[-\frac{V}{V_O}\left(\frac{S - S_U}{S_0}\right)^M \right] \tag{3.5}$$

where V is the sample volume and S_U is the lower limit stress, known as the proof stress (it must be exceeded to cause failure, but if the product is not subjected to proof testing it is set to zero), and V_O is the sample volume used for measuring the Weibull statistics. The exponent M is the Weibull modulus, an indirect measure of the inherent flaw population.

If N tests are performed to measure strength, then the fracture strength results are ranked in ascending order to link probability of failure to the stress. The samples containing larger flaws fail earlier to give a lower strength. Accordingly, to analyze the data Equation (3.5) is rearranged as a plot of the double logarithm of $1/(1-F)$ versus the logarithm of S. The slope is used to calculate M, assuming S_U is zero. At least 40 samples are recommended for determination of the distribution. A narrow range of failure strengths is desirable, giving a high Weibull modulus. This implies consistent sized flaws. The material appears stronger as the test volume V decreases, since the probability of a large flaw decreases. This says that small objects are always stronger when compared to large objects.

From a quality of fit standpoint, the Weibull distribution gives the best representation for variables related to fracture. This is evident using the WC-2TaC-6Co fracture strength data in Fig. 3.1 [5]. The cumulative probability of fracture is taken from several test samples, showing the probability of fracture versus stress level. In this plot the solid line represents the Weibull distribution and the dashed line is the normal distribution. For fracture strength, the Weibull distribution is the best fit to the data.

In contrast, for mass or size variations, the normal distribution is the better fit to data. An example is plotted in Fig. 3.2 for the cumulative mass distribution of an automotive component (from about 713 to 719 g) during 1 day of production. The

Fig. 3.1 Plot of the fracture strength distribution in terms of the cumulative failure probability up to 100 %, versus the fracture strength for a hard particulate composite consisting of WC-2TaC-6Co [3]. The individual measurements are given by the *square* symbols and both a Gaussian (normal) and Weibull distribution fit are plotted, with the best fit from the Weibull distribution

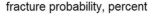

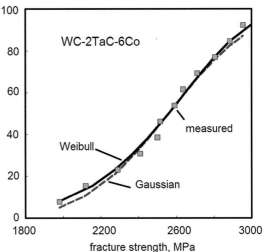

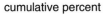

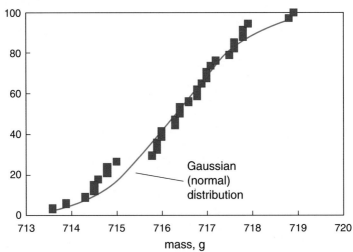

Fig. 3.2 Cumulative distribution in component mass for an automotive engine bearing fabricated from mixed particulates during 1 day of production. The Gaussian, or normal, distribution is shown versus the measurements, with a median mass of 716.3 g

line is the cumulative distribution fit to the individual points showing the percentage with a mass below each value (0 % are below 713 g while 100 % are below 719 g). Normal distributions apply when variation arises from random, unrelated effects. The effect of flaws typically arises in strength, fatigue life, impact toughness, fracture toughness, and ductility. For those properties, a high Weibull modulus indicates few defects. Audits of manufacturing operations rely on the Weibull modulus to assess process control.

Tests for Composition and Phases

Composite chemical analysis falls into two bins, corresponding to different magnification scales. Bulk chemistry ignores the individual phases and provides informative on the overall composition. Microchemical analysis allows determination of the individual phase composition. Several approaches are available for both [6].

An important parameter in chemical analysis is the "scale of scrutiny." A small scale of scrutiny, smaller than the grain size, gives the chemistry of a single phase. For example, in a mixture of aluminum and silicon carbide, microchemistry will record either 100 % aluminum or 100 % silicon carbide at different points in the composite. Many repeated and random tests, recording the frequency of each phase, are a means to assess the bulk chemistry. This is an inefficient means to determine if the ratio of ingredients is in specification.

In contrast, a bulk approach to measuring chemistry might use X-ray diffraction (XRD). A test sample is exposed to a monochromatic X-ray beam. The resulting diffraction pattern is used to assess the phases and the relative quantities [7]. For each material, a characteristic pattern of diffracted X-rays provides information on the crystal structure useful in identification of the phases. X-ray diffraction does not give grain size or homogeneity information, but the relative intensity of each phase is proportional to the composition. Computerized analysis identifies which phases are present and their relative proportion, at least near the component surface. Generally, X-ray diffraction analysis is best in situations where there is at least 5 vol% of each phase.

Most bulk chemistry measurements destroy a small piece of the test sample. For example combustion is the common means for measuring carbon, oxygen, nitrogen, and sulfur content. Atomic absorption (AA) requires the dissolution of the test sample into solution for injection into a combustion flame to analyze the optical emission spectra. X-ray fluorescence (XRF) provides chemistry based on X-ray emissions from the sample heated to a temperature that that induces fluorescence. The input energy generates X-ray emissions that are characteristic of the quantum shells for an atom. The quantum energy release is unique to each atomic species, allowing identification of the atoms and their relative abundance. A popular variant relies on an inductively coupled plasma arc to promote fluorescence. Fortunately these chemical analysis techniques rely on very small samples.

Microchemical techniques measure composition in small spots using lasers or electrons to stimulate the selected region. Sample damage is negligible so these are considered nondestructive. By coupled microscopy and analysis, it is possible to pinpoint each phase for analysis, and moving the stimulation beam around provides lateral composition information [8]. Auger electron spectroscopy relies on an electron beam to stimulate near-surface emissions for analysis. Scanning Auger microscopy coupled with slow mass removal by sputtering provides depth profiling, generating x-y-z composition information. It is tedious, expensive, but often necessary to solve lingering problems. Because the Auger electron energy is low, the depth from which emission occurs is small, just a few atoms deep or under 5 nm. Spatial resolution is good to 1 μm and the detection level is good to less than 1 at.%.

Auger electron spectroscopy is especially powerful in the analysis of fracture surfaces to determine interface chemistry. For example, detection of segregation identifies if processing changes are required to remove any segregation. Auger analysis penetrates a few atoms into the surface, so chemical identification of impurities responsible for brittle fracture is possible. Sputtering is a means to remove surface atoms selectively by bombarding the surface with ionized gas, such as argon. Auger spectroscopy is repeated after incremental sputtering to build a layer-wise chemical picture to identify segregation.

Scanning electron microscopy (SEM) is a popular tool for examining phases and is widely employed in this book. The SEM relies on a focused electron beam to sweep over the sample surface. That beam generates backscattered electrons (ricochets), secondary electrons (electrons dislodged from atoms in the sample), and X-rays (associated with quantum state jumps in the sample atoms). The SEM relies on high voltage (up to 25,000 V is common) to accelerate electrons that are focused into a narrow beam. Magnetic lenses do the focusing and move the beam over the sample surface. Detectors for the electrons and X-rays are coordinated with the beam motion to generate visual outputs of intensity versus beam location on the sample. The result are exceptional images with depth of field and magnification far superior to optical microscopy. Magnification is adjustable up to 10,000 and even 100,000-fold. The higher magnifications are more demanding in terms of instrument design and stability. As evident in Fig. 3.3, the images are three-dimensional.

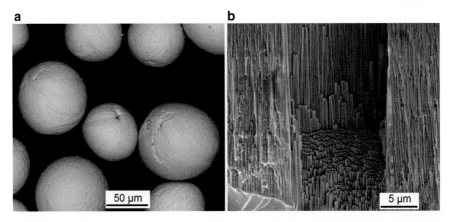

Fig. 3.3 Scanning electron microscopy provides considerable insight into structure, as evident in these two pictures; (**a**) is a nearly spherical powder with subtle differences in phases and defects in the particles and (**b**) is the fracture surface for a nanoscale polymer composite with anisotropic properties

Shown in these two pictures are particles and the fracture surface of a consolidated nanoscale composite.

The energy emissions stimulated by the sweeping electron beam are useful for chemical analysis. Tiny samples areas are possible, since the spot size ranges down to 1 μm^3. The emitted X-ray spectrum has chemical information based on energy dispersive spectroscopy (EDS) where the X-ray energy identifies which elements are present. Electron diffraction in the transmission electron microscope (TEM) is a means to extract crystal structure. Micro-chemical spatial resolution spreads to a diameters of 1 μm even though the electron beam is smaller. This is due to electron energy cascades spreading inside the sample. Generally micro-chemical analysis using electron stimulated X-rays is less accurate for low atomic number species, such as Li, Be, B, C, and such. Typically 1 wt% concentrations are needed for detection. Variants in the SEM allow for topographic contrast from the surface, compositional contrast (higher atomic number atoms reflect or backscatter more electrons), voltage and electron channeling contrast. Usually the SEM relies on energy as the discriminator in energy dispersive spectroscopy, but a related approach is electron probe microanalysis based on the X-ray wavelength. The wavelength route is more expensive but more sensitive to low concentrations.

Secondary ion mass spectroscopy relies on an ion gun shooting oxygen, cesium, gallium, or argon ions at the sample surface. The energetic ions impact with sufficient velocity to dislodge atoms from the sample surface. Those emitted atoms tell the local chemistry and are analyzed using mass spectroscopy. It is applicable to all elements, reaching to the range of one part per million. Stimulated volumes are from 100 to 1000 μm^3.

Density and Porosity Tests

Density is mass divided by volume, expressed in g/cm^3 or kg/m^3 or sometimes given as a ratio to the theoretical density. Specific gravity is the ratio to water, where the density of water is near 1.0 g/cm^3 at room temperature (at 20 °C pure degassed water has a density of 0.997 g/cm^3).

Composites consisting of mixed phases and the theoretical density for the mixture is calculated using the inverse rule of mixtures shown below. This relation assumes no solubility or reaction between phases. However, sometimes that the density differs slightly from the theoretical value, possibly due to contamination. Accurate determination of density is a means to sense fabrication or composition problems. Porosity is calculated using measured density and its departure from theoretical density. Pores degrade most properties, especially if the pores are large compared to the grain size. As examples, the following guides provide property comparisons to show how 15 % porosity reduces properties;

- thermal conductivity loss is 25 %,
- elastic modulus loss is 40 %,

- strength loss is 50 %,
- ductility loss is 65 %,
- fracture toughness loss is 70 %.

The theoretical density for a particulate composite corresponds to zero porosity. Pore-free density is calculated using the inverse rule of mixtures. For a two-phase composite, with phases designated A and B, the corresponding mass fractions are W_A and W_B with theoretical densities of ρ_A and ρ_B. The pore-free composite density ρ_C is given as,

$$\rho_C = \frac{1}{\frac{W_A}{\rho_A} + \frac{W_B}{\rho_B}} \qquad (3.6)$$

Density is relatively easy to measure on simple geometries, such as cylinders or rectangles, based on mass and dimensions. Dividing the measured density by the theoretical density gives the fractional density, f. Any departure from the theoretical density indicates porosity, defects, impurities, or chemical reactions.

Porosity comes in two forms. Closed pores sit in isolation and are not in communication with the exterior atmosphere. Open pores are connected to the surface and allow for gas exchange between the sample interior and exterior. Closed pores correspond to low porosity levels, typically below 5–10 % porosity.

Porosity is reported as a percentage or fraction of the component volume. Fractional porosity ε is calculated from the actual density ρ and theoretical composite density ρ_C,

$$\varepsilon = 1 - \frac{\rho}{\rho_C} = 1 - f \qquad (3.7)$$

where f is the fractional density. Although several approaches exist for measuring density, by far the most common is based on the ancient water immersion approach.

Archimedes Technique

For materials over 1 g/cm^3 the water immersion approach is most useful. This is also known as the Archimedes technique. It involves a sequence of mass measurements as schematically outlined in Fig. 3.4. First, the component is weighed dry (M_1). Steps are required to avoid water intrusion into surface pores that might alter the weight. Impregnating the sample with oil is a typical means to avoid this problem. The sample is placed in oil and evacuated to remove trapped air bubbles. Upon return to atmospheric pressure the oil flows to fill the pores. Excess oil is removed, but not the oil in the pores. The sample weight is again measured with the impregnated oil (M_2). If there are no surface pores, then oil impregnation is skipped since these two measurements are the same.

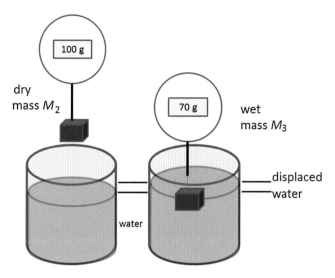

Fig. 3.4 This diagram shows the sequence of mass readings involved in the Archimedes approach to measuring density. The sample may first need to have pores sealed via oil impregnation, and then is weighed. Subsequently the mass is measured while immersed in water. The displaced water corresponds to the sample volume with a buoyancy force that reduces the immersed mass. The Archimedes density is calculated from the initial mass and displaced volume, the latter is calculated from the immersion mass change

Next the component is immersed in water to displace its volume and the immersed mass is measured (M_3). The displaced water volume contributes to a weight reduction. A wire is used to suspend the component in the water and its mass M_W is independently measured in water at the same level of immersion. The composite density ρ_C in g/cm^3 is calculated from the mass determinations as follows:

$$\rho_C = \frac{M_1 \rho_W}{M_2 - (M_3 - M_W)} \tag{3.8}$$

where ρ_W is the water density (in g/cm^3). Although we think of water at 1 g/cm^3, accurate density determinations include the slight change in water density with temperature as,

$$\rho_W = 1.0017 - 0.0002315\,T \tag{3.9}$$

with T being the water temperature in °C. When performed under careful conditions, the density is measured to six significant digits. The measurement error is about $\pm 0.001\,\%$. In routine testing situations the measured density is more typically only accurate to about $\pm 0.1\,\%$. As one option, water can be used as the impregnation fluid instead of oil, but care is needed to avoid evaporation out of the pores.

Pycnometry Technique

Pycnometry relies on gas to measure sample volume. The usual gas is helium because it is an ideal gas. Consequently, the ideal gas law is applied to calculate sample volume from pressure measurements, assuming constant temperature. For the pycnometer density, a test sample is placed in a chamber of known volume. The chamber is evacuated to remove all gas. A second calibration chamber of known volume is used to exchange gas via an interconnecting valve. Based on calibrated pressure-volume relations, the pressure after equilibration is used to determine the volume of the test sample. With an independent measure of sample mass, the density results. A less accurate approach relies on granular material instead of gas, but with a similar idea.

Gas pycnometry is applied to loose powder since gas easily infiltrates into the pores between the particles to give the true volume. It is less accurate for bulk samples. As illustrated in Fig. 3.5, the test relies on a chamber of known volume V_S containing the sample of unknown volume V. Initially, the test chamber is at a pressure P_1 and the calibration chamber is initially evacuated. The calibration chamber with volume V_C is connected to the sample chamber. After opening the connecting valve the pressure equilibrates in both chambers at P_2. Applying the ideal gas law gives,

$$P_1 (V_S - V) = P_2 (V_S - V + V_C) \tag{3.10}$$

Accordingly, the powder or component volume V is extracted,

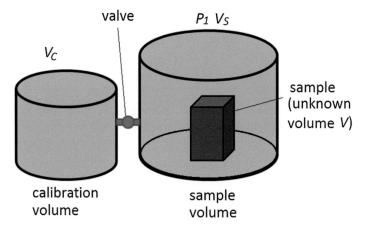

Fig. 3.5 Gas pycnometry is a means to measure sample volume using calibration and sample chambers, where pressure and volume are related, allowing extraction of the sample volume from pressure readings. A combination of sample mass and sample volume give the density. Usually helium is used for this test

$$V = V_S + \frac{V_C}{(P_1/P_2) - 1} \tag{3.11}$$

The independently measured component mass divided by this volume gives the pycnometer density. Pressure measurements are precise to about 0.05 %, so the technique is less accurate when compared to the Archimedes approach. Generally, pycnometry has a variability of $\pm 0.2\,\%$.

Hardness

Other than density, hardness is the most common property test. It is an effective means to quickly characterize a material with minimum damage, so it is a widely employed quality control tool. Hardness reflects the resistance to surface penetration by a loaded indenter [9]. Many variants exist, but generally the larger the indentation the softer the material; hard materials resist penetration and form small impressions. The tests differ in indenter design, applied load, and exact measurement—depth, width, or length of the indent. Because indentation deforms and works the material, hardness depends on a mixture of attributes that include crystallography, atomic bonding, yield strength, and elastic–plastic flow properties. Unlike destructive tests for strength or ductility, hardness indents are small and hardly noticeable.

Hardness scales include Rockwell (HRA, HRB, HRC, ...), Knoop (KHN), Vickers (HV or VHN), and Brinell (BHN or HB). Other scales exist, but most composites are covered by these scales. Figure 3.6 contrasts these tests. Conversion tables allow approximate comparisons between the scales, but the conversions are usually based on steels so they are less accurate for other materials.

Brinell Hardness

In the Brinell test, a spherical 10 mm diameter ball is pressed into the flat surface of a test sample. After a hold time of 10 to 30 s (longer for softer materials), the load is relaxed and the indentation diameter is measured. A hard material produces a small indentation. Accordingly, the apparent resistance to indentation is used to calculate the Brinell hardness number *BHN* (also designated *HB*) as follows:

$$BHN = \frac{2\,P}{\pi\,D^2 \left[1 - \sqrt{1 - (d/D)^2} \right]} \tag{3.12}$$

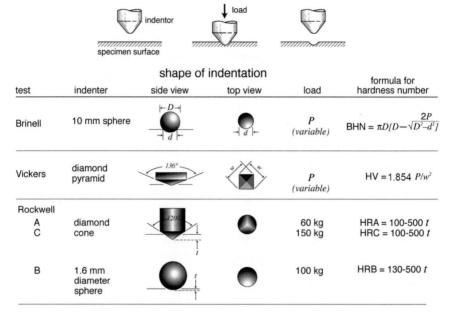

Fig. 3.6 An overview of common hardness tests showing the indenter, measured parameter, typical load, and calculation procedure

where P is the test load, D is the ball diameter (10 mm), and d is the impression diameter. In the typical test the test force F is 3000 kgf (29,000 N). The impression diameter d is measured in mm. The units are force over area, similar to strength. Compared to other scales, a value of 245 BHN is about the same as 100 HRB (Rockwell B) or 24 HRC (Rockwell C) or 250 HV or VHN (Vickers Hardness Number).

Vickers Hardness

The Vickers test is used to measure microhardness. The indent is small and requires a microscope to measure it's size. The small indenter allows testing of individual phases in the composite or evaluation of hardness versus position using small displacements between indents. This idea is extended to nanoscale indentation tests, using lower loads to reach impressions of just 1 μm.

In the Vickers test, hardness depends on the size of an indent from a 136° diamond pyramid pressed under a preset load P. After the load is removed, a microscope is used to measure the size of the impression. The average diagonal of the impression w is leads to the Vickers hardness number HV in kgf/mm^2 as follows:

$$HV = 1.854 \frac{P}{w^2} \tag{3.13}$$

The load P is in kg and w is the diagonal of the indent in mm. The reading is expressed in either of two units. One is HV, corresponding to units of kgf/mm^2, equivalent to 10^6 kgf/m^2. The second is determined by multiplying HV by the acceleration of gravity (9.8 m/s^2) to obtain stress, expressed as MPa or GPa (roughly 1000 HV = 1 GPa). The latter units are convenient for comparison with strength data.

Softer materials allow larger indentations, corresponding to a lower hardness. The applied load influences the hardness reading, especially for lighter loads, so sometimes the applied load is reported with the hardness (1, 2, 5, ... 100 kgf). For example 290 HV5 is one way to designate the hardness and load, with the 5 indicating the load in kg.

Knoop is another microhardness test. It is typically applied to brittle, fragile, or thin samples. Like other tests, hardness on the Knoop scale relies on the load divided by the impression area.

Rockwell Hardness

The Rockwell scales employ larger indenters. There are several scales, but the three most popular with respect to particulate composites are the A, B, and C scales. Note there about 30 related tests denoted as D, E, F, G, 15 T, 45 N, and so on. The measured hardness is reported as HRA, HRB, or HRC, to indicate which test is used—for example 45 HRC gives the value and the scale. The three common variants use either a 120° diamond cone with a 0.2 mm radius spherical tip (A, C, or D) or a ball with a 1.58 mm diameter (B, F, or G). Table 3.1 lists some of the specifics on the popular A, B, and C scales.

To make a measurement, the indenter is pressed into the sample surface in two steps. An initial force is applied to seat the indenter (say 98.1 kgf), followed by a higher force to create a deeper impression (say 981 kgf). The initial seating force is maintained for up to 3 s and the first indenter depth is recorded. The increase in force to the final, higher level occurs next and that higher force is maintained for 2–6 s. There is elastic relaxation when the additional force is removed. However, the initial seating force is still applied, so a second depth reading is made after a

Table 3.1 Rockwell A, B, and C hardness tests

Scale	N	Indenter	Initial force, N	Test force, N
HRA	100	Diamond cone	98	588
HRB	130	1.6 mm ball	98	981
HRC	100	Diamond cone	98	1471

short stabilization period. The Rockwell hardness number is calculated from the
change in depth as:

$$HR = N - \frac{t}{S} \tag{3.14}$$

Here t is the permanent increase in penetration depth in mm from the initial to final
loading, S is set to 0.002 mm, and N is 100 for the A and C scales, and 130 for the B
scale. Other Rockwell scales exist for polymers and other soft materials. All of the
details are included in automatic testing devices.

Tests for Mechanical Properties

Mechanical properties are widely reported for engineering materials [10]. This is
especially true for particulate composites. Elastic properties are measured nonde-
structively. Strength, fatigue, impact, and toughness tests are destructive. An array
of material attributes arises from these tests as used in evaluating structural materials. The main properties are generated via uniaxial loading in tensile and bending
tests, supplemented by fatigue and fracture tests. It is possible to adapt the test for
elevated temperatures or corrosive environments, or other environmental conditions to ensure the test is relevant to the application.

Nondestructive Tests

Nondestructive tests leave the material unchanged and the most common focus is
on elastic properties. The speed of sound in a solid u varies with density ρ and
elastic modulus E nominally as,

$$u \approx \sqrt{\frac{E}{\rho}} \tag{3.15}$$

Poisson's ratio ν enters into the calculations, where,

$$E = u^2 \rho \, \frac{(1 + \nu)(1 - 2\nu)}{1 - \nu} \tag{3.16}$$

There are two ways to use these relations. One way is to measure the density and
longitudinal sound velocity to extract the elastic modulus, assuming Poisson's ratio.
The other way is to measure the sound velocity (for a known material and elastic
properties) to assess the density for comparison with the theoretical density. The

calibration velocity u_O is known for full density material, allowing extraction of the fractional density f via the measured velocity u as follows:

$$u = u_O \sqrt{f} \tag{3.17}$$

One approach relies on a single transducer for both sending and receiving pulses. If the sample thickness is known, then the time from sending a pulse to receiving the return echo gives the velocity as thickness divided by half that time (the pulse travels twice the thickness). An alternative is to use through transmission with different sending and receiving transducers. At the same time, it is useful to measure the shear waves to extract Poisson's ratio. For particulate composites the usual assumption is a homogeneous body, so orientation effects are ignored.

Similar ideas provide a means to detect defects. Ultrasonic sound waves with frequencies up to 100 MHz are introduced in the material using a ceramic piezo-electric transducer. Transmission waves pass through the component, as illustrated in Fig. 3.7. The transmitted signal has embedded information on the component quality, including cracks, voids, and inclusions. Inclusions larger than 20 μm and voids larger than 50 μm are detectable if the constituent phases are smaller in size. The technique is best for thin sections; as the section thickness increases, the ability to detect cracks decreases. Ultrasonic inspection is a qualification step in accepting high-value, life-critical components for aircraft, oilfield, and biomedical applications.

Strength Tests

Tensile testing is the main approach to assessing mechanical properties [11]. In the tensile test the material sample is pulled to failure. Yielding is associated with the onset of plastic deformation, that is the onset of permanent strain. The stress to give 0.2 % plastic strain is the common basis for determining the yield strength

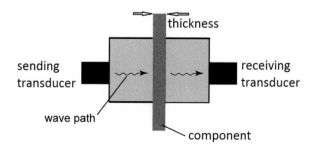

Fig. 3.7 Ultrasonic waves provide a means to measure for defects and determine elastic modulus. A through transmission approach is illustrated here, where the time delay between sending and receiving transducers allows extraction of elastic modulus. Dispersion in the signal provides information on defects, voids, or cracks

(implying a small permanent deformation). Engineering stress is given as load divided by initial cross-sectional area. The peak stress encountered up to failure is the ultimate tensile strength. The ultimate strength is not a safe design parameter so general practice is to discount the yield or ultimate strength by a safety factor, potentially allowing loading to say only 50 % the material capability.

All materials give higher strength for smaller samples—thus the frequently quoted "stronger than steel" statement. In reality even steel is stronger as the sample size decreases. The size effect on strength arises because small samples have limited defects; the smaller the sample size the lower the chance of including strength limiting defects. The Weibull relation in Eq. (3.5) includes this test volume effect.

For brittle solids the yield and ultimate strengths are the same, but for ductile solids significant deformation occurs prior to failure, measured by the elongation to failure. High ductility materials, such as pure metals and many polymers, exhibit 20 % and even 80 % permanent stretch prior to failure. Usually particulate composites exhibit less ductility while ceramics fracture prior to permanent deformation.

Compression tests for strength assume the composition is brittle. The peak load is measured up to failure. The strength is calculated from that failure load and the cross-sectional area. The two most common tests are illustrated in Fig. 3.8, although several variants are in use. Both tests rely on axial-symmetric geometries. One approach applies force perpendicular to the top and bottom faces of a simple right-circular cylinder. It is necessary to use squat cylinders (short in height compared to the diameter) to avoid buckling. An alternative is to machine a catenary compressive shape to concentrate the stress at the center. Likewise, sharp edges are beveled to avoid stress concentration, so the catenary geometry gives a higher strength when compared to the right-circular cylinder. In both cases the test geometry is crushed

Fig. 3.8 Two geometries used for compression strength testing. The simple cylinder tends to deliver a slightly lower strength versus the catenary cylinder. However, the catenary geometry requires machining and is used less frequently

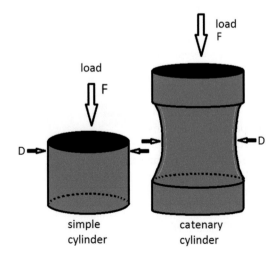

simple cylinder

catenary cylinder

Fig. 3.9 Strength in a
brittle sample is measured
by applying a fracture load
to the diameter. This is also
known as the Brazilian
strength test. A variant
called the theta test relies on
a disk with two "D" shaped
holes to concentrate the
stress in the center; the
sample looks like the Greek
θ

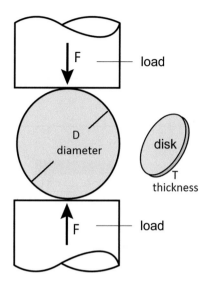

by orienting the flat faces perpendicular to loading platens. Crush or compressive
strength σ is calculated from the fracture load F and central sample diameter D,

$$\sigma = \frac{4F}{\pi D^2} \tag{3.18}$$

Due to the geometry role on crushing, data should report the test type. For brittle
composites, the compressive strength is up to tenfold higher compared to the
strength measured in tension.

An alternative compressive test relies on a cylindrical sample, with force applied
on the curved edge. Often this is a thin sample is sketched in Fig. 3.9 in what is
termed the Brazilian test. Compression is applied using parallel platens. Loading
increases up to the point of fracture, which tends to occur by a crack propagating
along a 45° diagonal across the disk. The fracture load F, disk diameter D, and disk
thickness T are used to calculate strength σ,

$$\sigma = \frac{2F}{\pi D T} \tag{3.19}$$

For particulate composites, typical samples are 5–7 mm thick and the diameter is
two to four times the thickness, often in the 20–25 mm range. In concrete testing the
same idea is applied, but the sample is much longer, up to 250 mm.

Bending tests are widely used to measure strength of brittle materials that exhibit
trivial plastic deflection prior to fracture. Test variants include both rectangular and
disk shapes. Most common is three-point bending, as diagrammed in Fig. 3.10. It
relies on a rectangular bar put into a bending moment to induce fracture on the
lower face where maximum tension exists. Commonly called the transverse rupture

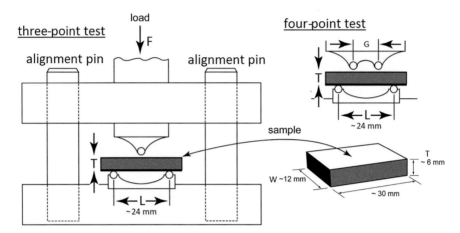

Fig. 3.10 The most popular strength test for low ductility materials is the transverse rupture test, also called the modulus of rupture. Two variants are illustrated here where a rectangular sample is loaded either with three or four line contacts. The four-point test delivers higher fracture strength

strength (TRS) it is also known as the modulus of rupture (MOR). Fracture strength σ is calculated from the maximum load F, lower support span length L between the two lower supports (not the sample length), sample width W, and sample thickness T as follows:

$$\sigma = \frac{3\,F\,L}{2\,W\,T^2} \tag{3.20}$$

Variation occurs in the absolute dimensions, but the size is scaled as 1-2-4; for example, $T = 6$ mm, $W = 12$ mm, and $L = 24$ mm. Smaller samples give higher strengths, so for this reason the 3-6-12 mm combination is also in use. Also, corner chamfers help boost the strength by eliminating stress concentrations, but usually this amounts to just 2 % higher strength. Should the midpoint deflection before failure exceed 4 % of the initial thickness, then the calculated strength is invalid (for 6 mm thickness this limit is a deflection before fracture of 0.24 mm).

The TRS is significantly lower when compared to compressive strength. Over a broad array of composites, the ratio of transverse strength to compressive strength is about 25 %. When comparing tensile strength σ_{TEN} and transverse rupture strength σ_{TRS}, the latter tends to be 60–80 % higher; the exact ratio depends on the Weibull modulus M from Eq. (3.5) as follows:

$$\sigma_{TRS} = \sigma_{TEN} \left[2\,(M+1)^2 \right]^{\frac{1}{M}} \tag{3.21}$$

For low ductility materials, the transverse rupture strength is typically 1.6 times the tensile strength, corresponding to a Weibull modulus near 13. In practical terms, this matches a typical scatter in measured strength near ± 10 % of the mean value.

For example, if the mean strength is 250 MPa, then the expected data range would be 225–275 MPa. Since the TRS is sensitive to defects, higher readings are possible by polishing and beveling the sample, enabling test strength gains by up to 1000 MPa. Thus, some particulate composites reach up to 3500 MPa using small samples, beveled edges, and polished faces, but are only 2500 MPa without these extra treatments.

A related transverse rupture test relies on two upper loading rods to spread the loading over a larger volume. This four-point loading is also sketched in Fig. 3.10. The four-point transverse rupture strength uses a modified variant of Equation (3.20) with the span decreased from L to $L - G$, where G is the gap between the two upper loading points $(L > G)$,

$$\sigma = \frac{3 F (L - G)}{2 W T^2} \tag{3.22}$$

Nominally the measured strength by the four-point test is about 14 % higher than that measured using the three-point test.

In some studies, bending tests are applied to disk shaped samples. A sample is supported on its outer rim and pressed at the center to induce fracture. As illustrated in Fig. 3.11, a variant relying on a ball to load the sample center. Other forms rely on flat faced punches to load the center.

Tensile testing is the foremost means for measuring elasticity, strength, and ductility. Samples for tensile testing are either rectangular or cylindrical in cross-section. Two sample geometries are illustrated in Fig. 3.12. Flat bars with rectangular cross-sections are easy to fabricate, but tend to give lower properties since the corners concentrate stress; machined round bars avoid stress concentrations. Typically round bars are employed for testing high ductility materials. The highest strengths come from polished round samples. A common geometry is 12.5 mm in diameter and 100 mm in length (with 50 mm of constant cross-section in the center). For the flat sample, the overall length is near 100 mm, with a central cross-section nominally 6 by 6 mm. Larger or smaller dimensions are in use, with the length scaled to the thickness or diameter: generally the cross-section should be 4× the thickness.

Fig. 3.11 A ball on disk strength test that is similar to the transverse rupture test. One variant uses a punch impressed into the sample, generally with a small thickness

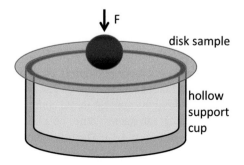

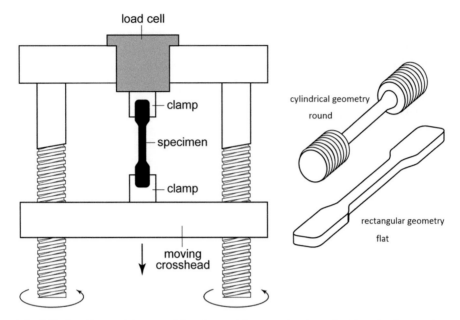

Fig. 3.12 Ductile materials are traditionally tested for strength using uniaxial tension in a moving screw-type machine. Two test geometries are illustrated here. The machined round cylindrical sample provides a higher strength. The sharp edges on the flat geometry induce stress concentrations that initiate fracture at a lower ductility and stress

Several parameters derive from a tensile test, including elastic modulus, yield strength (onset of permanent deformation), ultimate tensile strength, elongation to fracture, and reduction in area. Engineers recognize a variability in properties comes from adjustments to the sample size, geometry, surface finish, and rate of loading. If high properties are sought, such as for advertising literature, then small, circular samples with polished surfaces, tested at slow strain rates are optimal [12]. However, advertising values are not good for design, so it is best to perform testing using geometries, loading rates, sample size, temperature, and surface conditions reflective of the application conditions.

Fatigue Tests

The fatigue endurance strength is a measure of the cyclic stress that induces failure at ten million repeated loading cycles. Fatigue fracture occurs at a stress level below the yield strength captured in the compression or tension tests.

A common fatigue test relies on a bending moment applied to a rotating sample, as illustrated in Fig. 3.13. The test sample is a round bar that is placed in a bending moment such that the outer surface experiences compression and tension on each

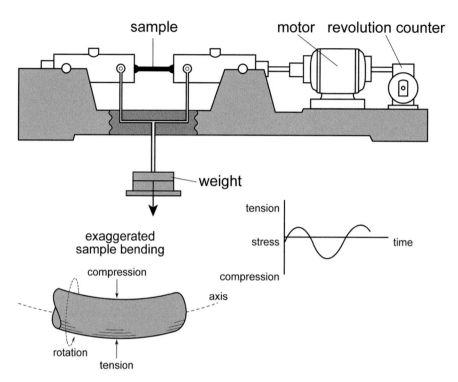

Fig. 3.13 A means to measure fatigue life is with a rotating bending beam test. The round sample undergoes compression and tension on each rotation. The bending load is varied to adjust the peak stress and the motor drives rotation until failure. Repeated tests determine the S-N fatigue curve based on the number of cycles to failure versus the applied stress

rotation. The motor causes rotations up to 10,000 RPM, while a counter measures the number of cycles to failure. Adjustments are made to control the weight or peak stress, and the corresponding number of cycles to failure is measured. Repeated samples subjected to an up and down stress search protocol eventually isolates the stress associated with 50 % failure at ten million cycles. The results are plotted as the S-N curve (S = stress, N = number), where the fractures are recorded in terms of the number of cycles N at the applied stress S. The survival stress at ten million cycles is the generally accepted fatigue strength or fatigue endurance strength. It is normally 30–70 % of the fracture strength; a first guess for the fatigue endurance strength is 50 % of the ultimate tensile strength. However, composites are sensitive to interactions between the two phases and the fatigue strength tends to be lower than such a first guess value.

Tensile stresses are most severe in causing fatigue failure, so compressive loading gives the longest life. The higher the applied tensile stress, the shorter the fatigue life. For example, WC-6Co with a compressive strength of 6.5 GPa fails at 6.0 GPa when cycled 1000 times and 4.4 GPa when cycled ten million times. Like other mechanical property tests, several factors influence the fatigue endurance strength, including sample size and surface finish. It is common practice to collect data using highly polished samples, but this is not reflective of many service situations. As with most testing, it is best to test under conditions relevant to the application.

Impact Tests

Impact toughness is a measure of the energy required to fracture a test sample. Like other mechanical property tests, several variants exist. Use of a Charpy bar with dimensions of 10 by 10 by 55 mm is most common, but other tests are adopted by various industries [13]. Using a swinging pendulum, the energy consumed to fracture the bar is measured in terms of energy lost over normal frictional losses in the test machine. The energy is in J and the sample is 1 cm by 1 cm (10 mm by 10 mm) in cross-section, giving units of J/cm^2. This is sometimes termed impact strength or impact toughness, where formally units are J/cm^2, reflecting the fracture energy divided by the unnotched cross-sectional area.

Two test geometries are is illustrated in Fig. 3.14, both the notched and unnotched variants. The notched Charpy bar is most common, with a 2 mm deep $45°$ notch at its center. To measure the fracture energy, the hammer is lifted to the starting height and allowed to swing through the sample. As more energy is consumed in fracturing the sample, the follow-through swing reaches a lower height. This lost energy is the impact toughness. Composites are notch sensitive,

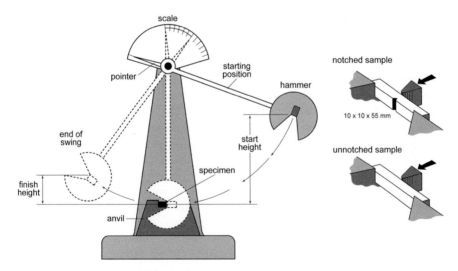

Fig. 3.14 Impact toughness involves fracture induced by swinging a pendulum through the test sample. The energy lost due to fracture is recorded. Various test geometries are in use, with the V-notch Charpy bar being the most common. For brittle materials an unnotched sample is possible, but the impact energy values are not comparable to standard Charpy results

so at times the unnotched variant is employed. Unnotched data should not be interchanged with notched data. For example, in a study on stainless steels, the notched fracture energy was 10 J/cm^2 and unnotched fracture energy was 272 J/cm^2. The unnotched geometry is larger in cross-sectional area and lacks the stress concentration from the notch; it obviously appears superior even though it is the exact same material with a different test configuration.

Fracture Toughness

Fracture toughness measures the resistance to crack propagation and is most important to understanding catastrophic failures [10]. Devices serving life-critical roles, such as biomedical implants or aircraft landing gears, are designed from high fracture toughness materials. The desire is to ensure resistance to rapid crack growth and sudden failure. High fracture toughness materials provide safety in service as compared with low fracture toughness materials—thus the frequent use of high fracture toughness stainless steel for surgical tools versus low fracture toughness ceramics. The latter are harder, but the former resists fracture. In this case the stainless steel tool will bend, deform, or crack, but still will resist failure in service. The ceramic on the other hand might unexpectedly shatter.

To measure fracture toughness, a structure with a pre-existing crack is stressed until the crack grows [14]. The stress to extend the crack depends on the square-root of the crack size, leading to fracture toughness units of MPa√m. Most ceramics are low in fracture toughness, typically below 15 MPa√m. On the other hand some particulate composites are more than 100 MPa√m. Several metals are even higher.

The measure of fracture toughness is reported in the parameter K_{Ic} which stands for –

stress intensity K
in tension I (the I implies the first case which is tension)
over the critical limit for crack growth c.

The K_{Ic} parameter enables prediction of the stress that will cause device fracture for any flaw size. Since pores are essentially small cracks, residual porosity lowers the measured fracture toughness. The methods used for determination of fracture toughness is advanced for several composites, especially cemented carbides [1, 2, 6, 10, 15]. As a material property, fracture toughness varies from about 0.2 MPa√m for ice to 4 MPa√m for alumina to more than 100 MPa√m for many steels.

Traditional fracture toughness testing relies on a pre-cracked sample with a crack of length A, such as sketched in Fig. 3.15. Sharp cracks generated by fatigue loading are the most detrimental. Tensile stress is applied via loading pins to open the crack, and the stress needed to induce crack growth is recorded. Fracture toughness K_{Ic} is calculated from the applied tensile stress for crack extension σ as,

Fig. 3.15 The compact
tension sample is used to
measure fracture toughness
for ductile materials. A
notch is machined into the
test geometry and a fatigue
crack is generated at the
root of the notch. Testing
involves determination of
the stress needed to advance
the crack versus the crack
size A

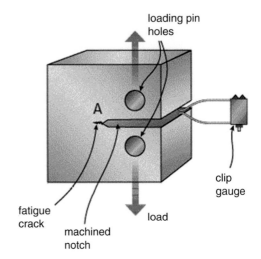

$$K_{Ic} = Y \sigma \sqrt{\pi A} \tag{3.23}$$

The geometric parameter Y depends on the sample width, crack shape, and crack placement; it is often near unity. About eight variants of the test are in use, all seeking to measure the stress needed to induce rapid crack advance. The reported values are usually accurate to about ± 1 MPa√m.

In composites there are many models for how the crack propagates, including interface failure, second phase deformation, phase cracking, second phase toughening, and crack blunting. The details are the subject of many models [1, 2, 4, 6]. In spite of several efforts, generally the models have a poor ability to predict measured behavior. Part of the difficulty arises from factors such as residual stresses, inhomogeneity, and low interface strength. As an example, data for Al-15Al$_2$O$_3$ at a grain size of 1 μm gives an elastic modulus of 200 GPa, yield strength of 469 MPa, and 1 % fracture elongation. Via different predictive models the fracture toughness would be expected to range from 3 to 8 MPa√m, but the experimental value is 19.5 MPa√m. This is a case where experiment is the only valid measure.

It is costly to form fracture toughness test samples with sharp cracks. An alternative in wide use induces cracks using microhardness testing, making for faster measurements. Materials with high fracture toughness resist cracking at the sharp tips of the Vickers hardness indent. But for brittle materials the microhardness indent induces cracking, especially for materials below about 20 MPa√m. About 30 variants of extracting fracture toughness from microhardness testing are in use, but they are largely derivatives on the idea initially called the Palmqvist fracture toughness.

In the Palmqvist approach, the fracture toughness K_{Ic} is estimated from the Vickers microhardness HV, crack resistance R, testing load P, and C determined by the crack length measured from the center of the indent while A is the crack

Fig. 3.16 The indentation
route to measuring fracture
toughness is applied to
lower ductility materials,
where the Vickers indent is
applied to generate cracks.
The crack length is a
measure of the fracture
toughness; short cracks
imply a high fracture
toughness

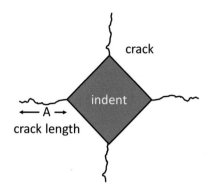

length, measured from the indenter tip as sketched in Fig. 3.16. The analysis gives
the fracture toughness as follows:

$$K_{Ic} = J \, [HV \, R]^{1/2} \tag{3.24}$$

the parameter J is adjusted to convert from the units used for hardness (kg/mm^2 or
MPa if multiplied by 9.8 m/s^2) and to correct for test details such as the hardness
load F and crack length measurement approach, while R is calculated using an
average for the crack length A as follows:

$$R = \frac{F - F_O}{4 \, A} \tag{3.25}$$

with F_O reflecting the threshold load for the onset of cracking. For brittle materials
the term F_O is often set to zero.

The relation is simplified to the following approximation:

$$K_{Ic} = 6.2 \left[\frac{9.8 \, HV}{4 \, A} \right]^{1/2} \tag{3.26}$$

where HV is the Vickers hardness number using 50 kgf loading, reported in kg/mm^2,
9.8 is the gravitational acceleration in m/s^2, and A is the mean crack length. It is
important to pay attention to the units, since indents are usually in mm, and loads are
given both in kg and kgf.

A semi-empirical model further includes the elastic modulus E and total crack
length C measured from the center of the indent,

$$K_{Ic} = 0.016 \, \frac{F}{C^{3/2}} \left[\frac{E}{HV} \right]^{1/2} \tag{3.27}$$

Repeated tests conclude the indentation based fracture toughness values have
typical measurement errors about $\pm 20\,\%$ of the measured value, especially for

brittle materials. Accordingly, fracture toughness based on microhardness testing is only nominal and should not be relied upon for design of life-critical components.

Magnetic Property Tests

Particulate composites are involved in both soft and hard magnets. Hard magnets are permanent magnets, characterized by magnetization retention after the polarizing field is removed. Soft magnets are the opposite and desirably are only magnetized when subjected to a polarizing field. They demagnetize when the field is removed. Both magnets are widely used in electrical devices. For example, permanent magnets are valuable to direct current motors while soft magnets are valuable for alternating current transformers. Flexible magnets, such as those used to mount pictures on a refrigerator, are permanent magnet composites; magnetic response arises from an iron oxide powder and flexibility comes from a polymer binder.

Magnetic properties are measured using an applied field to sense response parameters. A schematic plot of induced magnetization versus applied magnetic field is given in Fig. 3.17. Induced magnetization is delivered from a copper coil wound around the test sample. On the left is the applied field generated from the voltage supply. The responds is measured on the right in terms of the induced magnetic field. The adjustable load varies the applied field. As the sample magnetizes, the induced magnetization in the primary loop creates current flow in the galvanometer. The induced magnetic behavior is sensed in a secondary loop. The strength of the induced behavior is measured at a variety of applied stimulation levels until the material reaches saturation.

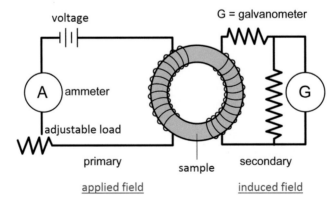

Fig. 3.17 Magnetic testing relies on a toroid sample geometry where current is applied to the primary loop and the magnetic response is measured in the secondary loop. The degree of induced magnetization is recorded versus the applied field. Several parameters are extracted from this test to characterize the magnetic response

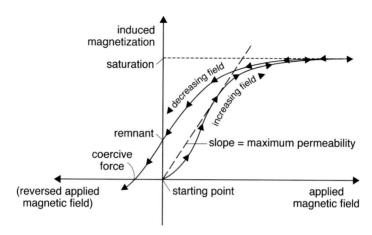

Fig. 3.18 A schematic of the initial magnetization for a magnetic material. The plotted trace gives induced magnetization and shows how with field reversal the material has a remnant magnetization that is not removed until a reverse coercive force is applied. Soft magnets should have little retained magnetization, but hard magnets have a desirably high retained magnetization

An example plot of applied and induced magnetization plotted during testing is shown in Fig. 3.18. A typical sample is 50 mm in outer diameter, 40 mm in inner diameter, and 4 mm thick. Primary and secondary coils are wound around the toroid for delivery of the applied field and measurement of the induced magnetic field. About 600 turns of insulated copper wire makes up the primary loop, and 150 turns are on the secondary loop. After the first magnetization, a hysteresis curve becomes apparent, with the area between the curves representing energy loss on each field reversal. Several response parameters are extracted from the test. These include initial permeability, saturated magnetization, remnant magnetization, and coercive force. The remnant value corresponds to the magnetization without an applied field, while the coercive force reflects the reverse field needed to remove all magnetization. The initial rate of magnetization is the tangent of the magnetization curve, giving the maximum permeability.

In Fig. 3.18, only one quadrant is shown since the magnetic fields can be reversed to give a similar curve in the lower left quadrant. As show, as the applied field is reduced from saturation, the reverse trace does not follow the loading trace. For a soft magnetic material, the desire is minimal remnant and coercive points, but for a permanent magnet the desire is the opposite. For many materials, magnetization is anisotropic, meaning the response depends on sample crystal orientation. In a polycrystalline material the different favorable magnetic orientations average out, so only average properties are a concern.

As temperature increases, permanent magnets lose the ability to remain magnetic. The temperature corresponding to loss of magnetism is the Curie temperature. It depends on purity and for example occurs at 771 °C (1044 K) for pure iron.

Tests for Thermal Properties

Thermal properties include melting behavior, heat capacity, thermal expansion, and thermal conductivity. Advanced instruments enable rapid and automated measurement of these properties [16]. Scanning temperature techniques, where heating is at a constant rate, say 10 °C/min, provide a means to identify temperatures where melting, phase changes, volatilization, or reactions occur. The common tests are the following:

- differential thermal analysis (DTA)—compares temperatures between the test sample and a standard during heating to determine temperatures where events such as melting and phase transformation occur—it is performed using one furnace and two sample cups, one containing the standard as sketched in Fig. 3.19,
- differential scanning calorimetry (DSC)—compares heat flow between a standard and a test sample during heating, it allows quantification of endothermic and exothermic reactions, such as crystallization or reaction,
- dilatometry—sometimes termed thermomechanical analysis (TMA), it measures dimensions versus temperature and is commonly applied to thermal expansion measurement; it is also used for sintering studies and for identification of phase transformation temperatures,
- thermogravimetric analysis (TGA)—measures the mass change on heating from events such as polymer decomposition, outgassing, oxidation or reduction, or

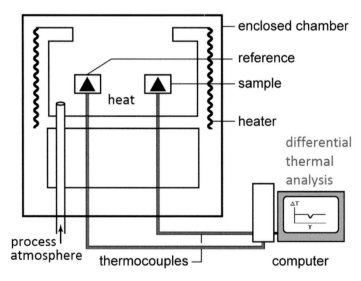

Fig. 3.19 Thermal analysis involves controlled heating to record changes versus temperature. Differential thermal analysis and differential scanning calorimeter involve heating a test sample and reference in the same chamber. Differences in heat flow or reactions result in temperature lags or accelerations corresponding to phase change, melting, or other events

reactions with the process atmosphere—sample mass is recorded versus an inert standard (platinum or alumina) during constant rate heating,
- thermal conductivity analysis (TCA)—quantifies the thermal diffusivity heat flow versus temperature to track the onset of sintering or to measure thermal conductivity—at high temperatures a laser is used to induce a thermal spike,
- evolved gas analysis (EGA) and related techniques such as residual gas analysis (RGA), mass spectroscopy (MS), or gas chromatography (GC), are means to measure the chemical species evolved during heating, by determining the temperature of the reaction and the compounds being liberated.

It is common to combine techniques in a single instrument, leading to a signature on heat flow, reactions, decomposition, phase changes, or thermal property variations.

Thermal shock resistance is the ability to undergo rapid heating and cooling without fracture. In particulate composites the difference in thermal and elastic properties between phases can contribute to low thermal shock resistance, especially if the two phases are very different. Thermal shock resistance R_T improves with strength σ and thermal conductivity K, but degrades with a high thermal expansion coefficient α, and elastic modulus E, roughly as follows:

$$R_T \simeq \frac{\sigma K}{E \alpha} \tag{3.28}$$

Material properties vary with temperature, so the protocol is to rely on material properties taken at the lower use temperature. Materials with a low thermal expansion coefficient, high thermal conductivity, and low elastic modulus are more resistant to thermal shock, especially when subjected to small temperature changes. Thermal shock is often reported as the maximum temperature change possible without fracture. However, it decreases as the number of oscillation cycles increases. This relation between the number of cycles and temperature difference is known as thermal fatigue.

Dilatometry measures dimensional change during heating and cooling. It is a tool that extracts thermal expansion, reaction, and sintering behavior versus temperature. Figure 3.20 illustrates a vertical dilatometer operating with a single push-rod. The push-rod rests on the sample and transmits dimensional information out of the heated region to an external sensor. Thermal expansion causes the sample to enlarge on heating while sintering causes the sample to shrink. Chemical reactions or phase changes also cause dimensional changes. The output signals are collected versus temperature, time, or atmosphere. A horizontal dilatometer relies on spring loaded push-rods to ensure sample contact. The spring motion leads to a small changes in applied stress with extension or shrinkage. Improved accuracy comes with two push-rods, one resting on the test sample and the other touching a calibration standard, providing data referenced to the comparative standard. Laser micrometry is a contactless alternative that allows measurements as close as ± 0.1 µm on a 10 mm length using a horizontally scanning laser beam.

Fig. 3.20 A sketch of a
vertical dilatometer. A
specimen inside the heated
furnace is touched by a
probe that transmits length
information versus
temperature or time during
heating and cooling. The
test is useful for measuring
thermal expansion
coefficient, phase
transformations, sintering
shrinkage, or decomposition
reactions. Protective
atmospheres are allowed

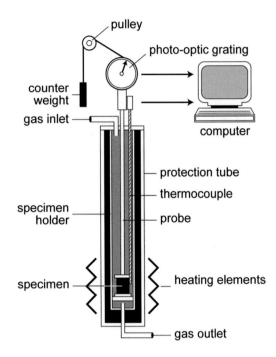

For thermal management applications, thermal expansion and thermal conductivity are the two important properties. Most insulators are highly porous to reduce heat flow. On the other hand, dense copper and aluminum provide high thermal conductivities. Several particulate composites are designed to deliver high thermal conductivity with a low thermal expansion coefficient.

Electrical Property Tests

Electrical properties range from insulator to conductor attributes. Metallic products are conductive, characterized by resistivity, or its inverse conductivity [17]. Conductive composites are employed in electrical systems for switching and sliding contacts. For example, combinations such as tungsten and copper provide a high conductivity composite with arc erosion and wear resistance. In those applications, arc erosion resistance is measured using repeated contact opening and closing with sufficient voltage to induce arcing on each cycle. The test measures the contact mass loss versus number of cycles [18]. The arc erosion resistant composites are also used in welding electrodes and electro-discharge machining tools.

Insulator composites are found in lighting, home appliance, and microelectronic devices. Conductivity is reduced by residual pores, so these composites might intentionally include porosity. Roughly, conductivity λ depends on the volume fraction of conductor phase V_C (assuming the conductor phase is connected) as follows:

$$\lambda = \frac{V_C \lambda_O}{1 + \psi \left(1 - V_C\right)^2} \tag{3.29}$$

where ψ represents the sensitivity to the nonconductive phase and λ_O is the conductivity of the full density conductor phase. This relation is effective for many particulate composites over a wide range of percolated conductor phase contents.

Electrical conductivity testing relies on a four-contact measurement technique, with two outer contacts providing current flow and two inner contacts measuring the corresponding voltage drop. Handheld units provide for quick field tests. In Ohm's law, resistance is given by voltage divided by current. Resistivity is a material property that depends on resistance as normalized to the sample cross-section area and length. Unlike resistance, resistivity and its inverse, conductivity, are geometry independent. For many materials, conductivity relies on electron motion, so electrical and thermal conductivity are proportional. Exceptions are ceramics, such as aluminum nitride, where thermal conduction is high in spite of the material being an electrical insulator.

Other electrical properties encountered in particulate composites include attributes specific to the application, such as thermoelectric behavior, dielectric constant, or piezoelectric response. Thermoelectric composites are compounds of Bi, Te, Ga, Ge, and Se that provide cooling with electric current flow. Dielectric materials are electrical insulators that polarize in an electric field as a means to store energy, such as in a capacitor, such as in a polymer filled with oxide mineral particles. Likewise, piezoelectric composites rely on ceramic particles dispersed in a polymer.

Environmental Degradation, Corrosion, and Wear Tests

Long-term survival in an application requires resistance to corrosion, tarnish, oxidation, and related avenues of attack; passivity in the use environment is desired. Resistance to environmental attack is quantified in terms of parameters such as mass loss or gain over time per unit of component surface area. The ability to resist corrosion is specific to the material, but depends on the preparation, test environment, and test details. In other words, environmental attack is sensitive to several factors, so it is not easily generalized. Further, residual pores accelerate environmental attack. Accordingly, testing is required using conditions relevant to the use environment to avoid unexpected failures. One classic example was in heat insulation used on steam lines. The three phase composite pipe coating reportedly caused corrosion of the stainless steel pipes during service. After extensive unsuccessful reformulation to cure the problem, the corrosion was traced to chlorine ions introduced by sweating installation crews during installation. All of the reformulations failed to include this factor, leading to many false cures that subsequently failed in service.

A simple test for corrosion is via immersion in the test fluid. Sample mass and surface area are measured. After exposure to the test fluid the mass is measured again. Mass change is normalized to the exposure time and sample surface area, leading to quantitative corrosion units of mass loss per unit area per unit time, or $g/(m^2\ h)$. For a composite, the attack usually is preferential against one phase. As that phase reacts and dissolves, the absence of fresh material on the surface slows the corrosion rate over time. Electrochemical tests are used to determine the corrosion current versus an applied voltage. Current flow is recorded over a range of applied voltages to determine the rate of corrosion versus the voltage. In service, voltages arise from contacting materials of different compositions.

While corrosion tends to remove mass from the test material, tarnish and oxidation add mass to the surface. The change in mass over time and exposed area is used to quantify the resistance. Often the growing reaction film slows continued attack, leading to an asymptotic film thickness if the film remains intact.

Friction tests are performed to quantify sliding wear [19]. A low friction coefficient is desirable to reduce wear damage. Wear testing involves intentional mass removal using forced abrasion against a moving, opposing test wheel. One common wear test is sketched in Fig. 3.21. Abrasive sand is fed between the stationary test sample and the moving rubber wheel, causing progressive sample

Fig. 3.21 Wear testing involves compressing a test sample against a rotating rubber wheel that is laden with abrasive sand. The wear is quantified by the loss of sample mass over time or based on the number of revolutions

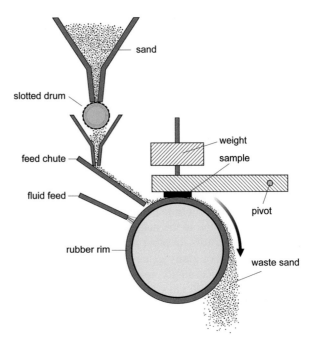

Fig. 3.22 Example data on wear rate, in terms of volume loss per distance traveled, versus composite hardness. These results are for a composite consisting of aluminum, silicon, and titanium diboride. The latter is quite hard so the higher TiB$_2$ content samples are more wear resistant

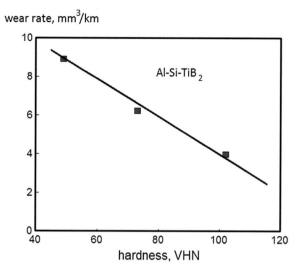

mass loss. Usually the test is performed to a predetermined number of revolutions by the rubber wheel (1000 or 10,000), then the test is stopped to measure sample mass loss. Elevated temperature options are used for testing wear related to engines, combustion equipment, mining, or drilling applications. Variants include pin-on-disk, and wheel-on-disk geometries; all rely on measuring mass change over time.

Wear resistance is quantified by the time to remove mass of over several rotations. Harder materials resist wear, as evident in Fig. 3.22 using Al-Si-TiB$_2$ data. Here wear is given by the volume removed versus sliding distance. As more titanium diboride (TiB$_2$) is added to the aluminum-silicon matrix, the hardness increases and wear decreases. Archard's law relates this behavior based on the wear volume V, load F, sliding distance y, and hardness HV as follows:

$$V = \frac{c\,y\,F}{3\,HV} \tag{3.30}$$

where c is a dimensionless wear constant. This relation is not accurate for high loads. Wear further depends on the interaction of the test material with its environment.

Biocompatibility

Medical devices require biocompatibility testing [20]. Biocompatible composites resist corrosion and tissue reaction over long exposure times. Further, the materials do not release toxic substances nor do they irritate contacting tissue.

An array of parameters are measured to ensure inert performance. Testing for biocompatibility starts with standard evaluation tests for mechanical properties and corrosion, but extends to a host of laboratory tests and eventually animal tests [21]. Typical tests include the following:

- tissue irritation—determined by implanting the material under animal tissue,
- cytotoxicity—death of cells from exposure to the material,
- thrombogenicity—determination of accelerated blood clotting from the material,
- systemic toxicity—animal tests to determine the concentration needed for death, often reported as the LD50 or similar number to indicate the lethal dosage for 50 % of the animal population to die,
- occupational exposure—set safe limits to the daily and annual exposure.

Implants, such as artificial heart valves, knees, or hips, require the exhaustive testing. It is best to collaborate with laboratories skilled in these tests to ensure results relevant to the application. Most typically the results must be repeatable and sufficiently well documented to withstand careful scrutiny by peer reviewers.

Human response to any substance involves a natural variation between individuals. For this reason caution is required to ensure a broad range of tests and test conditions over a large population to identify any problems. Clinical tests require time and much expense, but are required for device acceptance.

Microstructure Quantification

Property measurements are best coupled to microstructure quantification to understand the controlling parameters. Considerable expertise exists in materials characterization. This section explains microstructure characterization tests based on quantitative analysis using digital image analysis tools [22].

Sample Preparation

The first task is to select the sample for analysis. The higher the magnification, the smaller the test volume, so care is required to ensure the examined structure is representative of the overall structure. The frosting on a cupcake is not reflective of the taste for the whole structure. A random cross-section is best, unless some particular feature is of concern. Fracture surfaces are associated with regions of weakness, so if the analysis is focused on fracture problems, then examination of the fracture surface is mandatory.

Assuming a random section is desired, the first step is to slice the component to remove a small region. For optical microscopy that slice is mounted in a resin to allow grinding and polishing. The slicing process induces damage to the sample that is removed by a sequence of grinding steps. For example, pores are filled with epoxy resin to avoid smearing or filling with polishing grit. Progressively smaller diamond, silicon carbide, or alumina grit are employed, ending with a flat polished surface. Differences in color, hardness, or etching response give contrast in the image. Alternatives include vapor deposits that stick preferentially to one phase or thermal treatments to induce surface reactions. Without some difference in surface characteristics, the composite simply appears featureless. In the case of electron microscopy, especially scanning electron microscopy, conductivity helps avoid charge accumulation. For this reason a slight coating of conductor (silver, gold, or platinum) is deposited on the surface. Sputtering is an effective means to deposit thin layers to avoid charging that would distort the electron the image.

The image used for analysis ideally has no influence from the sample preparation steps. An example of a multiple phase composite structure after light etching is given in Fig. 3.23. This structure involves a predominant phase of tool steel (major phase), with a boron-rich phase (light color), hard refractory metal carbide (dark), and a few pores (black). Inspection finds cracks in the structure formed during fabrication. Such images help understand properties via the phase morphology.

Optical Imaging

Optical imaging is the most widely used characterization route. There are two options—reflection or transmission. These same options are available for electron microscopy. Transmission microscopy is used for samples that are not opaque where they are thinned sufficiently that light passes through. Interaction of the

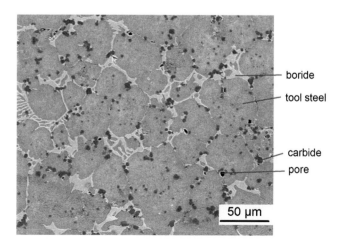

Fig. 3.23 Microstructure evident on a polished and etched composite. This structure is predominantly tool steel (major phase) with a boride phase surrounding the tool steel grains. The dark phase is a metal carbide. A few pores are evident as black spots [courtesy of T. Weaver]

Fig. 3.24 The idea behind
a reflected light microscope,
where sample viewing and
illumination share a
common axis. Images are
generated by differences in
the sample surface
topography

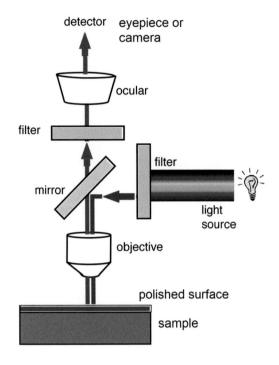

light with the structure provides information on the phases; an ideal technique for pathological tests. Most composites are opaque so microscopy relies on reflected light. The key concept is sketched in Fig. 3.24. The illumination beam is directed to the sample surface, where it interacts with the color or topography to alter the reflected light. That reflected light is captured by lenses and forms an image. That image is analyzed for amount, size, connection, homogeneity, and spacing of each phase [22]. The capability is limited by the wavelength of light, so usually 0.5–1 μm or larger features are recorded.

Options for extending the analysis include ultraviolet or infrared light, polarized light, laser light (monochromatic), and various combinations. Three-dimensional images are constructed using progressive polishing, photography, and polishing, in what is termed serial sectioning [23]. The images can be electronically melded into a three-dimensional representation. That reconstructed image provides information on phase spacing and phase connections. In addition, the microstructure and properties of the individual phases are suitable for finite element analysis to explain deformation and fracture [24]. Largely such findings show the importance of microstructure homogeneity with respect to maximization of properties.

Electron Imaging

As with optical imaging, electron imaging involves transmission and reflection modes. The devices are arranged around an electron source, accelerating mechanism, focusing device involving magnetic lens, sample manipulation, and means to capture the image. Transmission electron microscopy (TEM) requires a thin sample such that the accelerated electron beam passes through the sample prior to magnification. Interactions with the structure provide images that probe to the atomic scale. Differences in chemistry, crystal structure, or defect structure are evident in the micrographs. Figure 3.25 is a TEM micrograph at the junction of two phases. On the left and right are pure tungsten with body-centered cubic crystal structure and a grain boundary at the bottom of the image. The central "V" region is a face-centered cubic phase of nickel-tungsten-iron. The dihedral angle defining the three phase junction is approximately 60°. Dislocation are evident in the face-centered cubic region. At this magnification there is no interfacial film evident.

The diffraction of electrons from different crystals leads to options in TEM imaging. For example, one preferred crystal structure can be illuminated, such as the TiC crystals in Fig. 3.26. Additionally, X-rays are generated by the electron-atom interaction. Because of quantum mechanics, a precise energy level is associated with those X-rays to provide chemical identification. Transmission electron microscopy is a powerful imaging tool, but it is expensive to employ since the preparation of the required thin sample requires care. As evident in the micrographs, the imaged area is small, so it is not useful for analysis of large features.

The scanning electron microscope (SEM) is a favorite in imaging surfaces. Electrons are generated and accelerated to provide an electron beam that is focused with magnetic lens. As the beam scans over the sample surface several emissions occur, including electrons that bounce off the surface (backscatter), emission of internal electrons knocked out of their atomic orbitals (secondary), and release of

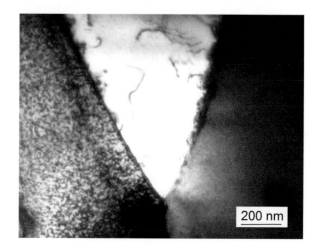

Fig. 3.25 Transmission electron micrograph of the Y junction of two grains and a second phase, with dislocations in the central region with interfaces and a grain boundary evident. The left and right are body-centered cubic, and the central wedge is face-centered cubic [courtesy A. West]

200 nm

Fig. 3.26 Bright field
transmission electron
microscopy of a TiC
particle in a tool steel
matrix. This image is
generated by electron
diffraction with the carbide
[courtesy J. Smugeresky]

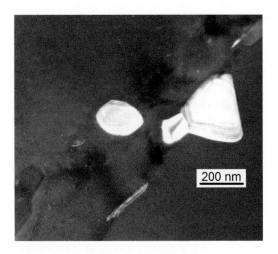

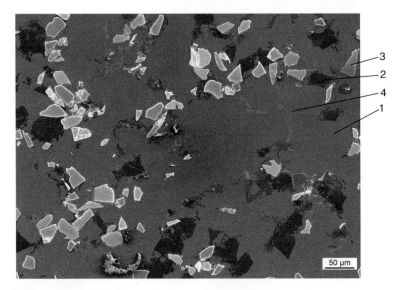

Fig. 3.27 Scanning electron microscope image of a composite consisting of four phases [25]. The
regions are as follows: (*1*) aluminum alloy matrix, (*2*) silicon grain, (*3*) silicon carbide grain, and
(*4*) copper-aluminum alloy [courtesy W. Li]

X-rays associated with quantum shifts of electron cascades between orbitals. The
latter is energy dispersive spectroscopy (EDS). A key to imaging is synchronization
of the electron beam position with display of the detected intensity.

The SEM images show phase contrast in the composite, such as pictured in
Fig. 3.27. This is a composite of mixed Al, Cu, Mg, Si, and SiC powders [25]. The
image corresponds 20 vol% SiC mixed with the other powders, compacted at
400 MPa, and vacuum sintered at 590 °C (863 K) for 90 min. Four different phases

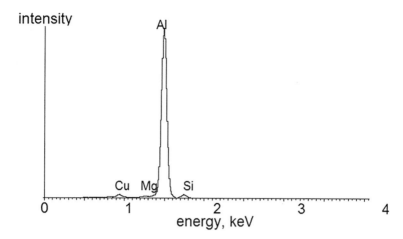

Fig. 3.28 The energy dispersive X-ray spectrum from region 1 in Fig. 3.27, showing intensity of X-rays versus energy in keV, with the peaks identified as Cu, Mg, Al, and Si corresponding to the alloy

are imaged. Using the electron beam of 1 μm diameter allows each phase to be individually analyzed. An example spectrum is plotted in Fig. 3.28 for region 1, corresponding to the Al-Si-Cu-Mg alloy matrix. Region 2 is almost pure silicon. Region 3 is silicon carbide. Region 4 is Al-Cu eutectic phase (nominally 22 wt% Cu). Note the small cracks are associated with the silicon and silicon carbide phases, indicating an undesirable fabrication cycle that damaged these brittle phases.

Scanning electron microscopy provides information on the material and structure in a fast and easy approach [26]. It is widely used for particles since they require almost no preparation. Unlike optical microscopy, with limited depth of field, the SEM easily captures the three-dimensional particle image. Further, scanning electron microscopy is useful in analyzing fracture surfaces; fractures follow the weak path, so SEM examination helps identify what aspect of the microstructure is weakest.

Other Imaging

The list of imaging options is large [6–8]. Optical and electron microscopy are the major approaches, but alternatives include microscopy imaging based on acoustic, low energy electron, atomic force, or electron tunneling. These various approaches rely on interactions with the test material, via some delivered energy. Some of these imaging techniques probe to the atom scale if required, especially in examining interfaces [27]. However, the advanced techniques require specialized equipment and trained operators, so assistance is required in moving to the novel imaging concepts.

Standards and Specification Bodies

Government research laboratories are proactive in setting standards for composition, testing procedures, and typical properties [28–31]. Industry standards arise from a thoughtful approval process. The industry standard starts with a geographic-based trade association. For example, companies located in Japan, USA, or Germany first gather to agree to a material, composition, or testing standard. They meet, discuss needs, and collect test data at a neutral site. Further test development may be needed to remove variations, such as from test fixture alignment. Round robin tests send samples to different labs. As data and protocols emerge, votes are taken on how the tests should be performed and how to present the data. Once a standard exists for a specific country-industry combination, then it is a candidate for a country-wide standard. Once adopted in a country, such as the USA (ASTM), the standards from different countries are melded into a global standard, such as those administered by the International Standard Organization (ISO). The process is slow due to the number of steps and organizations involved.

This is not the only route to industry standards. Large customers often set limitations or specifications related to their application. Further, insurance companies issue standards, government bodies pass specifications, while health and safety agencies regulate many substances, and in some cases different local rules exist. Confused? Of course, and for this reason data need to clearly define the testing procedures.

Comments on Accuracy and Precision

Accuracy indicates how close a parameter is to the specified quality. Precision indicates the repeatability of a measurement. These two attributes are illustrated by the bullet hole patterns on the targets in Fig. 3.29. A high precision is evident by closely clustered bullet holes. A high accuracy corresponds to the bullet holes being close to the center. Usually accuracy is associated with the specification—how close is good enough. The larger the number of significant digits used to specify a value then the greater is the required measurement accuracy. On the other hand, precision traces to instrument calibration, operator training, and standards for performing measurements.

Industries collaborate to set measurement standards, thereby allowing focus on determining how close the material or device is to the specified attribute. Several factors hinder agreement in any measurement. One factor is that each sample is

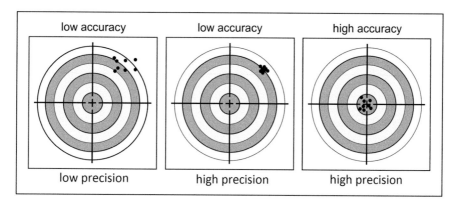

Fig. 3.29 Contrast of accuracy and precision in terms of how eight shots at a target are clustered. Accuracy relates to how close the shots are to the specified quality represented by the center of the target, while precision is characteristic of the repeatability represented by how closely clustered shots

slightly different, so there is inherent variation. This is illustrated by a study on 2084 production lots of steel. The yield strength showed 8.8 % coefficient of variation (standard deviation divided by mean), and the ultimate tensile strength showed 8.7 % coefficient of variation. However, the ductility as measured by the fracture elongation exhibited 16 % coefficient of variation. Note these were tests conducted by the same team, same test protocol, and same instruments.

Another factor comes from the individuals and test machines performing the measurements. Round robin testing is a means to quantify and reduce these variations. This involves lots of test material being shipped to several labs to quantify both the inherent material variation and the differences between sites. As an example of round robin testing, seven firms performed tensile tests to measure ultimate strength on a composite. The inherent material variation led to a coefficient of variation inside each firm that averaged 6 %. When the tensile test strengths were compared between firms, there was an additional 2 % variation. Beyond inherent material variations, the next two factors were test equipment and personnel. For similar compression strength tests, the variation of 12 % depended on personnel performing the test.

Property variations have several sources. One is the inherent difference in test materials, and the magnitude of the variation depends on the property measured. Differences between test sites, individuals, devices, and techniques add to the scatter. Accordingly, test accuracy for some properties introduced in this chapter are typically as follows:

- Density ±1 %
- Hardness ±4 %
- Strength ±8 %
- Ductility ±20 %

- Biocompatibility $\pm 20\%$.

Although parameters such as fracture toughness and fatigue strength have not been studied as extensively, most likely they are in the $\pm 10\%$ range.

Study Questions

3.1. The derivation of Equation (3.11) for sample volume based on gas pycnometry relies on ideal gas behavior. Show how this equation arises based on two chambers, each with known volume, with the first chamber filled to pressure P_1 and the second chamber containing the test material initially evacuated to zero pressure. The pressure after allowing the two chambers to equilibrate is P_2.

3.2. In the case of a composite with density below 1 g/cm^3, conjecture how the Archimedes approach must be modified to measure density.

3.3. A copper-alumina composite, consisting of 80 wt% copper (copper density of 8.96 g/cm^3 and alumina density of 3.96 g/cm^3), is formed by mixing and consolidating powders. The consolidated Archimedes density is 6.62 g/cm^3. What is the fractional porosity?

3.4. A firm selling a new particulate composite claims a fracture toughness 17 MPa$\sqrt{m}$ and Vickers hardness of 1360 HV. Based on a quick examination of existing data, is this anything special?

3.5. A new technique akin to computer tomography is used to create composition versus depth profiles on small areas of test samples. It relies on the output from secondary ion mass spectroscopy to record composition as the sample is being eroded. Where in particulate composite analysis might such a tool be of most value?

3.6. The Mohs hardness scale is popular in geology, where talc is ranked 1 and diamond is ranked 10. Create a relation between the Mohs scale and the Vickers scale, first graphically and then via a mathematical expression.

3.7. Hardness is often called a mechanical property test along with strength. It rumored the Vickers hardness (expressed in MPa by multiplying the HV by 9.8 m/s^2) is about three times the tensile yield strength (in MPa). Tabulate the yield strength and Vickers hardness for five particulate composites and determine if this is a good approximation (some data are included in the applications section of this book).

3.8. A Navy specification for a porous composite used to insulate steam pipes requires the transverse rupture strength to be proportional to the density. Comment on the merits of such a specification.

3.9. Dental amalgam is a particulate composite composed of toxic substances, yet it remains in widespread use? How is it possible to use mercury in this application?

3.10. Wear life is critical to the survival of hard particulate composites used in oil and gas exploration drills. However, most of the testing is performed in the field. What is the reason laboratory tests cannot replace expensive field evaluations?

3.11. Proponents of the four-point transverse rupture test like the higher values obtained for strength versus the three-point test. However, the three-point test is favored in practice. What is the practical issue that favors using the three-point test?

3.12. The water immersion density is performed at room temperature, but in some facilities room temperature varies. Over a reasonable range of temperatures, how much error arises from ignoring fluctuations in room temperature?

3.13. A component is specified with 10 dimensions at a coefficient of variation of 0.1 % and two dimensions at 0.05 %. To improve fit, it is decided to move to four dimensions specified at 0.05 %. If the dimensional budget is to be held constant, what should be the new tolerance on the remaining eight dimensions?

3.14. A composite is tested using three-point bending using 40 samples, giving 50 % failure at 350 MPa or less, and 90 % failure at 390 MPa or less. Estimate the Weibull modulus.

3.15. In developing a rocket engine composite there is concern for thermal shock. On large rockets, the engine is first cooled using liquid hydrogen prior to combustion. Estimate the temperature change required to define the thermal shock parameters relevant to this application.

3.16. Most all consumer products are covered by industry standards, and some of the common ones are CE, UL, ASTM, IEEE, and ASME. Even cardboard boxes are marked for ISO or ASTM compliance. Identify a product that is marked with at least one industry standard and comment on the specific intent of that standard.

3.17. Two aluminum—silicon carbide composites are offered for an application that has sensitivity to thermal shock. Which of these is more resistant to rapid temperature change?

	Strength MPa	Thermal conductivity W/(m °C)	Thermal expansion $10^{-6}/°C$	Elastic modulus GPa
Al-SiC-type A	395	150	7	206
Al-SiC-type B	450	180	10	192

3.18. Strength testing gives a rupture strength that is half that anticipated for the composite. Inspection of the fracture surface reveals the sample had a large 1 mm diameter void below the surface not visible prior to testing. What do you think the protocol should be for reporting such data?

3.19. Certain purchasing standards allow for random failures, sometimes denoted as the acceptable quality level. If 10,000 items are produced and 200 are tested for strength, and 10 are below the lower specified strength, then at an acceptable quality level of 1 % this lot should be rejected. Do you agree or is there some other recommendation?

3.20. Draw a target equivalent to Fig. 3.29, but for the missing situation of low precision and high accuracy.

3.21. Thermal conductivity of 0.08 W/(m °C) is reported for foamed calcium-silica-glass composite at 8 % density at 90 °C (363 K). Compare this conductivity to pure water and pure air. What would be the consequence if the ceramic insulator was saturated with water?

References

1. P.K. Mallick, *Composites Engineering Handbook* (Marcel Dekker, New York, 1997)
2. N.P. Bansal, *Handbook of Ceramic Composites* (Kluwer Academic Publisher, New York, 2005)
3. M. Evans, N. Hastings, B. Peacock, *Statistical Distributions*, 2nd edn. (Wiley, New York, 1993)
4. D.J. Green, *An Introduction to the Mechanical Properties of Ceramics* (Cambridge University Press, Cambridge, 1998)
5. U. Engel, H. Hubner, Strength improvement of cemented carbides by hot isostatic pressing (HIP). J. Mater. Sci. **13**, 2003–2012 (1978)
6. D.B. Miracle, S.L. Donaldson (eds.), in *Composites*, ASM Handbook, vol. 10. (ASM International, Materials Park, 2001)
7. E.N. Kaufmann, *Characterization of Materials*, vol. 2 (Wiley-Interscience, Hoboken, 2003)
8. E. Bauer, *Surface Microscopy with Low Energy Electrons* (Springer, New York, 2014)
9. J.J. Gilman, *Chemistry and Physics of Mechanical Hardness* (Wiley, Hoboken, 2009)
10. K.K. Chawla, *Composite Materials Science and Engineering*, 3rd edn. (Springer, New York, 2013)
11. P. Han (ed.), *Tensile Testing* (ASM International, Materials Park, 1992)
12. I.A. Ibrahim, F.A. Mohamed, E.J. Lavernia, Particulate reinforced metal matrix composites—a review. J. Mater. Sci. **26**, 1137–1156 (1991)
13. P. Suri, B.P. Smarslok, R.M. German, Impact properties of sintered and wrought 17-4 PH stainless steel. Powder Metall. **49**, 40–47 (2006)
14. R. Spiegler, S. Schmauder, L.S. Sigl, Fracture toughness evaluation of WC-Co alloys by indentation testing. J. Hard Mater. **1**, 147–158 (1990)
15. D. Han, J.J. Mecholsky, Fracture analysis of cobalt-bonded tungsten carbide composites. J. Mater. Sci. **25**, 4949–4956 (1990)
16. K.H. Kate, R.K. Enneti, S.J. Park, R.M. German, S.V. Atre, Predicting powder-polymer mixture properties for PIM design. Crit. Rev. Solid State and Mater. Sci. **39**, 197–214 (2014)
17. D.S. Mclachlan, K. Cai, G. Sauti, AC and DC conductivity based microstructural characterization. Int. J. Refract. Met. Hard Mater. **19**, 437–445 (2001)
18. X. Wang, H. Yang, M. Chen, J. Zou, S. Liang, Fabrication and arc erosion behaviors of Ag-TiB$_2$ contact materials. Powder Technol. **256**, 20–24 (2014)
19. R.L. Deuis, C. Subramanian, J.M. Yellup, Dry sliding wear of aluminum composites—a review. Compos. Sci. Technol. **57**, 415–435 (1997)
20. J.R. Davis (ed.), *Handbook of Materials for Medical Devices* (ASM International, Materials Park, 2003)
21. D.K. Pattenayak, V. Mathur, B.T. Rao, T.R.R. Mohan, Synthesis of titanium composite bio-implants. Int. J. Powder Metall. **40**(2), 55–62 (2004)

22. R.T. DeHoff, F.N. Rhines, *Quantitative Microscopy* (McGraw-Hill, New York, 1968)
23. A. Tewari, A.M. Gokhale, Application of three-dimensional digital image processing for reconstruction of microstructural volume from serial sections. Mater. Charact. **44**, 259–269 (2000)
24. Z. Shan, A.M. Gokhale, Digital image analysis and microstructure modeling tools for microstructure sensitive design of materials. Int. J. Plast. **20**, 1347–1370 (2004)
25. F. Findik, J.K. Thompson, A. Antonyraj, S.J. Park, R.M. German, Mechanical and physical properties of titanium and silicon carbide containing mixed powder sintered aluminum, *Advances in Powder Metallurgy and Particulate Materials* (Metal Powder Industries Federation, Princeton, 2007), pp. 10.103–10.113
26. M.S. Kumar, P. Chandrasekar, P. Chandramohan, M. Mohanraj, Characterisation of titanium—titanium boride composites process by powder metallurgy techniques. Mater. Charact. **73**, 43–51 (2012)
27. J.C. Williams, J.W. Nielsen, Wetting of original and metallized high-alumina surfaces by molten brazing solders. J. Am. Ceram. Soc. **42**, 229–235 (1959)
28. P. Klobes, K. Meyer, R.G. Munro, *Porosity and Specific Surface Area Measurements for Solid Materials*, Special Publication 960-17 (National Institute of Standards and Technology, Gaithersburg, 2006)
29. R.G. Munro, *Data Evaluation Theory and Practice for Materials Properties*, Special Publication 960-11 (National Institute of Standards and Technology, Gaithersburg, 2003)
30. S.R. Low, *Rockwell Harness Measurement of Metallic Materials*, Special Publication 960-5 (National Institute of Standards and Technology, Gaithersburg, 2001)
31. A. Jillavenkatesa, S.J. Dapkunas, L.S.H. Lum, *Practice Guide Particle Size Analysis: Particle Size Characterization*, Special Publication 960-1 (National Institute of Standards and Technology, Gaithersburg, 2001)

Chapter 4
Property Models

Property models require information on the composition and attributes of the constituent phases. Depending on which parameter is being addressed, the models range from linearly additive forms to models that require microstructure and interface information. Static properties, such as density, hardness, and thermal conductivity are modeled accurately. However, properties sensitive to interface strength, such as ductility and fracture toughness, often require laboratory tests to adjust for uncertain attributes. Many models involve empirical aspects related to interface behavior.

Introduction

This chapter assembles property models for particulate composites. Most of the models start with inputs of composition and constituent phase properties. These are often satisfactory for predicting properties. For improved accuracy, many formulations require additional features such as the grain size, microstructure homogeneity, or interface strength. The interface effect is especially important to mechanical properties that involve deformation. Interfaces can be weak or strong, leading to significant changes to composite conductivity, toughness, strength, and ductility [1, 2]. While a high concentration of hard phase creates a harder composite, often the maximum strength is at an intermediate concentration. For example, in a study using 0.1 μm WC and 1 μm Al_2O_3 mixed powders consolidated by 5 min of spark sintering at 70 MPa and 1800 °C (2073 K), the following observations were made with respect to composition effects [3]:

- peak hardness occurred at 100 % WC
- peak strength occurred at 4 vol% Al_2O_3
- peak fracture toughness occurred at 20 vol% alumina.

The lower strength alumina blunted crack propagation. Although strength declined with more than 4 vol% alumina, still it improved facture toughness up

© Springer International Publishing Switzerland 2016
R.M. German, *Particulate Composites*, DOI 10.1007/978-3-319-29917-4_4

to 20 vol%. At concentrations over 20 vol% alumina, a connected skeletal network formed that provided an easy fracture path in the weaker phase.

The linearly additive rule of mixtures is a starting point for modeling composition effects on properties [4]. The constituent volume fractions are used to weight the relative properties, and porosity is one possible phase. If we assume two phases, then the two volume fractions sum to unity, $V_2 = 1 - V_1$. Each phase provides an inherent property level, P_1 and P_2 (for example elastic modulus), so the proportional combination provides an estimate of the composite property P_C as [4, 5],

$$P_C = V_1 P_1 + V_2 P_2 \tag{4.1}$$

This linear rule of mixtures based on volume fraction produces a straight line connecting the two end compositions, as plotted in Fig. 4.1. The approach assumes the composite behavior is intermediate between that of the constituents; effectively weighted average based on relative volume of each phase.

Although easily created, linear composition models are not always accurate. Related constructions allow for curvature to the property versus composition plot. One empirical model relies on two adjustable parameters that reflect interface effects,

$$P_C = P_2 \frac{1 + N U V_1}{1 - N V_1} \tag{4.2}$$

Both U and N are varied to best fit the composite property data and boundary conditions, but such an approach only gives curvature and fails to model systems

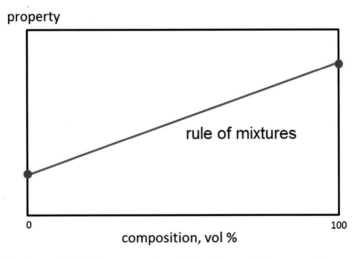

Fig. 4.1 The linear rule of mixtures on a volumetric basis is a model for composite properties that assumes a straight line connecting the two terminal phases. The properties of intermediate compositions are on this line

with maxima or minima at intermediate composition. The intent of introducing curvature is to allow for some possible interaction between the phases. The conjecture is properties are nonlinear with composition due to residual strains or changes in microstructure. Nonlinear models might improve data fitting, but there is no predictive sense to the fitting parameters.

In some cases, a composite property reaches a maximum or minimum at an intermediate composition. Such behavior requires models involving interaction terms, interface quality, or microstructure data as detailed later in this chapter. Common microstructure features are grain size, phase connectivity, and homogeneity. With the more complicated models it is a classic give and take decision: Is the improved predictive accuracy justified by the additional data and effort needed to determine additional input parameters? Essentially, is it easier to simply measure performance or is the effort needed to measure microstructural parameters justified? Usually the response is to perform the experiment rather than contrive a model that might still require trials to determine adjustable parameters.

Inherently, particulate composites are homogeneous. Likewise, the models ignore defects, such as phase agglomeration and residual stresses, by assuming idealized microstructures. However, porosity is a frequent concern that must be added to behavior models.

The presentation in this chapter starts with some easily modeled parameters and moves to progressively more involved properties. Complexity arises in treating deformation and fracture characteristics, such as fracture toughness.

Density

The theoretical density of mixed phases follows the volumetric rule of mixtures. For two phases of volume fractions V_1 and V_2, the sum $V_1 + V_2$ is unity; the composite theoretical density ρ_C is estimated as follows:

$$\rho_C = V_1 \rho_1 + V_2 \rho_2 \qquad (4.3)$$

where ρ_1 is the theoretical density of phase 1, and ρ_2 is the theoretical density of phase 2. The composite theoretical density is in kg/m^3 or more conveniently g/cm^3.

Most composition formulations are based on weight fractions, so it is typical to calculate the theoretical density using weight fractions. On the weight basis this leads to an inverse rule of mixtures. Again, for two phases designated as 1 and 2, with the mass of the first being W_1 and the being W_2, the mixture theoretical density is obtained by dividing the total mass by the total volume. To do this realize the total mass W_T is,

$$W_T = W_1 + W_2 \qquad (4.4)$$

The volume of each phase is the mass divided by the phase density,

$$V_1 = \frac{W_1}{\rho_1} \tag{4.5}$$

and

$$V_2 = \frac{W_2}{\rho_2} \tag{4.6}$$

The total volume and total mass are added for the constituents; hence, the composite theoretical density ρ_C is given as the total mass divided by the total volume,

$$\rho_C = \frac{W_1 + W_2}{\frac{W_1}{\rho_1} + \frac{W_2}{\rho_2}} \tag{4.7}$$

Reducing to mass fractions X, where $X_1 = W_1/(W_1 + W_2)$ and $X_2 = W_2/(W_1 + W_2)$, gives the generalized form as follows:

$$\frac{1}{\rho_C} = \sum \frac{X_i}{\rho_i} \tag{4.8}$$

This equation extends includes any number of phases. What is calculated is the estimated pore free density. Since pores have no mass but take up volume, the density falls in direct proportion to the volume fraction of pores.

Example calculations of density versus composition are illustrated in Table 4.1 for W-Cu composites [6, 7]. The difference between measured and calculated density is small, within the experimental error.

Hardness

Following density, a common property reported for particulate composites is the hardness. Models linking hardness to composition have several levels of complexity. The simple approach is a linearly additive approach based on the volume

Table 4.1 Density comparison for tungsten-copper composites [6, 7]

Wt% Cu	Vol% Cu	Measured density g/cm^3	Predicted density g/cm^3
0	0	19.3	19.3
10	19	17.3	17.3
20	35	15.6	15.7
30	48	14.1	14.3
40	59	13.3	13.2

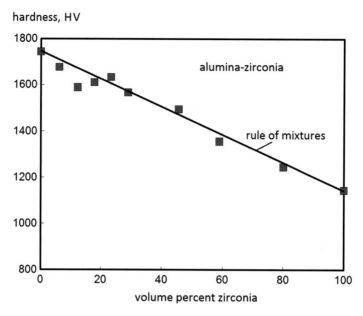

Fig. 4.2 Hardness for alumina-zirconia particulate composites across the full composition range. The experimental data [9] are compared with the volumetric rule of mixtures using the Vickers hardness scale, HV

fraction and hardness of the constituent phases [8]. The composite hardness H_C is estimated in terms of constituent phase hardness H and volume fraction V as,

$$H_C = V_1 H_1 + V_2 H_2 \tag{4.9}$$

where the subscripts designate each phase. As a demonstration of hardness variation with composition, Fig. 4.2 plots microhardness data versus composition across the alumina-zirconia composite system [9]. No substantial chemical interaction occurs between these phases. Thus, Eq. (4.9) is a good representation over the whole composition range in spite of a substantial hardness difference.

The linearly additive rule is accurate for many composite systems where there is no chemical interaction between phases, including Al-SiC, W-Cu, Mg-ZrO$_2$, and Mo-Al$_2$O$_3$ [5, 6, 10–12]. The microstructure role is ignored, but considerable variation in strength, ductility, and toughness arise from changes to the spatial relations between phases.

Contiguity is a microstructure parameter based on the fractional perimeter of a grain in contact with similar chemistry grains. A contiguity of unity is expected for a pure, dense solid. In practice, contiguity is varied by thermal treatments or chemical additives to adjust the interfacial energies. To allow for such variations, hardness models extent to include contiguity C in a modified expression [13, 14]:

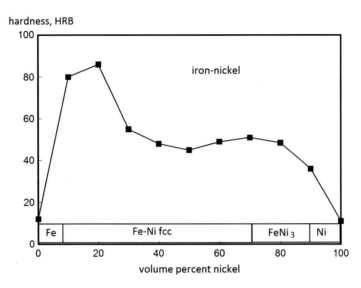

Fig. 4.3 Hardness (HRB scale) versus composition for Fe-Ni mixed phases [15]. The interactions between the constituents produce an involved hardness variation in contrast with that anticipated from the rule of mixtures, which fails to predict maximum values at intermediate compositions

$$H_C = V_1\, C_1\, H_1 + V_2\, C_2\, H_2 \qquad (4.10)$$

For a dense, two-phase composite $V_1 = 1 - V_2$ and $C_1 = 1 - C_2$. The contiguity modified model is accurate for cemented carbide WC-Co compositions. Composite hardness is a function of the carbide and cobalt hardness, composition, and microstructure contiguity.

If the phases exhibit solubility, then the hardness prediction is more complicated. For example, sintered mixtures of nickel and iron powders produce a hardness maximum near 20 vol% Ni, as plotted in Fig. 4.3 [15]. At high temperatures this system homogenizes, but at lower temperatures four phases form depending on composition. These phases are body-centered cubic Fe, face-centered cubic Fe-Ni alloy, ordered intermetallic FeNi$_3$, and face-centered cubic Ni. Approximate composition ranges are marked on the bottom of the hardness plot. A composition consisting of mixed nickel and iron powders produces different phases depending on composition. Each phase differs in hardness, leading to dramatic changes across the composition range depending on the chemical interactions. In this system the links between hardness and composition are complicated even by processing details. Care is needed to properly account for such interactions.

Microstructure scale is another factor, commonly included by grain size or grain spacing. For a constant composition, grain size G is proportional to the ligament size or mean separation between grains λ. Both parameters link hardness to microstructure since larger grains correspond to larger ligament sizes. The microstructure scale role involves an inverse square-root character known as the Hall–Petch relation [12, 16],

$$H_C = H_O + \frac{K_G}{\sqrt{G}} \qquad (4.11)$$

For a fixed composition, H_O represents the hardness associated with the composition at a large grain size and K_G is the grain size sensitivity parameter. For example, in a study of WC-11Co composites H_O is 1330 HV, K_G is 24 HV-μm$^{1/2}$ when grain size G is measured in μm. An alternative relies on the ligament size λ between grains, with a similar form,

$$H_C = H_O + \frac{K_\lambda}{\sqrt{\lambda}} \qquad (4.12)$$

where K_λ reflects the experimentally determined mean separation sensitivity. Figure 4.4 plots the relation between hardness and ligament size (inverse square-root) for WC-Co composites, showing agreement with the relationship given as Eq. (4.12) [14].

Other factors influence measured hardness, including residual strain (from deformation), porosity, and grain connectivity. Deformation is generally associated with dislocation generation and entanglement, leading to a power law relation between the composite hardness without strain H_O and after a deformation strain of ε;

$$H_C = H_O \, \varepsilon^m \qquad (4.13)$$

Fig. 4.4 Hardness for WC-Co compositions plotted versus the inverse square root of the microstructure spacing, known as the ligament size or mean free path, measured in mm [14]; the units on the horizontal axis are (mm)$^{-1/2}$

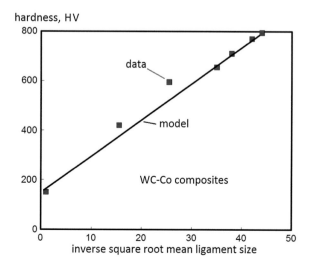

The exponent m depends on the material; for example ferrous alloys are characterized by m near 0.13.

Porosity reduces hardness. Over many systems, the porosity effect on composite hardness H_C is modeled as follows [17]:

$$H_C = H_O \, exp[-B\,(1-f)] \tag{4.14}$$

Again H_O reflects the expected composite hardness without porosity, f is the fractional density relative to theoretical (so $1-f$ is the porosity), and B is a material parameter providing a measure of the sensitivity to porosity. Two plots illustrate this relation. The first is for ductile stainless steel, plotted in Fig. 4.5 [18]. The sensitivity parameter B is 4 in this case. The second plot is for brittle composites of ruby (Al_2O_3-Cr_2O_3) with 6 vol% titanium, with up to 30% porosity, plotted in Fig. 4.6 [19]. In the second case the model corresponds to a sensitivity B of 3.2. Both cases exhibit agreement with Eq. (4.14).

The several effects lead to many expressions, depending on what parameters are known. It is not implied that all factors need to be included, but when models are constructed to explain composite hardness data, these factors might be considered:

- constituent phase hardness, H_1 and H_2
- composition, V_1 and V_2
- intersolubility
- contiguity, C_1 and C_2
- grain size, G

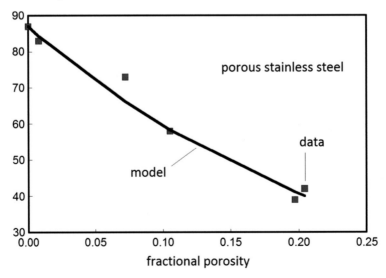

Fig. 4.5 A plot of hardness versus porosity for stainless steel, illustrating the agreement of the model given by Eq. (4.14) when compared to data for a ductile stainless steel [18]

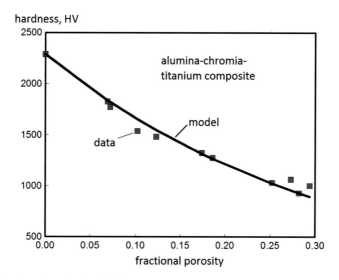

Fig. 4.6 Hardness for a brittle alumina-chromia-titanium composite showing the Vickers hardness decline with increasing porosity and the fit based on Eq. (4.14) [19]

- mean grain separation, λ.
- porosity, $1 - f$

Each situation provides additional adjustments. It possible to assemble all of the factors into a master model. Most typically a cascade of models is appropriate from constituent hardness and composition, to the influences dictated by the specific situation.

Elastic Modulus

The elastic modulus is also known as Young's modulus. It the stiffness of a material since it relates deformation strain to the applied stress. It is typically extracted from the loading portion of a tensile test. Elastic behavior is fully recoverable, meaning that when the stress is removed the strain relaxes back to zero. For crystalline materials, elastic modulus is isotropic and simply calculated from the slope of early stress-strain curve. For some materials elastic modulus varies with crystal orientation and even strain, but for particulate composites the assumption is that these variations are small, giving a single elastic modulus because of the large number of grain orientations contained in a composite test sample. Elastic modulus is a widely tabulated engineering property and treated by several calculation approaches in particulate composites [4, 5, 12, 20–22].

Fig. 4.7 Elastic modulus for composites consisting of silica added to polyimide, comparing measured values with the behavior expected from the volumetric rule of mixtures

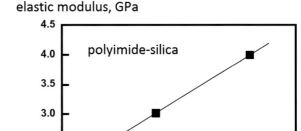

Models for the elastic modulus variation with composition start with a relation attributed to Einstein, a linear form applicable to low second phase concentrations [20],

$$E_C = E_1 \left(1 + 2.5\,\theta\,V_2\right) \tag{4.15}$$

Where E_C is the composite elastic modulus, E_1 is the major phase or matrix elastic modulus, θ is the sensitivity factor, and V_2 is the volume fraction of second phase. As a demonstration, Fig. 4.7 plots the measured elastic behavior for polyimide loaded with varying silica contents. In this case the sensitivity factor θ is about unity. Equation (4.15) is applicable to low concentrations of second phase, usually up to 20 vol%. Curiously, this model disregards the properties of the second phase and is only concerned with the amount of second phase.

Equation (4.15) has been modified to include a volume fraction squared term [22],

$$E_C = E_1 \left(1 + 2.5\,V_2 + 14.1\,V_2^2\right) \tag{4.16}$$

where the intent is to allow for interactions between the second phase grains. Again, the properties of the second phase are ignored for dilute concentrations, giving a slightly nonlinear behavior for low values of V_2.

As interaction terms are included, an adaptation of Equation (4.2) arises for composites with low concentrations of second phase [23]. The empirical behavior is represented as follows:

$$E_C = E_1\,\frac{1 + N\,U\,V_2}{1 - N\,V_2} \tag{4.17}$$

Although empirical, the sense is U relates to second phase grain shape and N relates the second phase elastic modulus. More involved relations for dilute second phase content composites include Poisson's ratio ν_1 of the major phase [20],

$$E_C = E_1 \left[1 + \frac{15 V_2 (1 - \nu_1)}{(1 - V_2)(8 - 10 \nu_1)} \right] \qquad (4.18)$$

Generally behavior is not sensitive to grain size. Trial data with systematic changes in grain size find a trivial effect, the variation is about the same as experimental scatter. For example, in glass filled epoxy, the elastic modulus at 30 vol% glass is 6.9 GPa using 1 μm spheres, and 6.6 GPa using 12 μm glass spheres [4, 24]. The mean value over a wide size range was 6.9 GPa. However, when the glass grain size reaches into 10 nm the elastic modulus changes due to the high interfacial area. Likewise, the interfacial adhesion between phases has no measurable impact on elastic modulus.

For composites with high second phase contents, the properties of each phase impact the elastic modulus. Classically, elastic modulus dependence on composition is treated using two models—isostrain or isostress. For these, the assumption is that either the two phases are carrying the same strain, but vary in stress, or the two phases are carrying the same stress, but vary in strain. The rule of mixtures assumes isotropic, homogeneous behavior with the composite elastic modulus E_C set by the relative volume fraction V and elastic modulus of each phase by the isostrain relation:

$$E_C = \sum V_i E_i \qquad (4.19)$$

where E_i is the elasticity and V_i is the volume fraction of phase i. Note $\sum V_i = 1$.

The second isostress relation employs the inverse rule of mixtures as follows:

$$\frac{1}{E_C} = \sum \frac{V_i}{E_i} \qquad (4.20)$$

Equation (4.19) assumes the strain is the same in all phases while Eq. (4.20) assumes the stress is the same in all phases. Most particulate composites fall between these two estimates, as demonstrated in Fig. 4.8. This plot for Al-SiC composites compares data with the isostrain and isostress predictions [5, 12, 25, 26]. The composite behavior is intermediate between the two extremes, with slight curvature over the composition range.

At 40 vol% SiC, the microstructure consists of dispersed SiC grains in a soft Al matrix. The elastic modulus is 69 GPa while the SiC is 450 GPa. Equation (4.19), predicts the composite elastic modulus at 221 GPa, while Eq. (4.20) predicts 104 GPa. The measured value is 166 GPa, almost what is given by the average from the two models at 163 GPa. This is common, where the particulate composite is intermediate between the two model predictions. To reinforce this sense, Fig. 4.9

Fig. 4.8 Elastic modulus versus composition for various Al-SiC composites [5, 12, 25, 26]. The plot compares experimental behavior with the upper and lower bound estimates from the isostrain and isostress formulations; Eqs. (4.19) and (4.20), respectively

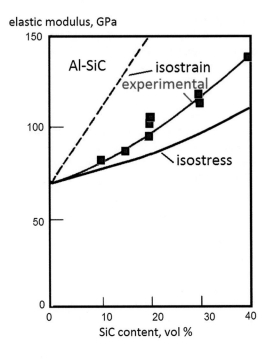

Fig. 4.9 Elastic modulus for alumina-zirconia composites across the full composition range [9]. The *square symbols* are the experimental measurements, the *solid black line* is the predicted behavior by averaging the isostress and isostrain predictions

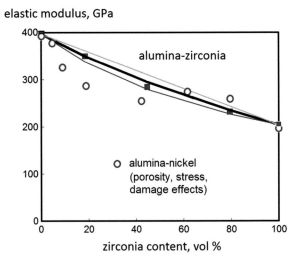

is a plot of elastic modulus for alumina-zirconia composites over the full composition range [9]. Also plotted are alumina-nickel data where porosity, residual stress, and grain damage adjustments are used to reduce the predictions at high alumina contents. Along with the experimental data are the expected values from averaging Eqs. (4.19) and (4.20). Test data from Cu-Al$_2$O$_3$ composites find the average of Eqs. (4.19) and (4.20) is within 3 % of the measured value. The average of the isostress and isostrain is a good first approximation.

A combination of models is the best representation for particulate composites. Other models require microstructure information, such as phase contiguity [5, 14, 23, 26]; however, such modifications often provide marginal gains in model accuracy.

A factor with recognized impact on elastic modulus is residual porosity. Porosity can be included in Eqs. (4.19) and (4.20) by adding a pore phase with zero elastic modulus. Various other relations are offered to discount the predicted elastic modulus with respect to porosity, including power law relations such as,

$$E_C = E_O f^Y \tag{4.21}$$

where f is the fractional density, E_O is the calculated elastic modulus for full density, and Y is an empirical exponent between 3 and 4. Other models are applied to the special cases when the pores are closed. The models are similar to ideas already developed in this chapter, with a few variants as follows [17]:

$$E_C = E_O \, exp[- \, B \, (1-f)] \tag{4.22}$$

$$\frac{1}{E_C} = \frac{1}{E_O} + C \, \frac{1-f}{f} \tag{4.23}$$

$$E_C = E_O \, exp\left[- \, b \, (1-f) - c \, (1-f)^2\right] \tag{4.24}$$

In these models E_O represents the elastic modulus without porosity. The symbols are the same as previously defined, with the constants B, b, C, and c determined by fitting to experimental data. Generally Eq. (4.22) provides the best fit. The last expression, Eq. (4.24), is accurate for high porosity levels, but pore shape becomes a factor at more than 80 % porosity, when the elastic modulus is just 1 % of the full density value.

Since the shear and bulk moduli are related to the elastic modulus, essentially all of the above models can be formulated to predict bulk modulus or shear modulus from the constituent properties and composition.

Poisson's Ratio

Models for Poisson's ratio in particulate composites are generally absent. As with other properties, a proportional rule is a starting point. A complication arises if there is some interaction between phases. Otherwise the assumption is a linear variation with composition. As an illustration, Fig. 4.10 gives data for WC-Co compositions

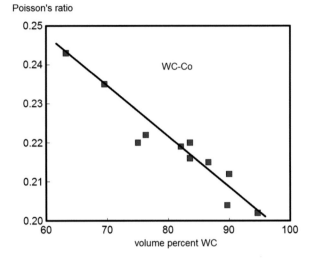

Fig. 4.10 Poisson's ratio versus composition for WC-Co composites and a linear fit to the behavior [27]

ranging from 3 to 25 wt% Co (5–37 vol%) with a linear fit [27, 28]. Other models add considerable complexity, leading to high complexity with diminishing merit.

The Poisson's ratio ν_C for a composite is degraded by porosity. It is treated as a linear function of density as follows [17]:

$$\nu_C = \nu_O \left(1 - \omega \left(1 - f\right)\right) \qquad (4.25)$$

with ν_O being the full density composite Poisson's ratio, f is the fractional density, and ω being a pore sensitivity factor. The pore sensitivity factor for a ductile composite is generally below 1. Another model relies on an exponential relation to fractional density f,

$$\nu_C = \nu_O \, A \, exp(-f) \qquad (4.26)$$

Where A is an empirical factor. In critical analysis, data from Cu-Al$_2$O$_3$ composites tends to show the rule of mixtures is reasonably accurate; thus, Eq. (4.19) with Poisson's ratio substituted for elastic modulus, is the best form for broad composition ranges.

Strength

A first estimate of strength is possible using hardness. As a rule of thumb, the tensile strength is about three times the Vickers hardness. Figure 4.11 plots 36 different particulate composites to show this correlation [5, 7, 12–14, 20, 21, 25]. The included composites range from epoxy-glass to mixed carbides (WC-5Co-1TaC). The correlation between Vickers hardness and strength is 0.95, or very significant.

Fig. 4.11 Correlation of strength and hardness from 36 different particulate composites [5, 7, 12–14]. Although scattered, the key sense is that the strength in MPa is approximately three times the HV hardness

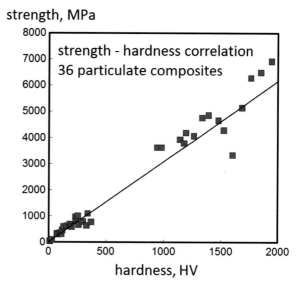

The low hardness materials rely on tensile results, but for the high hardness, easily fractured materials, rely on the higher compressive strength.

High hardness materials are poor in tension due to low fracture toughness. Thus, tensile strength is low while compression strength is high. For example, WC-15TaC-10Co-5TiC has a hardness of 1476 HV, tensile strength of 2380 MPa, and compressive strength of 4666 MPa. The latter value is used in this plot. For ductile and tough composites, such as Zn-50SiC, the hardness is 114 HV and tensile strength is 310 MPa, which is almost the same as the compression strength of 330 MPa.

A first estimate of the composite strength σ_C comes from the Vickers hardness H_C,

$$\sigma_C = 3\,H_C \qquad (4.27)$$

Under tension the interface role is dominant. Interface area in the composite increases with a smaller grain size. Accordingly, the tensile behavior depends on several factors, including the grain size, interface area, and microstructure. A demonstration of the complexity is given in Figs. 4.12 and 4.13. The first figure gives stress-strain curves for polyurethane reinforced with 60 μm glass beads from 12 to 58 vol% [29]. The reinforcement decreases the strain to failure, but lacks any systematic influence on peak strength. On the other hand, for the polypropylene-calcite composite in Fig. 4.13, strength decreases with the addition of all but the smallest particles.

The phase morphology influence stress transmission leading to an abundance of strength models [5, 23, 25, 30]. Depending on the number of phases, phase size, and connectivity, there might be strengthening, softening, or premature debonding. The

Fig. 4.12 Tensile test
results for four
polyurethane-glass
composites
[29]. Comparison of
strength shows a variable
response as the glass
content increases, but
ductility declines
systematically

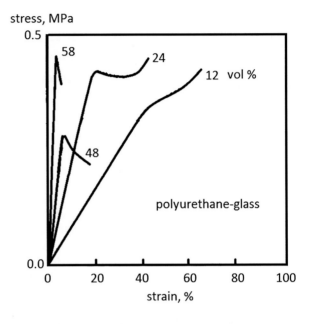

Fig. 4.13 Tensile strength
of polypropylene-calcite
composites versus the
volumetric calcite content
for four different particle
sizes of spherical calcite
grains

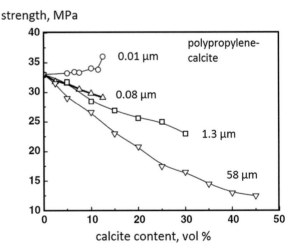

relative behavior shifts with composition and grain size ratio, leading to much variation [31]. Table 4.2 tabulates tensile strength and elongation to fracture for W-Ni-Fe composites, with Ni:Fe = 7:3 on a weight ratio. The data are for varying tungsten contents [32]. Note the strength is nonlinear with composition.

Particulate composite strength models include an interface cohesion term χ [20, 33],

Table 4.2 Strength and ductility for W-Ni-Fe composites [32]

Composition, wt% W	Tensile strength, MPa	Fracture elongation, %
88	907	35
90	925	31
93	986	23
95	917	18
97	888	12

$$\sigma_C = \sigma_1 \left(1 - 1.21 \, V_2^{\frac{2}{3}}\right) \chi \tag{4.28}$$

The mechanical properties of the second phase are absent, only the first phase strength σ_1 and second phase volume fraction V_2 are considered. Interface cohesion is adjustable from about 0.2 to 1.0. One difficulty is the assumption that all composites are weaker than the strongest constituent, which is not always the case. Indeed, load transfer between phases is a dominant parameter that depends on processing, mainly in how wetting and interdiffusion takes place between phases.

Strength trials over a range of compositions are important to developing and optimizing particulate composites. In practice, the behavior is unpredictable as illustrated by the data in Table 4.3 [1, 5–13, 25, 29–36]. Collected here are strength ratios (composite strength as a ratio to the pure matrix strength) for several systems at 20 vol% second phase. Cases included in the table show both strengthening and weakening. When the ratio is near unity, the implication is that the second phase is replacing the matrix, providing neither benefit nor detriment. No specific pattern emerges in the data. The wide variation reflects the combined roles of interface, phase connectivity or contiguity, and the relative properties of the two phases.

Decreased ductility also impacts strength, especially for composites where work hardening occurs prior to fracture [6]. During tensile testing a low strain fracture implies less work hardening and a lower tensile strength. Large gains are possible in systems that form interpenetrating networks during consolidation. Further, strength declines with a loss of homogeneity [20, 37]. On the other hand, when one phase is isolated, the composite shows degraded strength. Thus, three variants are possible:

1. *beneficial*—the composite is stronger since the second phase forms an interpenetrating network to distribute loading,
2. *neutral*—the second phase has no strength impact,
3. *degradation*—the composite is low in strength since the second phase acts like porosity to reduce load bearing area.

Consequently, reliable composite strength models are missing. As factors such as microstructure, residual stress, or processing history are added, there is a progressive gain in the models, but there are no general strength models as yet.

It is possible to increase strength using a small grain size. The behavior follows the Hall–Petch relation [38],

Table 4.3 Strength ratio for composite versus pure matrix [1, 5–13, 29–36]

Major phase	Second phase	Vol%	Strength ratio[a]
Al	AlN	15	0.9
Al	Si	20	1.2
Al	TiB_2	20	1.8
Al alloy	Al_2O_3	20	1.3
Al alloy	B_4C	20	1.3
Al alloy	SiC	20	1.6
Al_2O_3	ZrO_2	20	0.6
Al_2O_3	Mo	16	1.0
Al_2O_3	Ni	20	0.5
Be	BeO	20	1.2
Co	WC	24	1.5
Cu	TiB_2	20	1.2
Fe alloy	TiB_2	20	1.2
Epoxy	Glass	20	1.1
Glass	Ni	20	0.7
HfC	$MoSi_2$	20	1.0
Mg	B_4C	20	2.8
Mg alloy	SiC	20	0.8
Mg alloy	B_4C	20	1.1
Mo	TiC	25	2.7
Polyester	$CaCO_3$	20	0.7
Polyimide	SiO_2	20	0.9
Polypropylene	$CaCO_3$	20	0.8
Polystyrene	Glass	20	0.9
Polyurethane	$BaSO_4$	20	1.0
SiC	ZrO_2	15	1.4
Si_3N_4	TiN	20	0.9
Stainless steel	Al_2O_3	20	0.3
Stainless steel	Cr_2C_3	20	0.6
Stainless steel	TiB_2	15	1.6
Stainless steel	TiC	20	0.5
Stainless steel	TiN	20	0.6
Ti	(TiB+TiC)	22	1.5
Ti alloy	TiC	20	1.0
Ti alloy	TiB	20	1.4
TiAl	TiB_2	20	1.6[b]
TiB_2	Fe	20	0.5
TiB_2	ZrO_2	20	1.9
TiC	Ni	20	4.3
Ti(C,N,B)	ZrO_2	20	0.5
W	Cu	20	1.0
W	(Ni+Fe)	20	1.7

(continued)

Table 4.3 (continued)

Major phase	Second phase	Vol%	Strength ratio[a]
WC	Co	20	4.5
WC	Ni	20	4.1
Zn alloy	SiC	20	0.8

[a]Ratio of composite strength to major phase strength under same conditions
[b]Test data at 600 °C

$$\sigma_C = \sigma_O + \frac{K_G}{\sqrt{G}} \tag{4.29}$$

where G is the grain size and K_G is a sensitivity parameter. For example in Al-50SiC composites, with either 13 or 165 μm grains, the tensile strength was 590 MPa for the smaller grain size and 390 MPa for the larger grain size. Similar to hardness, strength varies with the inverse square-root of the microstructure scale. The intergrain spacing is an alternative parameter used in some studies [14, 39].

Grain shape is important to load transfer between phases. Irregular grains induce stress concentrations that reduce strength when compared to rounded grains. In some cases an irregular grain shape reduces strength by 50 %. An elongated grain, although impractical in some instances, provides more interface area for load transfer. Conceptually elongated grains can double the strength over that of a spherical grain. Ellipsoidal grains give a minor strength gain over spheres.

For many composites, interface failure occurs prior to failure of either phase. Cracks formed at the interface lead to significant strength loss. Analysis of fractures shows the cracks form in the interfaces between closely spaced hard grains. As the grain size decreases the induced early failure leads to less deformation prior to fracture, possibly with more strength. Each composite has a different behavior, so a generalized model is not possible. Instead, experimental data are fit with multiple parameter models that include adjustable interface terms, but there is no predictive basis [40].

Porosity is a factor that lowers composite strength σ_C. In one case 13.5 % porosity reduced tensile strength by 43 %. Several models incorporate the role of porosity. A simple power law relation is useful, where,

$$\sigma_C = \sigma_O \, k f^P \tag{4.30}$$

where σ_O is the strength of the composite at full density, in the same condition with regard to grain size and thermal treatment, k is effectively a stress concentration factor that varies with test geometry and processing details, and P is exponent between 3 and 6. High values for P indicate large stress concentration effects at the pores. In such a case the fracture surface shows an abundance of pores, indicative of cracks bridging between pores. The high sensitivity to pores often leads to justifiable demands for full density.

Another simple model to explain composite strength σ_C variation with fractional density f follows an exponential behavior similar to that shown earlier for hardness, as follows [17]:

$$\sigma_C = \sigma_O \, exp[- a \, (1-f)] \tag{4.31}$$

where σ_C is the composite strength, discounted from the full density composite strength σ_O, with a sensitivity factor a near 6.

For strength, the list of considerations in modeling behavior is long, including -

- constituent phase strength, σ_1 and σ_2
- composition, V_1 and V_2
- contiguity, C_1 and C_2
- grain size or other microstructure scale parameter, G
- grain shape with long, slender rod-shapes prove best, but rounded or spherical grains are superior over irregular grains
- microstructure homogeneity
- residual strain, ε
- porosity, $1-f$
- interfacial adhesion.

Failure Distributions

Often strength is limited by processing defects, especially in low ductility composites loaded in tension. A statistical measure of the spread in fracture strength is taken from the Weibull distribution fit to fracture strength data [41]. The Weibull distribution model assumes failure depends on the defect population. Large defects cause fracture at a low stress. The defect population depends on sample size. Accordingly, small test samples give higher strengths, simply because the flaw size and number of flaws is inherently limited in a small sample.

The statistical distribution for failure is in the form of an exponential distribution function; the failure probability F depends on the applied stress S and the characteristic strength of the composite S_O as follows:

$$F(S) = 1 - exp\left[-\frac{V}{V_O} \left(\frac{S - S_U}{S_O} \right)^M \right] \tag{4.32}$$

Letting $F(S)$ be the cumulative probability for failure, then the probability of surviving to a loading stress S is $1 - F(S)$. In Eq. (4.32), V is the actual volume, and V_O is the sample volume used in testing. The characteristic strength S_O corresponds to a 63 % fracture probability. The parameter S_U is the proof stress, indicating the minimum stress each component is subjected to without failure prior to service. For many situations, where no preloading is applied the proof stress is zero. Finally, M is the Weibull modulus, a measure of the strength dispersion. A high modulus indicates a narrow spread in strength and is associated with few manufacturing defects.

One consequence of this equation is seen in the common claim of being "stronger than steel". Such claims are based on handbook strength for steel, but compared to test results on tiny threads of a new material. Even steel is much stronger (about threefold) when tested as a tiny thread. So unless the test volume is large, comparable to the component volume, strength always appears high and is difficult to attain in practice.

The Weibull distribution is a valuable representation for strength variations in particulate composites. A strength distribution occurs around the characteristic strength, with a variation described by the modulus M. For typical manufacturing conditions, the Weibull modulus is about 10, and with care it reaches 20. Only in situations where everything is carefully controlled does it reach 25–35.

In life-critical applications, it is important to predict failure probability to avoid injuries. It is impossible to have zero failure probability unless proof testing is applied. Without proof testing, a failure probability of 10^{-9} is typical in life-critical situations. A high Weibull modulus is required, implying the fabrication process is carefully controlled at all steps and the applied stress is kept low even when using a high strength material. Likewise, most desirable are small components to minimize the volume effect in Eq. (4.32). Additionally, failure probability is reduced by proof testing. It is best to perform testing on specimens the same test volume as used in an application. For instance, a material with 850 MPa characteristic strength, proof tested at 400 MPa, and produced with a Weibull modulus of 25, can only be safely subjected to 730 MPa if the use volume is tenfold larger than the test volume.

To evaluate the strength distribution, usually at least 40 samples are tested. The fracture strengths are ranked from lowest to highest and the probability is associated with the relative progression. For 40 samples, the lowest fracture strength corresponds to 2.5 % failure probability; the next fracture corresponds to 5 % failure probability, and so on. Rearranging Eq. (4.32) gives a linear equation in terms of the inverse survival probability and stress, often ignoring the proof stress,

$$log\left[log\left(\frac{1}{1-F}\right)\right] = log\left(\frac{V}{V_O}\right) + M\,log\left(\frac{S}{S_O}\right) \qquad (4.33)$$

A plot of the double logarithm of inverse survival probability (left hand side) versus logarithm of fracture strength is fit with a straight line. Such a plot is illustrated in Fig. 4.14 for glass, alumina, and silicon nitride. The slope is the Weibull modulus M. To understand the relative failure probability, the plot is according to Eq. (4.33), labeled for the corresponding values of F. A least squares analysis is used to extract the two adjustable terms. Curiously the high strength silicon nitride comes from a self-composite, where large Si_3N_4 grains are used to strengthen a smaller grain size silicon nitride [42].

Fig. 4.14 Fracture statistics plotted according to Eq. (4.33) for extraction of the Weibull modulus based on least squares fit to the cumulative failure distribution, illustrated here for glass, alumina, and silicon nitride systems [32]

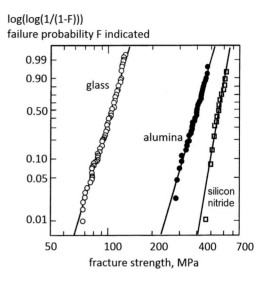

Ductility

Ductility is a moot point for brittle materials. A ductile matrix decreases ductility when a hard, brittle phase is added; thus, most metal matrix composites are low in ductility [1, 30, 43]. For example, Fig. 4.15 plots the declination in ductility with increased additive concentration (SiC or Al_2O_3) for composites based of aluminum and magnesium alloys [12, 25]. The behavior shows considerable scatter, but at about 15 vol% ceramic the ductility is less than 50 % of the matrix alloy.

Ductility is difficult to predict. For low ductility materials, both the ultimate tensile strength and fracture energy depend on the ductility. The lower the fracture elongation, the lower the tensile strength, meaning less net fracture energy. In the case of a poorly bonded interface, cracks nucleate and grow at low stresses, linking together to induce failure with low ductility.

A ductile matrix containing a low ductility second phase results in ductility loss as the second phase content increases. Starting with a matrix phase 1, the higher the second phase contiguity C_2, corresponding to the lower ductility phase, the lower the composite ductility ε_C. The following empirical relation fits a range of experimental data:

$$\varepsilon_C^B = \kappa \left(1 - C_2\right) \tag{4.34}$$

the parameter κ relates to the ductility of the full density, pure matrix phase 1, and B is a sensitivity parameter. For example, in trials with tungsten composites, B is 0.7.

Another situation corresponds to a brittle material with an added ductile second phase. For this situation, ductility varies with composition. An example is plotted in

Fig. 4.15 Scatter plot of
the ductility reported for
various metal-ceramic
composites (aluminum
alloy with alumina and
magnesium alloy with
silicon carbide)
[12, 25]. The ceramic
additions cause a loss of
ductility compared to the
parent alloy (relative
ductility is measured
ductility divided by the
alloy ductility). Modeling
this behavior is difficult due
to the wide variation in
interface quality

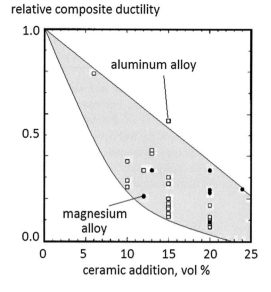

Fig. 4.16 Ductility
measured by the fracture
elongation for tungsten
composites with increasing
content of ductile alloy
[32]. Pure tungsten is brittle,
but composites with the
tough nickel-iron phase
(Ni-30Fe) significantly
increase ductility

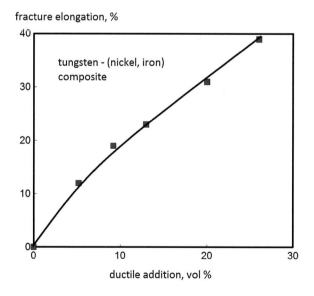

Fig. 4.16 [32]. This plot gives the fracture elongation versus ductile phase content
for a tungsten-based composite, where the ductile phase is a nickel-iron alloy. Over
the range from 0 to 25 vol% ductile phase, the fracture elongation increases
dramatically, due to excellent interface adhesion between the phases. The tough
phase addition significantly alters the otherwise brittle tungsten.

Work hardening means a material requires a progressively higher stress to continue deformation. Accordingly, the ultimate tensile strength is linked to the fracture ductility [44]. The exact behavior changes for each composite system, but once the behavior is understood it is possible to predict strength and ductility from each other [45].

Pores concentrate stress and act to initiate fracture, so ductility falls with incomplete densification. For example, in one system the fracture elongation at 10 % porosity is 15 % of the full density value. By 30 % porosity the relative elongation is 3 % of the full density value, effectively behaving as a brittle material. The ratio of porous composite ductility with respect to the full density ductility is the relative ductility δ. When compared under equivalent processing conditions, the relative ductility depends on the fractional density f,

$$\delta = \exp[-\alpha (1 - f)] \qquad (4.35)$$

where α is an experimental parameter. Other models link ductility to fractional density using polynomial relations, but these models overestimate ductility at high porosity levels.

Deformation and Fracture

Fracture behavior depends on several factors. A strong interface enables load transfer between phases while a weak interface results in premature failure with low strength and ductility [6, 12]. Accordingly, deformation and fracture models require assumptions on characteristics of the composite interface [1, 46]. Figure 4.17 sketches of several fracture path variations. The energy required to fracture the composite depends on the weakest path, implying fracture resistance increases as the crack interacts with the phases. If low ductility is sought, such as in frangible ammunition, then weak interfaces are desired to allow early fracture. More commonly toughening and strengthening are desired via strong interfaces.

High toughness is attained when the fracture path passes through both phases. Especially desirable is crack blunting. A homogeneous second phase generally eliminates easy fracture though one phase versus the other. Figure 4.18 compares heterogeneous and homogeneous Al-SiC microstructures [12]. Fracture preferentially jumps though the hard SiC grain clusters in the heterogeneous structure, reducing fracture strength [47, 48]. Residual porosity is another negative influence on fracture resistance.

Interface related properties are evident in the following:

1. impact toughness,
2. work hardening and plastic deformation,
3. strain rate hardening,
4. fatigue,
5. fracture toughness.

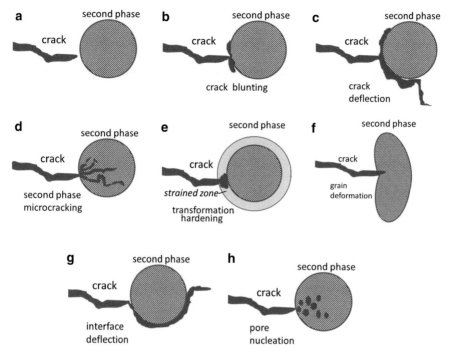

Fig. 4.17 Possible fracture path interactions with a second phase that add to the composite toughness by increasing the energy required for crack advance. Starting with the crack approaching the second phase in (**a**), the variations include; (**b**) crack blunting, (**c**) crack deflection, (**d**) crack propagation into the second phase with microcracking, (**e**) induced toughening due to induced transformation, (**f**) grain deformation and plastic working, (**g**) interface deflection, and (**h**) pore nucleation inside the second phase

Early information on the interface role is evident by fracture surface inspection using scanning electron microscopy. An example is evident in the contrasting images in Fig. 4.19. The images are of the fracture surface for the same composite, but with different heat treatments. The upper image corresponds to fracture with a weak interface, induced by impurity segregation during slow cooling. The material was brittle. The lower image corresponds to fast cooling to avoid impurity segregation. Without the weak interface, the strength, ductility, and toughness all improved.

Prediction of toughness variation with composition is not difficult. For example, alumina-zirconia system fracture toughness data are summarized in Table 4.4 [34]. The toughness depends on the amount and size of the zirconia phase. It is common to find toughness reaching a maximum at an intermediate composition. Additionally, fracture toughness varies with grain size, as illustrated by the plot in Fig. 4.20 for B_4C-10TiC composites [49]. The peak fracture toughness occurs near

Fig. 4.18 Comparative
micrographs of
heterogeneous and
homogeneous
microstructures of Al-SiC
composites [12]. The closer
spacing of the hard SiC
grains in the heterogeneous
structure leads to preferred
fracture through the
clustered regions

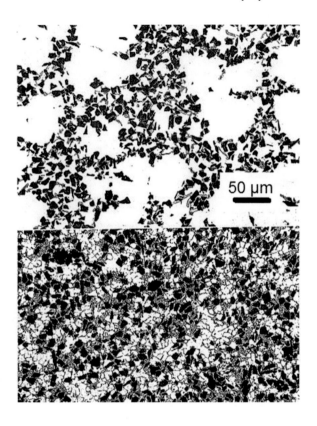

a grain size of 10 μm. For this system, fracture toughness declines with more TiC, in spite of higher hardness and strength; for example, falling from 4.1 MPa√m at 290 MPa to 2.1 MPa√m at 580 MPa.

Thermal softening, the loss of strength with heating, tends to increase ductility, but mostly lowers strength and lower toughness. Figure 4.21 plots the fracture toughness at elevated temperatures for an alumina-zirconia composite consisting of 29 vol% zirconia [50]. For this system, the zirconia was toughened using 2 % yttria. As an approximation, fracture toughness decreases in a linear manner with increasing temperature.

There are instances where composites increase fracture toughness during crack advance. These situations are usually aligned with discontinuous cracking, possibly when small cracks form but do not propagate.

Because of issues related to interface quality, microstructure, and homogeneity, accurate models for deformation and toughness prove elusive. However, indirect approaches using experimental data can be successful [46]. Properties such as strength, ductility, and hardness provide a basis for predicting fracture toughness. An example correlation is given in Fig. 4.22, plotting fracture toughness versus hardness for WC-Co compositions [13, 14, 39]. A high hardness implies a low fracture toughness. Further, since hardness and strength depend on microstructure

Fig. 4.19 Fracture surfaces for a composite with different heat treatment cycles to alter the interface. The *upper image* corresponds to slowly cooled material with low ductility due to impurity segregation to the interfaces, while the *lower image* corresponds to rapidly cooled material with high ductility since impurity segregation was avoided. Both images are backscatter scanning electron micrographs

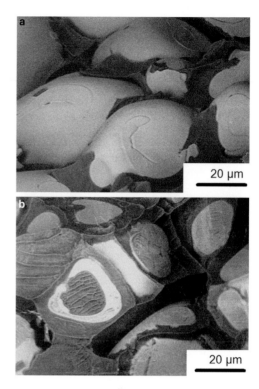

Table 4.4 Fracture toughness for alumina-zirconia composites [34]

ZrO$_2$ content vol%	ZrO$_2$ grain size μm	Fracture toughness K$_{Ic}$, MPa√m
0	–	4.9
10	1.25	6.1
10	1.75	6.8
15	1.25	9.8
20	1.25	5.1

features such as grain size, contiguity, and grain spacing, it is possible to correlate via regression the fracture toughness to these features. Experimentation is required to isolate the relations. Hardness is often a linear function of composition as illustrated in Fig. 4.2, but this is not true for fracture toughness. For example, Fig. 4.23 plots fracture toughness as a function of composition for alumina-zirconia composites [9, 13, 14, 39, 51]. In these data, the correlation between fracture toughness and strength is 0.000, meaning there is no relation. Thus, relying on one property, such as strength, to predict fracture toughness, is without merit except for special cases.

Porosity is detrimental to toughness; in one study the impact toughness fell by 50 % with 1 % porosity and fell by 90 % with 10 % porosity. Pores act as initial

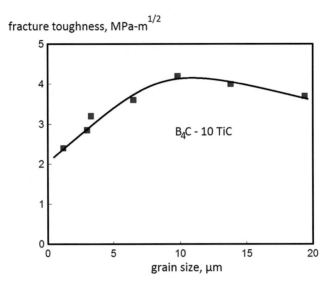

Fig. 4.20 Fracture toughness for B₄C-10TiC composites with intentional variations in grain size, showing a peak at about 10 μm [49]

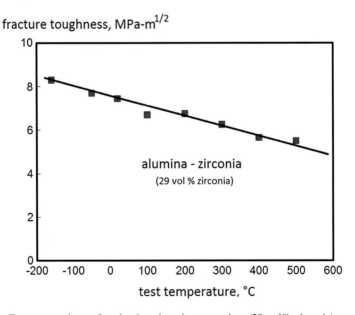

Fig. 4.21 Fracture toughness for alumina-zirconia composites (29 vol% zirconia) versus test temperature, demonstrating a linear decline as temperature increases, similar to that seen with hardness, strength, and elastic modulus [50]

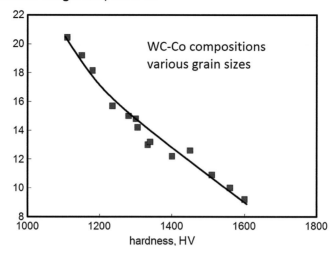

Fig. 4.22 The relation between hardness and fracture toughness for WC-Co composites of varying cobalt content and grain size. The relation is not quite linear

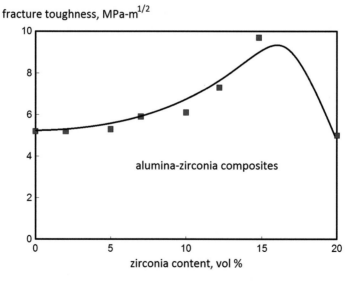

Fig. 4.23 Fracture toughness for alumina-zirconia composites over a range of compositions [9, 13, 14, 39, 51]. The increase and then decline in toughness with increasing zirconia content reflects a strong microstructure role. In this case hardness or other parameters are poor predictors for toughness

cracks and provide easy fracture paths. Consequently, many high performance composites mandate pore removal for life-critical applications. The larger and irregular pores are the most detrimental.

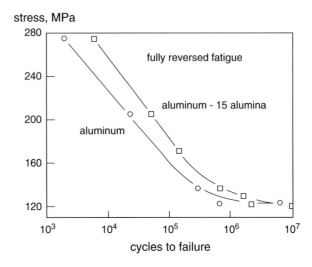

Fig. 4.24 Fully reversed fatigue data for aluminum 6061-T6 and the same alloy with 15 vol% alumina [12]. The failure stress and number of cycles show convergence to the same fatigue endurance strength

Fatigue behavior is difficult to model in particulate composites. Two phase structures generate differential strains on loading because of different elastic moduli. How these differential strains interact influences fatigue life. Failure occurs at the interface, possibly inducing early fracture. As shown by Fig. 4.24, the fatigue endurance strength for aluminum at ten million cycles is not improved by 15 vol% alumina addition [12, 52]; however, fatigue life is improved for composites at fewer cycles. Elevated temperatures reduce the fatigue endurance strength, mostly due to thermal softening [53, 54].

High Temperature

Thermal softening causes materials to lose strength when heated. The same is true for composites, in spite of second phase strengthening effects. As an example, Fig. 4.25 plots the temperature dependence of tensile yield strength for a magnesium alloy and that same magnesium alloy with 15 vol% silicon carbide [12]. Although the SiC retains strength to high temperatures, it is not able to offset thermal softening of the magnesium. The curves are nearly parallel, displaced by an average of 18 MPa; the composite is higher in strength, but shows similar softening. Strengthening also depends on grain size, so significant benefit arises from small second phase grains. Table 4.5 compares strength data for an aluminum alloy at two temperatures with three SiC grain sizes [1, 5, 12, 25]. At both 20 °C (293 K) and 300 °C (573 K) the highest strength is for the composite with the smallest second phase grain size. Thermal softening is less pronounced for the composites; the 300 °C (573 K) strength at 152 MPa is 35 % of the room temperature strength 428 MPa strength for the composite.

strength, MPa

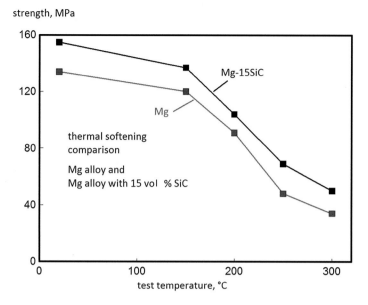

Fig. 4.25 Thermal softening behavior for a magnesium alloy and a composite of that same alloy with 15 vol% silicon carbide [12]

Table 4.5 Test temperature and grain size effects on Al-alloy strength with and without 15 vol% SiC [1, 5, 12, 25]

SiC content vol%	SiC size, μm	20 °C strength, MPa	300 °C strength, MPa
0	–	91	23
15	20	200	72
15	3	301	88
15	1	438	152

Simultaneous exposure to high temperature and stress induces creep. Composites can resist creep since the interfaces disrupt deformation. The activation energy is a measure of the temperature sensitivity, and composite creep generally has a higher activation energy indicative or improved creep resistance. Even so, the composite deformation strain rate increases with temperature or stress. Amorphous phases deform by viscous flow, while crystalline phases deform by diffusional flow. As a generalized treatment, the deformation strain rate $d\varepsilon/dt$ is a function of stress S and absolute temperature T as follows [53–55]:

$$\frac{d\varepsilon}{dt} = \frac{B}{V_2^P} S^N exp\left[-\frac{Q}{RT}\right] \tag{4.36}$$

where B is a collection of material constants, V_2 is the volume fraction of second phase, Q is the activation energy, R is the gas constant, and the exponent P is

Fig. 4.26 Steady state
creep strain rate plotted
versus the applied stress for
Al-30SiC at 375 °C (648 K),
illustrating the behavior
expected from Eq. (4.36)

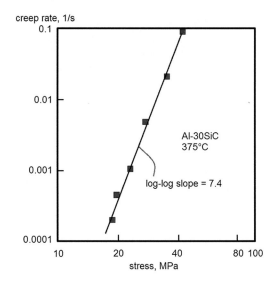

Fig. 4.27 Creep strain rate
data for alumina-zirconia
composites ranging from
20 to 80 vol% alumina,
tested at 1375 °C (1623 K)
at applied stress levels of
20–70 MPa [55]

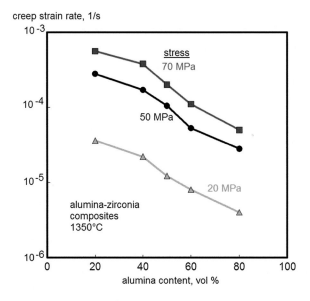

typically near 3. Typical behavior gives the stress sensitivity N ranging from 5 to
25. Some models embed the stress other forms, but still show a high sensitivity.
Figure 4.26 plots data for the creep deformation of Al-30SiC at 375 °C (648 K). The
strain rate and applied stress are on log-log scaling. The slope of 7.4 corresponds to
N in Eq. (4.36). Experiments at different test temperatures provide a means to
extract the activation energy [56]. So the key factors are composition, stress, and
temperature. The first two of these are evident in creep rate plot for alumina-
zirconia composites in Fig. 4.27 [55]. Three stress levels are included at 1350 °C
(1623 K); the creep rate declines with high alumina content and lower stress.

Composites with two phase connectivity are generally creep resistant. As an example, the NiAl-20TiB$_2$ system at 1027 °C (1300 K) exhibits one-third the creep of pure NiAl at the same temperature. Systems of hard reinforcement grains, such as in Al-SiC and NiAl-TiB$_2$, exhibit $N = 7$ with activation energies near 400 kJ/mol. The high activation energy indicates an inherent resistance to deformation.

Heat Capacity

Heat capacity is a linearly additive parameter, following the rule of mixtures. Pores have zero heat capacity (except for a trivial trapped gas role). For a dense, two-phase composite, consisting of mass fractions M_1 and M_2 with heat capacities C_{P1} and C_{P2}, the linear combination gives the composite heat capacity as C_{PC} as follows [57],

$$C_{PC} = \sum M_J \, C_{PJ} = M_1 \, C_{P1} + M_2 \, C_{P2} \qquad (4.37)$$

However, this equation typically underestimates behavior versus experimental values. To address the error, modifications include thermal expansion coefficient and bulk modulus to reflect heat stored as differential strains between phases [58]. Since energy is stored in the composite as strain, the modified models include the test temperature range.

The volumetric rule of mixtures is a reasonable basis for estimating composite heat capacity. The relation is the same form as Eq. (4.37), but with volume fraction instead of mass fraction,

$$C_{PC} = \sum V_J \, C_{PJ} = V_1 \, C_{P1} + V_2 \, C_{P2} \qquad (4.38)$$

with V_1 and V_2 being the volume fractions. The predicted behavior is closer to measurements, but tends to be slightly low as a demonstrated in Fig. 4.28 for W-Cu. The same data differ from the mass-based model of Eq. (4.37) by 10 % at the intermediate compositions.

To handle interactions between the phases, modeling efforts include properties such as the contiguity to account for different levels of second phase contact [59]. An empirical form includes interaction between phases as follows:

$$C_{PC} = (V_1 \, C_{P1} + V_2 \, C_{P2}) \, (1 + 0.2 \, V_1 \, V_2) \qquad (4.39)$$

This particular form assumes the particles are spherical. It is designed for polymer-particle composites (where the solid particles are not bonded) and for those systems provides an accurate fit to experimental data.

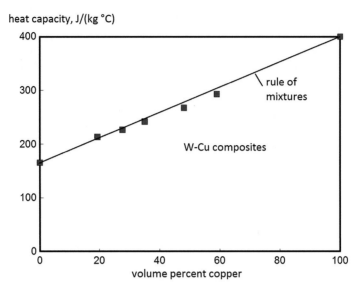

Fig. 4.28 Heat capacity versus composition for W-Cu composites, showing the agreement with the linear volumetric rule of mixtures

Thermal Expansion

Thermal expansion measures the change in component length on heating. It is the fractional length change ΔL normalized to the starting size L_O in proportion to the temperature change ΔT,

$$\frac{\Delta L}{L_O} = \alpha \, \Delta T \tag{4.40}$$

The thermal expansion coefficient α is a material parameter, often assumed to be constant. Effectively this give the elastic strain with heating. The thermal expansion coefficient is expressed as $10^{-6}/°C$ or ppm/°C or per K, where ppm is an abbreviation for parts per million). A low thermal expansion coefficient reflects strong atomic bonding, high elastic modulus, and high melting temperature. High thermal expansion phases have the opposite characteristics. Compare three metals at room temperature, Al, Ni, and W;

- Al—expansion   coefficient = 23.8   ppm/°C,   melting = 660   °C, elasticity = 70 GPa
- Ni—expansion   coefficient = 13.3   ppm/°C,   melting = 1453   °C, elasticity = 214 GPa
- W—expansion   coefficient = 4.6   ppm/°C,   melting = 3410   °C, elasticity = 400 GPa.

Often the thermal expansion coefficient increases slightly during heating, so it is not truly a constant. For example, between room temperature and 200 °C the thermal expansion coefficient of nickel increases from 13.3 to 13.9 ppm/°C.

For a composite, the thermal expansion coefficient depends on composition and microstructure, as well as defects such as cracks or pores [58–61]. When one phase is dispersed in the other, without connective bonds for the second phase, the volumetric rule of mixtures gives the composite thermal expansion coefficient α_C:

$$\alpha_C = \sum V_J \, \alpha_J = V_1 \, \alpha_1 + V_2 \, \alpha_2 \tag{4.41}$$

The latter form applies for two phases, where α_I is the thermal expansion coefficient of the first phase present at a volume fraction V_I, and α_2 is the thermal expansion coefficient of the second phase present at a volume fraction $V_2 = 1 - V_I$, assuming no pores. The volumetric rule of mixtures is illustrated in Fig. 4.29 for alumina-nickel composites at 200 °C (473 K). Some predictive accuracy gain is possible by adding differential strains associated with elastic interactions between the phases [62]. One form relies on assumption of equal dimensional change (no damage) for both phases, giving,

$$\alpha_C = \frac{\alpha_1 \, K_1 \, V_1 + \alpha_2 \, K_2 \, V_2}{K_1 \, V_1 + K_2 \, V_2} \tag{4.42}$$

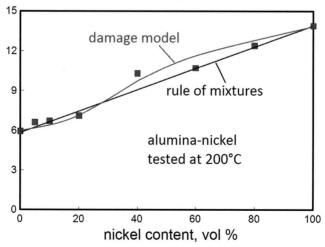

Fig. 4.29 Thermal expansion coefficient data for nickel-alumina composites tested at 200 °C (473 K). Two models are included with the data, with a straight line for the rule of mixtures but curvature to the model that includes damage from the elastic modulus difference

Here K is the bulk modulus and the subscripts correspond to the two phases. Other models assume dilute compositions, spherical particles, or require the shear modulus. One such model requires first calculation of the composite bulk modulus, K_C, which changes with composition, giving the following composite thermal expansion coefficient estimate:

$$\alpha_C = \alpha_2 + \frac{(\alpha_1 - \alpha_2)\left[\frac{1}{K_C} + \frac{1}{K_2}\right]}{\left[\frac{1}{K_1} - \frac{1}{K_2}\right]} \tag{4.43}$$

In composites where both phases are connected, differential strains arise between the phases that necessitate elastic deformation corrections. The high thermal expansion phase is retarded in expansion while the low thermal expansion phase is either stressed, delaminates, or cracks. These interactions require inclusion of elastic properties of the two phases as shown below:

$$\alpha_C = \alpha_1 - \frac{E_2}{E_1}\left\{\frac{3\,V_2(\alpha_1 - \alpha_2)(1 - \nu_1)}{2\,[1 - 2\,\nu_2 V_2 + 2\,V_2\,(1 - 2\,\nu_2) + (1 + \nu_2)]}\right\} \tag{4.44}$$

where ν indicates Poisson's ratio and E indicates elastic modulus, with the subscripts denoting the respective phases. For the nickel-alumina system, with experimental data plotted in Fig. 4.29, the measured composite thermal expansion coefficient at 40 vol% nickel is 10.2 ppm/°C. The volumetric rule of mixtures using Eq. (4.41) gives 9.2 ppm/°C, while Eq. (4.44) predicts 13.2 ppm/°C. Including microstructure damage lowers the estimated thermal expansion to 10 ppm/°C, but requires estimates on the degree of interface damage [63]. The thermo-elastic predictions over-estimate and the damage predictions under-estimate. Equation (4.21) often proves a reasonable balance between accuracy and complexity of estimating or measuring bulk modulus, interface damage, and phase connectivity.

Pores lower the composite mass, but provide no contribution to thermal expansion. Accordingly, porosity decreases the thermal expansion coefficient in proportion to the missing mass. To account for the solid expansion and contraction, the composite thermal expansion coefficient is modified,

$$\alpha_C = \alpha_{CO}\,f^{1/3} \tag{4.45}$$

where α_{CO} represents the full density thermal expansion coefficient for the composite and α_C is the value expected for a fractional density f. Pores in the composite remove atomic bonds that expand or contract with temperature change, but the underlying system expansion is unchanged.

Thermal Conductivity

Heat conduction is measured by energy transfer over time, depending on the thickness, area, and temperature gradient. The SI units are W/(m °C) or per K. Thermal conductivity of diamond is 30,000-fold larger than foamed ceramic insulators. Copper is an excellent heat conductor at 400 W/(m °C), while insulation is near 0.07 W/(m °C), about twice that of still air.

High thermal conductivity composites are important to thermal management applications, such as heat spreaders for microelectronic circuits. Composite materials applied to this balance a need for low thermal expansion and high thermal conductivity, leading to composites of copper-diamond, copper-tungsten, aluminum-silicon carbide, aluminum nitride-yttria, silver-invar, aluminum-diamond, silicon-diamond, aluminum-aluminum nitride, and copper-molybdenum. On the other hand, insulators for space vehicles, boilers, petrochemical plants, and ships rely on foamed, high-porosity, low thermal conductivity materials, such as calcium silicate-glass, alumina-zirconia, magnesia-glass, and silica-glass.

In a composite the conduction of heat and electricity depend on the constituent properties and the connectivity of the two phases. One phase is always connected, but the second phase might be either connected or isolated. As a simple model, conductivity is estimated by the relative amount and properties of each phase. Without microstructure data, an estimate of the composite thermal conductivity κ_C comes from the familiar volumetric rule of mixtures, assuming no pores:

$$\kappa_C = \sum V_J \, \kappa_J = V_1 \, \kappa_1 + V_2 \, \kappa_2 \qquad (4.46)$$

with κ_1 being the thermal conductivity of the first phase at volume fraction V_1, and κ_2 is the thermal conductivity of the second phase at volume fraction $V_2 = 1 - V_1$. It is possible to include porosity with near zero thermal conductivity (depending on the gas filling the pores). Only in structures over 50 % porosity is significant thermal conduction possible through the gas phase inside the pores. Likewise, a homogeneous composite is assumed, but if one phase is improperly dispersed or poorly bonded, then the models are inaccurate.

Several models exist, generally geared to filled polymer composites where only the matrix phase is connected [64]. Two examples are given below that only require information on the composition and constituent properties. Both models assume the second phase exists as spherical particles dispersed in the first phase,

$$\kappa_C = \kappa_2 \left(\frac{\kappa_1 + 2\,\kappa_2 + 2\,V_2\,(\kappa_1 - \kappa_2)}{\kappa_1 + 2\,\kappa_2 - 2\,V_2\,(\kappa_1 - \kappa_2)} \right) \qquad (4.47)$$

$$1 - V_2 = \left[\frac{\kappa_2 - \kappa_C}{\kappa_2 - \kappa_1} \right] \left(\frac{\kappa_1}{\kappa_C} \right)^{1/3} \qquad (4.48)$$

Additional terms might add elastic properties and interface characteristics. Even the simple assumption of spherical grains dispersed in a matrix phase becomes complicated if the grain shape is anisotropic.

A large grain size improves heat transport, so corrections are sometimes needed for grain size [65]. Interfaces interrupt heat transport, so more interface area is associated with a smaller grain size, pores, or impurity films. For example, in nanoscale W-Cu, the measured thermal conductivity is 30 % lower due to the large number of interfaces associated with the grains. Roughly, composite conductivity varies with the inverse grain size. Coated powders ensure improved microstructure homogeneity and higher conductivity. In a study of aluminum-diamond, a 50 vol% composite reached 490 W/(m °C) thermal conductivity when the diamond was coated with titanium prior to consolidation. This compares with under 300 W/(m °C) for equivalent composites prepared with uncoated diamond. The titanium coating ensures interfacial bonding to decrease heat dispersion in the composite.

Some models incorporate elastic effects by including elastic modulus along with contiguity. These models are predict the upper bound thermal conductivity. For example, in the tungsten-copper system, with 85 wt% W (72 vol%) the rule of mixtures predicted thermal conductivity is 238 W/(m °C). If connectivity is included to correct the model, the prediction is 223 W/(m °C). Experimental reports in ranking of lowest to highest are 136, 139, 167, 175, 191, 193, 200, 202, 203, 215, and 230 W/(m °C). Thus, these models provide good upper bound estimates, but many practical issues degrade actual behavior.

In many cases heat conduction and electrical conduction rely on electrons, so thermal conductivity and electrical conductivity are proportional. For those cases, it is possible to measure electrical conductivity to estimate thermal conductivity. However, compounds such as AlN, SiC, BeS, GaP, GaN, and BeO are exceptions, where phonons provide significant thermal conduction. These ceramics conduct heat but are poor electrical conductors.

Thermal conductivity decreases with porosity and usually is not sensitive to the pore shape. In the higher-density range, over about 70 % of theoretical density, thermal conductivity κ_C exhibits a linear dependence on fractional density f [17],

$$\kappa_C = \kappa_O \left(1 - a \left(1 - f\right)\right) \tag{4.49}$$

Here κ_O is the conductivity for the full density material and a is a sensitivity coefficient usually between 1 and 3, theoretically predicted at 2.5. A few isolated, spherical pores have little impact, but extensive porosity is associated with thermal insulators. An empirical relationship effective in modeling the density influence is given as,

$$\kappa_C = \kappa_O \frac{f}{1 + 11 \left(1 - f\right)^2} \tag{4.50}$$

The factor of 11 in Eq. (4.50) reflects a rounded pore shape.

Thermal Shock Resistance

Thermal shock resistance reflects a combination of properties. Characteristically a high thermal shock resistance indicates an ability to accept large and rapid temperature changes without fracture [66]. This resistance, designated T_R, depends on the temperature change along with the thermal expansion coefficient α_C, thermal conductivity κ_C, and elastic modulus E_C, as follows:

$$T_R = \frac{\sigma_C \, \kappa_C}{\alpha_C \, E_C} \tag{4.51}$$

where σ_c is the composite strength at the test temperature. The thermal shock resistance has units of N/s. Thermal shock resistance improves with a higher thermal conductivity and lower thermal expansion coefficient.

Another measure of thermal shock resistance is in the strength loss experience from a rapid cooling cycle. Plotted in Fig. 4.30 are strength data from an AlON-ZrN composite [67]. The indicated temperature change was repeated three times. The strength remaining after three cycles declines as the magnitude of the temperature change increases. Similar behavior is seen in terms of the number of cycles to failure for a given temperature change, a property related to thermal fatigue.

One problem with particulate composites lies in the differential internal strain arising on heating and cooling. The two phases differ in many properties, including thermal conductivity, heat capacity, thermal expansion, strength, and elastic modulus. Accordingly failure occurs in the weaker phase due to differential thermal stresses exceeding the strength of the interface or weaker phase [68].

Fig. 4.30 Evidence of thermal shock damage for an AlON-ZrN particulate composite based on the remaining strength after repeated quenching from various peak temperatures [67]. The elastic modulus also declined due to thermal shock damage

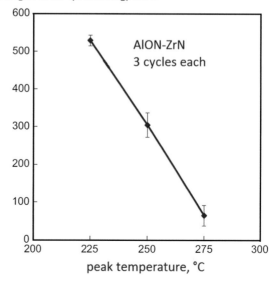

Thermal Fatigue

A structure subjected to cyclic heating and cooling will fail by thermal fatigue. The treatment for thermal fatigue is similar to mechanical fatigue where life is measured by the number of cycles to failure. Strength is lost after repeated cycles and eventually results in spontaneous failure [69]. Temperature fluctuations act similar to stress fluctuations; large temperature changes cause failure in fewer cycles.

Thermal fatigue is a problem in composites since the phases have different mechanical and thermal properties. Thus, temperature oscillations lead to damage accumulation and eventual failure. For example, Cu-Mo test data for a 65 °C temperature change using 25 mm sample size shows mean failure in 800 cycles. Pure copper or molybdenum easily withstand many more cycles without failure. An estimate on the number of cycles to failure N is given as follows:

$$N = A \, exp \left[\frac{\psi}{\Delta T \, L \, (\alpha_1 - \alpha_2)} \right] \qquad (4.52)$$

where A is an empirical constant, ΔT is the cyclic temperature change, L is a characteristic size (interface width), and α is the thermal expansion coefficient, with subscripts 1 and 2 denoting the two phases. The parameter ψ is a constant with dimensions of size. Actual failure is statistically distributed, so repeat tests are required to assess the distribution.

Thermal fatigue is a problem in repetitive operations. For example electric switches heat and cool with power cycles, as do automotive engines, jet engines, furnaces, solar cells, and welding equipment. In electronics the on-off cycle causes the components to heat and expand, then to cool and contract. Differences in thermal expansion coefficient and elastic modulus generate interfacial stresses, eventually causing failure.

Electrical Conductivity

Composite electrical conduction depends on constituents, composition, temperature, and microstructure, most especially phase connectivity [60, 70, 71]. Several additional factors degrade conductivity, including residual stress, impurities, and porosity. For composites, various cases are possible,

- composites consisting of two conductors,
- composites consisting of two insulators,
- composites consisting of mixtures of conductor and insulator,

  - conductor phase not connected,
  - conductor phase connected.

The easiest to model is the conductor-conductor composite. Here the volumetric rule of mixtures provides a first estimate of composite conductivity λ_C;

$$\lambda_C = V_1 \lambda_1 + V_2 \lambda_2 \tag{4.53}$$

where λ is the conductivity and V is the volume fraction (subscript C = composite, 1 and 2 = the two phases). Resistivity R is the inverse of conductivity λ, so the inverse rule of mixtures is employed to model resistivity. Equation (4.53) is effective for a wide variety of systems, such as copper or silver mixed with arc erosion or wear resistant metals such as Cr, W, or Mo [25, 72].

If both phases are nonconductors, then the composite is likewise nonconductive. Again, a volumetric rule of mixtures is appropriate. Experimental data are highly scattered, so it is difficult to conduct a critical test for the insulator-insulator combination.

The more difficult situation occurs with a mixture of conductive and nonconductive phases. If the conductive phase is isolated, as sketched in Fig. 4.31a, then the composite conductivity is low, like that of the insulator. The dispersed conductor gives a slight improvement in conductivity with increasing conductor content. Still the composite is an insulator. When the concentration of conductor is sufficient to form a percolated structure, as illustrated in Fig. 4.31b, a dramatic change in conductivity occurs over a relatively small composition range. Models for conductivity versus composition include two adjustable parameters related to microstructure. With experimental data, it is possible to solve for the composite resistivity R_C,

$$\frac{V_1 \left(R_C^Q - R_1^Q \right)}{R_C^Q + A\,R_1^Q} = \frac{V_2 \left(R_2^Q - R_C^Q \right)}{R_C^Q + A\,R_2^Q} \tag{4.54}$$

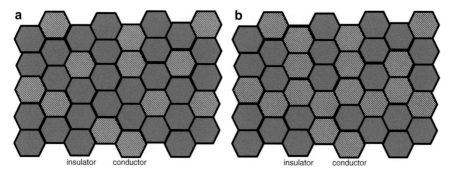

Fig. 4.31 A two-dimensional sketch of an isolated conductor. In (**a**) the conductive phase is not connected and the composite acts as an insulator, while in (**b**) the conductive phase is at sufficient concentration to form complete networks and the composite is conductive

Fig. 4.32 Electrical behavior of tungsten-zirconia composites over the range of possible compositions [73]. The logarithm of resistivity is plotted versus tungsten content and the model of Equation (4.54) is shown for comparison corresponding to a critical percolation concentration of 26 vol% tungsten

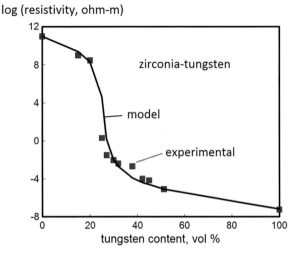

In this effective medium model, V_2 is the volume fraction of conductor, R is the resistivity (subscript C = composite, 1 = insulator, and 2 = conductor), Q is an adjustable parameter, and A is calculated from the conductor volume fraction f_P associated with the onset of the percolated conductor network, $A = (1 - f_P)/f_P$. This model is illustrated using data on the W-ZrO$_2$ system in Fig. 4.32 [73]. Tungsten is the conductor and zirconia is the insulator. For this case Q is 0.286 and f_P is 0.26. At low tungsten contents, the conductive phase is dispersed, resulting in a nonconductive composite. At the percolation limit, near 26 vol% tungsten in this case, connections form to switch the composite into a conductor. The conductivity increases about 10-orders of magnitude. At higher tungsten contents the conductivity continues to increase. Dramatic shifts occur at the percolation limit when the conductive network forms.

Grain size is a factor in electrical conduction, since grain boundaries and interfaces disrupt current flow. In trials with 30 vol% SiC in Ag, the electrical resistivity increased as the silicon carbide grain size decreased −46 nΩ-m at 12.5 μm, 54 nΩ-m at 5 μm, and 62 nΩ-m at 1.5 μm.

With respect to porosity, electrical conductivity is similar to thermal conductivity: Nonconductive pores reduce conductivity. The conductivity loss, in an otherwise conductive composite, increases with fractional porosity assuming dispersed pores:

$$\lambda_C = \lambda_O \frac{f}{1 + \chi (1 - f)^2} \qquad (4.55)$$

This form applies to less than 50 vol% porosity. The conductivity for the full density composite is λ_O. The coefficient χ expresses the sensitivity to pores. This curve applies to several materials, usually relying on relative conductivity (λ_C/λ_O) as a function of fractional density, with a typical value of $\chi = 11$. Figure 4.33 plots

relative conductivity

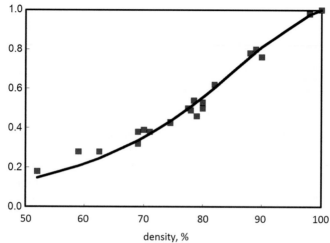

Fig. 4.33 Relative electrical conductivity of iron-copper composites versus the fractional porosity and the model fit given by Eq. (4.55)

electrical conductivity of iron-copper composites at different densities. The solid line is from Eq. (4.55). At high densities, electrical conductivity linearly increases with the elimination of pores.

Several other electrical properties are important to particulate composite applications [71]. One is arc erosion resistance, relevant to welding electrodes and electrical contacts. A high hardness and high melting temperature improve arc erosion resistance. Thus, refractory metals such as W and Mo are combined with conductors such as Ag and Cu. Dielectric behavior seems to be linear in changes with composition. Experimental scatter makes discrimination between models difficult. Complex calculations do not improve the fit to experimental data.

Magnetic Behavior

Magnetic properties of particulate composites scale with the magnetic particle content. This is encountered in filled polymers, where magnetic particles are added to a flexible polymer. The greater the magnetic particle ratio, the stronger the magnetic response. The response, such as magnetic saturation, is linear with magnetic particle volume faction.

Wear Properties

In practice, a wide range of wear mechanism are encountered, including adhesion, abrasion, sliding, and fretting, so for any application it is best to test under conditions relevant to the application. Sliding wear is quantified based on material loss as a function of time, with response depending on the hardness, sliding distance, and normal load. The wear rate depends on the coefficient of friction between the substrate and sliding component; a low coefficient of friction is desirable. Likewise, hard phases improve wear resistance when added to softer phases, such as the addition of ceramics to metals. This is established for composites such as Al-SiC as plotted in Fig. 4.34 [74]. As SiC is added to the aluminum alloy the increase in hardness reduces wear loss.

The Archard relation enables calculation of the wear behavior by assuming mass loss is via asperity removal. The asperities are the small surface bumps existing on a surface. For a circular cross-section wear is a function of the material parameters as follows:

$$M = \frac{W \, L \, K}{H_C} \, \rho \qquad (4.56)$$

Here M is the mass of material removed from the test composite, K is a wear constant that measures wear resistance, W is the load perpendicular to the surface causing wear, L is the total sliding length, ρ is the density of the composite, and H_C is the composite hardness (Vickers's hardness) assuming the opposing material is harder. Underpinning this model is a conjecture that hardness is proportional to

Fig. 4.34 Abrasive wear data for Al-SiC composites where the hardness and wear loss are expressed as functions of the SiC content [74]. Less wear occurs as hardness increases. Testing was performed using a standard test with 4.9 N load, 60 mm/s velocity over 240 m (800 cycles of 300 mm each)

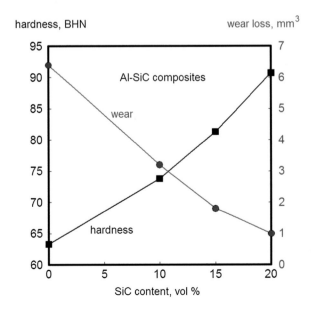

yield strength. Alternative expressions show a high sensitivity to hardness via an exponential relation,

$$M = M_O \, exp[-B \, H_C] \qquad (4.57)$$

The idea is to accentuate the sensitivity to hardness, where M_O and B are fitting parameters. The high sensitivity to hardness shows up in abrasive wear testing of various composites. Wear resistance increases tenfold from 400 to 1500 HV, and tenfold again from 1500 to 1850 HV. Obviously, improved wear resistance leads to composites containing diamond or cubic boron nitride.

Practical testing for wear resistance examine tool life, such as in metal cutting. The time to failure or number of components machined prior to failure are direct tests. One option is to build up a small grain size composite with dispersed grains using coated powders (hard particle at the core) [75]. Wear resistance decreases with porosity, so full density is most typical.

Sound Attenuation

Composites are employed to dampen sound waves in vehicles to remove noise in the passenger compartments. Elastic sound wave attenuation is improved by differences in elastic modulus between the phases. Pores with essentially no stiffness are effective in this regard [76]. An alternative is to rely on a composite with hollow particles. It is useful to employ a viscous phase to shift the frequency response in a sound dampening structure. Most of the models for sound attenuation by composites are empirical.

Comments on Property Models

Models for composite property variations with composition and constituent phase selection are favorite subjects for thesis research. Accordingly, many variants exist, but few are widely used. The linear volumetric rule of mixtures is dominant. As already noted, a property P (hardness, thermal conductivity, wear resistance, or such) and volume fraction V, with subscripts 1 and 2 to identify the two phases, combine in a linear form as follows:

$$P_C = V_1 \, P_1 + V_2 \, P_2 \qquad (4.58)$$

More phases are handled by additional terms. Equation (4.58) is effectively is a weighted average. It is applicable to several composite properties, such as hardness. A related form is the inverse rule of mixtures, where the composite property is given by the same parameters expressed as an inverse summation,

$$P_C = \frac{1}{\frac{W_1}{P_1} + \frac{W_2}{P_2}} \qquad (4.59)$$

This is based on weight fractions of the two phases as W_1 and W_2.

Interface quality is responsible for significant property variations. To handle the interface role, a generalized curve fitting relation is useful using two interaction parameters U and N, giving,

$$P_C = P_1 \frac{1 + N U V_2}{1 - N V_2} \qquad (4.60)$$

Note this model has no dependence on the properties of the minor phase, phase 2, but only depends on the volume fraction and phase interaction. Figure 4.35 plots Eq. (4.60) for a few different combinations of N and U. The response is curved with volume fraction, but intermediate between the two terminal phases. It is not useful in situations where a maximum or minimum occurs at an intermediate composition, such as the case in Fig. 4.23 for fracture toughness.

Clearly, modeling properties that vary with interface quality is a difficult task. This is especially true in modeling thermal [63, 77], deformation [43, 46, 53, 54], and toughness-ductility [2, 40, 78] behavior. Easy interface fracture significantly degrades properties. Figure 4.36 schematically illustrates the difficulty in modeling ductility and toughness. Both drawings show a composite structure, but the fracture path and energy to cause fracture depends on the interface. The upper case corresponds to a weak interface. In this situation the ligaments between grains are the source of strength, so by leaping between weak interfaces there is a significant decrement in fracture energy. On the other hand, the lower case is the same

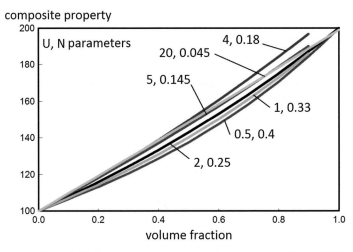

Fig. 4.35 Plots of Eq. (4.60) for various combinations of the parameters U and N. The composition effect is nonlinear, but the range of predictions is generally limited

Fig. 4.36 Two illustrations of fracture path change with interface strength. A weak interface allows the crack to propagate along the easy path with little energy consumption, resulting in low ductility and toughness. A strong interface forces the crack to propagate with greater energy consumption, leading to higher properties for the composite

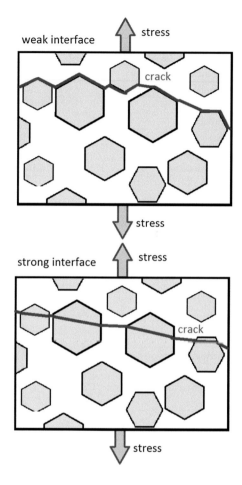

structure, but the high interface strength forces the crack to propagate though the second phase, typically with increased energy consumption. Interface quality is a major influence on toughness. Indeed, for the fracture micrographs shown in Fig. 4.19, the hardness is essentially the same (286 HV for slow cooling and 294 HV for fast cooling), but the impact toughness for the fast cooled composite is eightfold higher (81 versus 10 J/cm^2). It is a challenge in modeling composites to incorporate these factors into predictive scheme. The preferred approach is to test, and once data are collected, then a means is sought to explain the observation.

Properties of particulate composites degrade with porosity. For a porous composite P_C, the porosity effect depends on the full-density property P_O and fractional density f as follows [17]:

$$P_C = P_O \, exp[-B \,(1-f)] \qquad (4.61)$$

Table 4.6 Typical *B* parameters for porosity sensitivity in Eq. (4.61)

Property	Range of *B* values	Typical *B* value
Elastic modulus	2.5–5.4	3.7
Electrical conductivity	1.9–2.9	2.5
Thermal conductivity	2.5–4.5	2.5
Strength	4.0–7.0	5.1
Hardness	4.1–5.8	5.0
Poisson's ratio	0.4–0.5	0.5
Shear modulus	2.2–4.3	3.6
Bulk modulus	4.1–5.0	4.5

Table 4.7 Comparison of measured and predicted properties for Mo-Cu composites [79]

		Wt% copper		
Attribute		15	20	25
Volume %		16.7	22.2	27.5
Density, g/cm^3	Measured	10.0	9.9	9.9
	Predicted	10.0	9.9	9.9
Elastic modulus, GPa	Measured	290	–	–
	Predicted	283	270	258
Thermal expansion coefficient, ppm/°C	Measured	6.6	7.2	7.7
	Predicted	7.3	8.0	8.7
Thermal conductivity, W/(m °C)	Measured	175	186	185
	Predicted	169	181	191

where *B* is a parameter that measures the sensitivity to porosity. Conceptually the property might have a low sensitivity, such as for thermal expansion coefficient, or a high sensitivity, such as for fracture toughness. Table 4.6 provides values for *B* related to some engineering properties.

As a final comment on predictive relations, Table 4.7 illustrates a typical situation. It compares predictions with measurements for Mo-Cu composites using three compositions [79]. There are obvious differences, but overall the models are as accurate as the experimental measurements. Thus, models guide us to possible compositions that meet specific design goals, but the models cannot replace test data taken under conditions relevant to the application.

Study Questions

4.1. The linearized version of the Weibull strength distribution equation, where $\log(\log(1/1 - F))$ is plotted versus $\log S$, allows for least-squares solution to extract the characteristic strength and Weibull modulus. Show the detailed steps needed to arrive at this linear relation.

4.2. In a particulate composite the fracture ductility varies with the phase contiguity. If the two phases differ greatly in intrinsic ductility, where one phase is brittle and the other is ductile, explain how ductility might vary with contiguity?

4.3. A research group is trying to determine the porosity gradients in a structure. They plan to section the component and perform repeated microhardness tests over the cut surface at 10 mm spacing. Assume the component has a 300 HV at full density and typical measurement scatter is 5 HV. If the porosity is expected to vary up to 10 % within the component, is the suggested approach suitable for identification of low porosity gradients?

4.4. Aluminum with silicon carbide is a favorite particulate composite. A low stiffness aluminum alloy with 74 GPa elastic modulus and 0.33 Poisson's ratio is combined with silicon carbide grains with 401 GPa elastic modulus and 0.19 Poisson's ratio. The reported properties of the 15 vol% SiC composite are 98 GPa and 0.31. Are these realistic values?

4.5. For the composite described in Study Question 4.4, if the porosity is 5 %, what is the anticipated elastic modulus?

4.6. Trials determine yield strength for a composite varies with grain sizes as follows:

Grain size, μm	Yield strength, MPa
30	290
5	430
1	700

Check to see if this behavior agrees with the Hall–Petch relation?

4.7. If the creep rate depends on the inverse cube of the second phase volume fraction, suggest why particulate composites are not commonly employed in high temperature structures; what are some of the other limitations besides creep rate?

4.8. The isostress model for elastic modulus relies on a laminated structure with the two phases stacked as layers. Derive the isostress model starting with a two-dimensional layer stack, where the volume fraction is 50 vol% of each phase. Stress is applied parallel to the stacked layers. Show the calculation of the elastic modulus.

4.9. Repeat the calculation in Problem 4.7 for the isostrain case, again for a composite consisting of 50 vol% of each phase loaded in the perpendicular orientation.

4.10. Composites of 20 vol% calcium carbonate dispersed in polypropylene are tested for yield strength using 58 and 1 μm carbonate particles. The tensile strength is higher for the smaller particle size. Both composites are significantly lower in strength versus untreated polypropylene tested under similar conditions. What factor explains the particle size effect on strength?

4.11. Human bone is on average about 130 MPa in strength. If a titanium scaffold is designed to match bone in strength, what porosity would be appropriate using an alloy with 800 MPa full density strength?

4.12. New composites are reaching upward to four ingredients—for example $TiC+TiN+TiB_2+ZrO_2$. Comment on the difficulty anticipated in modeling strength or hardness of such a composite.

4.13. One of the issues with fiber reinforced composites traces to the Weibull distribution, which says inherently the fiber strength falls as its volume increases. Assume a Weibull modulus of 25 for a fiber with 400 MPa characteristic strength using 100 μm diameter fibers with 200 mm length. What is the design strength that would give 99 % survival probability if the fibers are 6 m long?

References

1. F.L. Matthews, R.D. Rawlings, *Composite Materials: Engineering and Science* (CRC Press, Boca Raton, 2008)
2. S.Y. Fu, X.Q. Feng, B. Lauke, Y.W. Mai, Effects of particle size, particle/matrix interface adhesion and particle loading on mechanical properties of particulate-polymer composites. Compos. Part B **39**, 933–961 (2008)
3. D. Zheng, X. Li, X. Ai, C. Yang, Y. Li, Bulk $WC-Al_2O_3$ composites prepared by spark plasma sintering. Int. J. Refract. Met. Hard Mater. **30**, 51–56 (2012)
4. Z. Fan. P. Tsakiropoulos, A.P. Miodownik, A generalized law of mixtures. J. Mater. Sci. **29**, 141–150 (1994)
5. K.K. Chawla, *Composite Materials Science and Engineering*, 3rd edn. (Springer, New York, 2013)
6. J.A. Belk, M.R. Edwards, W.J. Farrell, B.K. Mullah, Deformation behaviour of tungsten-copper composites. Powder Metall. **36**, 293–296 (1993)
7. Anonymous, *Cambridge Engineering Selector* (Granta Design, Cambridge, updated annually)
8. Y.G. Gogotsi, Review particulate silicon nitride based composites. J. Mater. Sci. **29**, 2541–2556 (1994)
9. J. Wang, R. Stevens, Review: Zirconia-toughened alumina (ZTA) ceramics. J. Mater. Sci. **24**, 3421–3440 (1989)
10. L. Xu, S. Wei, J. Li, G. Zhang, B. Dai, Preparation, microstructure, and properties of molybdenum alloys reinforced by *in-situ* Al_2O_3 particles. Int. J. Refract. Met. Hard Mater. **30**, 208–212 (2012)
11. R. Sadangi, D. Kapoor, T. Zahrah, in *Powder Metallurgy Processed Magnesium Matrix Composites*, Advances in Powder Metallurgy and Particulate Materials (Metal Powder Industries Federation, Princeton, 2015), pp. 7.113–7.124
12. D.J. Lloyd, Particle reinforced aluminum and magnesium matrix composites. Int. Mater. Rev. **39**, 1–23 (1994)
13. S. Luyckx, in *The Hardness of Tungsten Carbide-Cobalt Hardmetal*, ed. by R. Riedel. Handbook of Ceramic Hard Materials, vol 2 (Wiley-VCH, Weinheim, 2000), pp. 946–964
14. J. Gurland, A structural approach to the yield strength of two-phase alloys with coarse microstructures. Mater. Sci. Eng. **40**, 59–71 (1979)
15. T.Y. Chan, S.T. Lin, Sintering of elemental carbonyl iron and carbonyl nickel powder mixtures. J. Mater. Sci. **32**, 1963–1967 (1997)

16. J. Corrochano, M. Lieblich, J. Ibanez, On the role of matrix grain size and particulate reinforcement on the hardness of powder metallurgy Al-Mg-Si/MoSi$_2$ composites. Compos. Sci. Technol. **69**, 1818–1824 (2009)
17. R.W. Rice, *Porosity of Ceramics* (Marcel Dekker, New York, 1998)
18. A. Manonukul, N. Muenya, F. Leaux, S. Amaranan, Effects of replacing metal powder with powder space holder on metal foam produced by metal injection moulding. J. Mater. Process. Technol. **210**, 529–535 (2010)
19. S.A. Cho, F.J. Arenas, J. Ochoa, Densification and hardness of Al$_2$O$_3$-Cr$_2$O$_3$ system with and without Ti addition. Ceram. Int. **16**, 301–309 (1990)
20. S. Ahmed, F.R. Jones, A review of particulate reinforcement theories for polymer composites. J. Mater. Sci. **25**, 4933–4942 (1990)
21. S.D. Henry, C. Moosbrugger, G.J. Anton, B.R. Sanders, N. Hrivnak, C. Terman, J. Kinson, K. Muldoon, W.W. Scott (eds.), *Composites, ASM Handbook* (ASM International, Materials Park, 2001)
22. M.G. Phillips, Simple geometrical models for Young's modulus of fibrous and particulate composites. Compos. Sci. Technol. **43**, 95–100 (1992)
23. L.D. Wegner, L.J. Gibson, The mechanical behaviour of interpenetrating phase composites: I: modelling. Int. J. Mech. Sci. **42**, 925–942 (2000)
24. F.J. Guild, R.J. Young, A predictive model for particulate filled composite materials, part 1 hard particles. J. Mater. Sci. **24**, 298–306 (1989)
25. T.W. Clyne, P.J. Withers, *An Introduction to Metal Matrix Composites* (Cambridge University Press, Cambridge, 1993)
26. P. Kwon, C.K.H. Dharan, Effective moduli of high volume fraction particulate composites. Acta Metall. Mater. **43**, 1141–1147 (1995)
27. H. Doi, *Elastic and Plastic Properties of WC-Co Composite Alloys* (Freund Publishing House, Tel-Aviv, 1974)
28. H. Doi, Y. Fujiwara, K. Miyake, Y. Oosawa, A systematic investigation of elastic moduli of WC-Co alloys. Metall. Trans. **1**, 1417–1425 (1970)
29. F. Danusso, G. Tieghi, Strength versus composition of rigid matrix particulate composites. Polymer **27**, 1385–1390 (1986)
30. A. Mortensen, J. Llorca, Metal matrix composites. Annu. Rev. Mater. Res. **40**, 243–270 (2010)
31. D.K. Hale, Review: the physical properties of composite materials. J. Mater. Sci. **11**, 2105–2141 (1976)
32. B.H. Rabin, R.M. German, Microstructure effects on tensile properties of tungsten-nickel-iron composites. Metall. Trans. **19A**, 1523–1532 (1988)
33. M.B. Waldron, The production of cermets containing a relatively large amount of dispersed phase. Powder Metall. **10**, 288–306 (1967)
34. N. Claussen, Fracture toughness of Al$_2$O$_3$ with an unstabilized ZrO$_2$ dispersed phase. J. Am. Ceram. Soc. **59**, 49–51 (1976)
35. S.C. Tjong, Z.Y. Ma, Microstructural and mechanical characteristics of in situ metal matrix composites. Mater. Sci. Eng. **29**, 49–113 (2000)
36. E. Pagounis, V.K. Lindroos, Processing and properties of particulate reinforced steel matrix composites. Mater. Sci. Eng. **A246**, 221–234 (1998)
37. J. Segurado, J.L. Llorca, Computation micromechanics of composites: the effect of particle spatial distribution. Mech. Mater. **38**, 873–883 (2006)
38. V. Provenzano, N.P. Louat, M.A. Imam, K. Sadananda, Ultrafine superstrength materials. Nanostr. Mater. **1**, 89–94 (1992)
39. K. Jia, T.E. Fischer, B. Gallois, Microstructure, hardness, and toughness of nanostructured and conventional WC-Co composites. Nanostr. Mater. **10**, 875–891 (1998)
40. H.K. Lee, S.H. Pyo, Multi-level modeling of effective elastic behavior and progressive weakened interface in particulate composites. Compos. Sci. Technol. **68**, 387–397 (2008)
41. J.L. Chermant, A. Deschanvres, F. Osterstock, Factors influencing the rupture stress of hardmetals. Powder Metall. **20**, 63–69 (1977)

42. N. Hirosaki, Y. Akimune, M. Mitomo, Effect of grain growth of beta-silicon nitride on strength, weibull modulus, and fracture toughness. J. Am. Ceram. Soc. **76**, 1892–1894 (1993)
43. D.L. McDanels, Analysis of stress-strain, fracture, and ductility behavior of aluminum matrix composites containing discontinuous silicon carbide reinforcement. Metall. Trans. **16A**, 1105–1115 (1985)
44. X. Liu, G. Hu, A continuum micromechanical theory of overall plasticity for particulate composites including particle size effect. Int. J. Plast. **21**, 777–799 (2005)
45. J. Llorca, C. Gonzalez, Microstructural factors controlling the strength and ductility of particle reinforced metal matrix composites. J. Mech. Phys. Solids **46**, 1–28 (1998)
46. F.J. Humpherys, W.S. Miller, M.R. Djazeb, Microstructural development during thermo-mechanical processing of particulate metal-matrix composites. Mater. Sci. Technol. **6**, 1157–1166 (1990)
47. T. Christman, A. Needleman, S. Suresh, An experimental and numerical study of deformation metal-ceramic composites. Acta Metall. **37**, 3029–3050 (1989)
48. S. Li, D. Xiong, M. Liu, S. Bai, X. Zhao, Thermophysical properties of SiC/Al composites with three dimensional interpenetrating network structure. Ceram. Int. **40**, 7539–7544 (2014)
49. L.S. Sigl, Processing and mechanical properties of boron carbide sintered with TiC. J. Euro. Ceram. Soc. **18**, 1521–1529 (1998)
50. Z. Wang, S. Li, M. Wang, G. Wu, X. Sun, M. Liu, Effect of SiC whiskers on microstructure and mechanical properties of the $MoSi_2$-SiCw composites. Int. J. Refract. Met. Hard Mater. **41**, 489–494 (2013)
51. V. Naglieri, P. Palmero, L. Montanaro, J. Chevalier, Elaboration of alumina-zirconia composites: role of the zirconia content on the microstructure and mechanical properties. Materials **6**, 2090–2102 (2013)
52. C.C. Perng, J.R. Hwang, J.L. Doong, Elevated temperature, low-cycle fatigue behaviour of an Al_2O_3 P/6061-T6 aluminum matrix composite. Compos. Sci. Technol. **49**, 225–236 (1993)
53. Y. Uematsu, K. Tokaji, M. Kawamura, Fatigue behaviour of SiC-particulate reinforced aluminum alloy composites with different particle sizes at elevated temperature. Compos. Sci. Technol. **68**, 2785–2791 (2008)
54. J.D. Whittenberge, R.K. Viswanadham, S.K. Mannan, B. Sprissler, Elevated temperature slow plastic deformation of $NiAl$-TiB_2 particulate composites at 1200 and 1300 K. J. Mater. Sci. **25**, 35–44 (1990)
55. J. Wang, E.M. Taleff, D. Kovar, High temperature deformation of Al_2O_3/Y—TZP particulate composite. Acta Mater. **51**, 3571–3583 (2003)
56. A.V. Nair, J.K. Tien, R.C. Bates, SiC-reinforced aluminum metal matrix composites. Int. Metals Rev. **30**, 275–290 (1985)
57. K.H. Kate, R.K. Enneti, S.J. Park, R.M. German, S.V. Atre, Predicting powder-polymer mixture properties for PIM design. Crit. Rev. Solid State Mater. Sci. **39**, 197–214 (2014)
58. Y. Hirata, Representation of thermal expansion coefficient of solid material with particulate inclusion. Ceram. Int. **41**, 2706–2713 (2015)
59. B. Weidenfeller, M. Hofer, F.R. Schilling, Thermal conductivity, thermal diffusivity, and specific heat capacity of particle filled polypropylene. Compos. Part A **35**, 423–429 (2004)
60. H.A. Bruck, B.H. Rabin, Evaluation of rule-of-mixtures predictions of thermal expansion in powder-processed Ni-Al_2O_3 composites. J. Am. Ceram. Soc. **82**, 2927–2930 (1999)
61. R.M. German, A model for the thermal properties of liquid phase sintered composites. Metall. Trans. **24A**, 1745–1752 (1993)
62. A.A. Fahmy, A.N. Ragai, Thermal expansion behavior of two-phase solids. J. Appl. Phys. **41**, 5108–5111 (1970)
63. Y.Z. Wan, Y.L. Wang, H.L. Luo, G.X. Cheng, Effect of interfacial bonding strength on thermal expansion behaviour of PM Al_2O_3/copper alloy composites. Powder Metall. **43**, 76–78 (2000)
64. J. Jancar, A. Dianselmo, A.T. Dibenedetto, The yield strength of particulate reinforced thermoplastic composites. Polym. Eng. Sci. **32**, 1394–1399 (1992)
65. A. Boudeene, L. Ibos, M. Fois, E. Gehin, J.C. Majeste, Thermophysical properties of poly-propylene/aluminum composites. J. Polym. Sci. B **42**, 722–732 (2004)

66. Z.H. Jin, R.C. Batra, Stress intensity relaxation at the tip of an edge crack in a functionally graded material subjected to a thermal shock. J. Therm. Stresses **19**, 317–339 (1996)
67. N. Zhang, X.J. Zhao, H.Q. Ru, X.Y. Wang, D.L. Chen, Thermal shock behavior of nanosized ZrN particulate reinforced AlON composites. Ceram. Int. **39**, 367–375 (2013)
68. J.W. Zimmermann, G.E. Hilmas, W.G. Fahrenholtz, Thermal shock resistance of ZrB_2 and ZrB_2–30 % SiC. Mater. Chem. Phys. **112**, 140–145 (2008)
69. S.G. Long, Y.C. Zhou, Thermal fatigue of particle reinforced metal-matrix composite induced by laser heating and mechanical load. Compos. Sci. Technol. **65**, 1391–1400 (2005)
70. H. Tian, T.T. Liu, H.F. Cheng, Microstructural and electrical properties of thick film resistors on oxide/oxide ceramic—matrix composites. Ceram. Int. **41**, 3214–3219 (2015)
71. D.S. McLachlan, M. Blaszkiewicz, R.E. Newnham, Electrical resistivity of composites. J. Am. Ceram. Soc. **73**, 2187–2203 (1990)
72. X. Wang, H. Yang, M. Chen, J. Zou, S. Liang, Fabrication and arc erosion behaviors of Ag-TiB_2 contact materials. Powder Technol. **256**, 20–24 (2014)
73. S. Vivs, C. Guizard, C. Oberlin, L. Cot, Zirconia-tungsten composites: synthesis and characterisation for different metal volume fractions. J. Mater. Sci. **36**, 5271–5280 (2001)
74. C. Garcia-Cordovilla, J. Narciso, E. Lewis, Abrasive wear resistance of aluminum alloy/ceramic particulate composites. Wear **192**, 170–177 (1996)
75. F.E. Kennedy, A.C. Balbahadur, D.S. Lashmore, The friction and wear of Cu-based silicon carbide particulate metal matrix composites for brake applications. Wear **203**, 715–721 (1997)
76. G.T. Kuster, M.N. Toksoz, Velocity and attenuation of seismic waves in two-phase media: part I. Theoretical formulations. Geophysics **39**, 587–606 (1974)
77. C.W. Nan, R. Birringer, D.R. Clarke, H. Gleiter, Effective thermal conductivity of particulate composites with interfacial thermal resistance. J. Appl. Phys. **81**, 6692–6699 (1997)
78. D.M. Bigg, Mechanical properties of particulate filled polymers. Polym. Compos. **8**, 115–122 (1987)
79. J.L. Johnson, Opportunities for PM processing of metal matrix composites. Int. J. Powder Metall. **47**(2), 19–28 (2011)

Chapter 5
Constituents

Although many materials are available as building blocks for composites, still the 80-20 rule is in effect; most particulate composites are formed from relatively few constituents. These constituents have unique attributes, such as exceptional strength, toughness, or conductivity. This chapter focuses the selection of constituents via hierarchical analysis based on application requirements.

Leading Candidates

Composite design starts with the evaluation of existing materials and their properties. The initial intent is to determine what fits with the application needs. Material properties for the constituents are assembled in handbooks, database, web sites, trade association publications, and commercial programs. Several sources are listed as references [1–31]. Software programs are effective since the data collections include a broad range of candidates. Data from these resources are employed in this chapter to illustrate a diverse range of examples [15, 32–41]. It is immediately clear that thousands of material options are available.

In forming particulate composites, a few materials are used frequently. These tend to be materials with outstanding specific properties. Composites are built by considering properties, but also are limited by cost and availability. Materials used frequently tend to be lower in cost, and vice versa, materials lower in cost are used more often. If a monolithic structure satisfied the property needs for an application, then the only opportunity for a composite is by lower cost.

Plastics, steels, and concrete dominate engineering [1]. For the high volume applications, these nominally are in a cost ratio of 16:6:1, namely about \$2/kg, \$0.7/kg, and \$0.12/kg for plastic, steel, and concrete, respectively. The first two are frequently used in automobiles and the latter two are frequently used in construction.

© Springer International Publishing Switzerland 2016

R.M. German, *Particulate Composites*, DOI 10.1007/978-3-319-29917-4_5

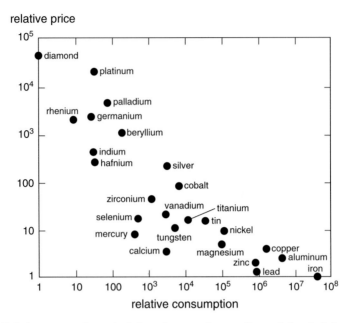

Fig. 5.1 Relative consumption and relative price per unit mass of several chemical elements. The relation between these two factors begs the question as to whether we use more of an element such as iron because it is lower in price, or is the low price enabling more consumption. Composites heavily rely on a third dimension, that being performance

Figure 5.1 is a plot that rationalizes how humans use various elements on the periodic table. The relative price is relative to iron, which is set as unity, and the consumption is relative to diamond. Thus the high volume iron systems are a price baseline at 1, with actual annual tonnage in 2015 exceeding 10^{12} kg/year. Prices for steels generally range from \$0.60 to \$3.00/kg depending on the shape, alloy, and heat treatment. On the other hand, diamond is the baseline for consumption at 1, with annual sales of \$100 billion/year. Diamond prices range from \$1250/kg for industrial diamonds to \$8500/kg for high clarity small stones. Gemstones are even more expensive, depending on the size, color, clarity, and cut, reaching up to \$20,000/g. These prices change daily as supply shifts. In the normalized plot of Fig. 5.1, the interplay between cost and consumption helps understand a key basis for constituent selection in building a composite.

The critical factor in selecting candidate constituents derives from the available properties. Particulate composites form unique property combinations. Materials with unique properties are attractive. This means favoritism toward constituents that are particularly outstanding in specific features:

high density	osmium, iridium, platinum, rhenium, tungsten-rhenium alloys, gold-platinum alloys, tungsten, gold, platinum-rhodium, uranium, tungsten alloys, tantalum-tungsten alloys
high stiffness	diamond, tungsten carbide, osmium, iridium, rhenium, boron carbide, silicon carbide, alumina, molybdenum disilicide, ruthenium, titanium carbide, titanium diboride, rhodium
high hardness	diamond, boron carbide, tungsten carbide, silicon carbide, boron compounds, vanadium carbide
high strength	diamond, tungsten, tool steel, osmium, martensitic stainless steel, graphite
high toughness	stainless steel, alloy steels, nickel-chromium alloys, palladium alloys, nickel-copper alloys, rhenium, gold-platinum, niobium alloys, rhodium, tungsten alloy, zirconium-niobium alloy, nickel-molybdenum alloys, tantalum alloys
high wear resistance	diamond, boron carbide, sapphire, titanium carbide, tungsten carbide, silicon carbide, vanadium carbide
high corrosion resistance (i.e. salt water)	nickel silver, zirconium alloys, zirconium carbide, zirconia, bronze, nickel-iron alloys
high temperature oxidation resistance	ruthenium, zirconium alloys, zirconium carbide, zirconia, nickel alloys, alumina, silica, mullite
high electrical conductivity	silver, silver alloys, copper, gold, copper alloys, aluminum
high thermal conductivity	diamond, silver, sterling silver, copper, silver alloys, copper alloys, graphite, beryllia, gold, aluminum
high thermal expansion	ethylene acrylic rubber, polyvinylidene chloride, styrene isoprene styrene copolymer, methyl acrylate, cellulose acetate, polyethylene
soft magnetic response	iron-nickel amorphous metal, iron-silicon, iron-cobalt alloys, nickel-iron alloys, nickel-molybdenum-iron alloys, ferrite compounds, iron, nickel, samarium-cobalt compound
hard magnetic response	iron-neodymium-boron, alnico, cobalt-samarium, iron-cobalt alloys, iron-nickel-cobalt-aluminum alloys
low density	polymeric foam (melamine, polyethylene, polyurethane, polystyrene, polypropylene), carbon foam, balsa wood, cork, silica and glass foam, calcium silicate, foamed concrete, wood
low stiffness	polyurethane foam, melamine foam, polyethylene foam, styrene isoprene styrene block copolymer,

	ethylene vinyl acetate rubber, acrylic rubber, natural rubber, nitrile rubber
low thermal conductivity	polyurethane foam, polyvinylchloride foam, ethylene acrylic rubber, balsa wood, polystyrene foam, cork, polyethylene foam, graphite foam, glass foam, calcium silicate foam, mullite foam
low thermal expansion	silicate glass, titanium silicide, graphite, silica, nickel-iron (Invar), diamond, lithium aluminosilicate, boron nitride
low flammability	aluminum hydroxide, antimony oxide
antimicrobial	silver, copper, brass, bronze.

In addition, fabricability is important, leading to frequent use of lower processing temperature phases since they are easier to handle—epoxy, aluminum, copper, bronze, and polymers such as polypropylene. Other factors might include optical clarity (glass) or opacity (graphite), flammability (magnesium) or inflammability (mullite), biocompatibility (titanium), or corrosion resistance (platinum).

Successful composites mix constituents from the "highest" or "lowest" categories to provide complementary properties. The effort is to improve on the property combination at a low cost. Most combinations are selected to avoid reactions during fabrication, but in some composites reactions occur during fabrication. Thermite consisting of mixed aluminum and iron oxide is one example. When this mixture is ignited the exothermic reaction heats to produce molten iron, useful for remote steel welding. Other mixtures provide portable heat sources, such as Ni+Al, Mn+S, or Ti+C. Once initiated the reactions provide portable heating, or localized bonding. For example, the Mn+S mixture is inert until ignited, but once ignited it produces MnS and heat, so it is useful for remote heating.

Constituent Selection Protocol

A hierarchical selection protocol is advocated for identify constituent phases in a particulate composite. The approach relies on sequential steps that compare candidate constituents against application requirements. A successful comparison leads to further detailed evaluation. Composites emerge that blend properties, where one phase fills a key requirement, and the other phase provides complimentary attributes.

An example of hierarchical down-selection is seen with organizations hiring to fill a vacant position. First, there is a vision of the ideal candidate in terms of education, experience, and personality, usually derived from the attributes of successful prior employees. Candidate resumes are compared to this vision as a means to evaluate for matches. After ranking resumes, factors are explored in interviews. Interviews evaluate intangible factors such as personality, attentiveness, punctuality, and clarity of communication.

In a similar sense, constituent phase selection for an engineering design involves a matrix of constituent properties held up against the application requirements. Intersections are sought to identify the best candidates. After *pro forma* analysis, there is still a need for experimentation and optimization, including assessment of secondary factors such as microstructure homogeneity or residual stress related to fabrication details.

The creation of a customized particulate composite starts with a statement of goals as captured in a definition document. This might be a brief statement or listing of goals written in the vision of an ideal solution. The definition document provides insight into what combination of attributes are needed for success. It avoids unrealistic or loose goals; such as the request for the lowest possible cost. Quantitative criteria are used, since what is deemed "good" or "slow" for one person is not universally perceived the same. Limits are needed, such as a fabrication cost below \$2 per unit or raw material cost below \$1/cm^3. A properly formulated definition document is short and quantitative. The key parameters are listed first followed in sequence by secondary and optional attributes.

The definition document is the basis for whittling down the candidate constituents in a sequence of evaluation steps, as outlined in Fig. 5.2. Relying on hierarchical criteria ensures the important considerations are given priority. A spreadsheet with logic cells is one option. It might include weighing factors, giving more weight to the most important attributes. A properly constructed definition gives evaluation criteria in decreasing order of importance with clear goals, boundary conditions, and constraints.

Constraints might include toxicity, health hazard, low cost, lack of reactivity, or avoiding prior patented compositions. Subjective criteria are avoided. For example, a "noncorrosive" criterion is not quantitative; instead published corrosion rates provide an acceptance standard. Exact criteria might specify a maximum 0.1 % corrosive weight change (allowing for rusting or dissolution) for 100 mm^2 surface area (7 by 7 by 1 mm thick sample) after 7 days in 600 mL of mildly agitated 3.5 wt% salt water at room temperature. On such a quantitative basis a large number

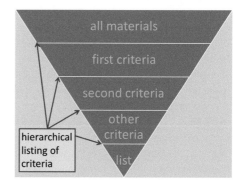

Fig. 5.2 Constituent selection involves a progressive down-selection process wherein the candidate materials are evaluated against the most important criteria first. Fewer candidates remain after each step in the hierarchical evaluation, leading to a short list of good candidates

of candidate materials pass, such as metals (silver, gold, platinum, palladium and their alloys), nickel-alloys (nickel, superalloys, high temperature alloys), most plastics, many rubbers, graphite, several copper alloys (brass, naval bronze), cobalt-chromium alloys, aluminum alloys, zirconium alloys, titanium alloys, most oxide ceramics, silicon carbide, silicon nitride, glass, and granite.

Illustrated Definition Document: Thermal Management

An example definition document is given in Table 5.1. This is for a thermal management materials intended to dissipate heat from microelectronic integrated circuits [32]. The dictates for success are arranged in decreasing importance, and in most cases the desired property is quantified.

The application arose as computer chips increased speed, complexity, and shrank in size. Semiconductors are slower and less conductive on heating, leading to premature failure. This problem arose in desktop computers, internet servers, land-based radar controllers, electrical power inverters, and base stations for cellular telephone networks. These applications were stationary, so component mass was not a concern; the lowest cost solutions were tungsten-copper, molybdenum-copper, and Invar-silver.

Subsequently, microelectronics emphasized portability, so low density materials were sought for computers, cellular telephones, and missile electronics. Likewise,

Table 5.1 Definition for microelectronic heat dissipation substrates

Attribute	Description	Measure
Thermal conductivity	As high as practical	Exceed 200 W/(m °C)
Hazards, toxicity	None	No recognized concerns
Material cost	As low as possible	Cost below \$0.5/cm^2
Shape availability	Flat wafers	Nominally 1–3 mm thick
Magnetic response	Nonmagnetic	No Fe, Ni, Co
Thermal expansion	Close to silicon	4×10^{-6} to 6×10^{-6} 1/°C
Melting temperature	Must survive soldering	Over 300 °C
Fabricated cost	As low as feasible	Maximum \$1/cm^2
Dimensional concerns	Flatness, roughness	Surface roughness <0.5 μm
Density	As low as possible	Maximum of 8 g/cm^3
Strength	Sufficient for assembly	Minimum 100 MPa
Assembly	Braze or thermal epoxy	Brazing most desirable
Thermal fatigue	No failure in on-off cycles	1500 cycles, −20 °C to 130 °C
Availability	Avoid exotic materials	At least 2 qualified vendors
Corrosion resistance	No reaction	Benign in humid air to 200 °C
Oxidation resistance	No reaction	No oxidation in air to 200 °C
Fracture toughness	Able to withstand dropping	Toughness over 4 MPa√m

automotive electronics in hybrid and electric vehicle systems embraced low density solutions. The final two solutions in use are;

- active heat dissipation, meaning heat dissipation is assisted by fluid flow through a small radiator especially designed for the electronics
- passive heat dissipation, meaning the heat spreader is in contact with ambient air and cools using convection.

The latter is a more demanding design, but is more reliable since there are no moving parts and avoids expensive repairs.

Illustrated Selection Protocol: Thermal Management

This section relies on a problem definition to illustrate constituent phase selection. The approach involves a matrix of application requirements compared to candidate constituent properties. According to Table 5.1, the initial concern is with thermal conductivity. A search shows several materials exceeding the target of 200 W/ (m °C). Purity and residual stress degrade conductivity, so only soft and pure phases are useful. An alphabetical listing of candidates and thermal conductivities, in W/(m °C), is as follows (simplified to show only major categories):

aluminum (Al) = 240 (unalloyed, annealed)
aluminum nitride (AlN) = 200 (polycrystalline, pure)
beryllia (BeO) = 265 (hot pressed, no additive)
beryllium (Be) = 215 (hot isostatically pressed, low oxygen)
brass (Cu-5Zn) = 235 (gilding metal, annealed)
calcium (Ca) = 200 (annealed)
copper (Cu) = 395 (oxygen-free, annealed)
diamond (C) = 400 (polycrystalline)
gold (Au) = 310 (pure, annealed)
graphite (C) = 370 (anisotropic, high in preferred orientation)
silicon carbide (SiC) = 210 (reaction bonded)
silver (Ag) = 415 (pure, annealed)
sterling silver (Ag-10Cu) = 365 (annealed).

No polymers pass this first hurdle. The remaining options include a few alloys, such as sterling silver—an alloy of silver and copper. Practically, thermal conductivity varies with purity, microstructure, and residual strain. For example, thermal conductivity of commercial silicon carbide ranges from 83 to 210 W/(m °C) depending on these factors. Hot isostatically pressed SiC is on the low end while reaction bonded SiC is on the high end. The best case is listed above. Pyrolytic graphite is anisotropic, amounting to a 200-fold difference in thermal conductivity with orientation, but that is ignored.

In terms of attributes, the next concern is toxicity or hazards. This eliminates beryllium and beryllia due to toxicity and calcium due to reactivity. Other criteria such as corrosion, oxidation resistance, and fracture toughness are not limitations for the remaining candidates. Copper and silver are higher density materials compared with the others.

A major consideration requiring specific attention is material cost. For a semi-conductor heat sink, the thermal behavior is quantified on the basis of unit area assuming 1–3 mm thickness. Assume 1 mm thickness to reduce mass. For large scale applications, the industry limit is $1/cm^2 (future designs are ranging to $8/cm^2). A material cost metric M_C allows a comparative evaluation of phases as follows:

$$M_C = C\frac{\rho}{100} \tag{5.1}$$

where ρ is the material density in g/cm^3 and C is the cost in $/kg; the denominator factor of 100 comes from the assumed 1 mm thickness and 1000 g/kg. In this application, the volume is constrained, so conversion from cost on a mass basis to cost per device basis is most fruitful. Cost fluctuations are ignored, realizing updates are possible using an internet search, especially for materials such as diamond, silver, gold, copper, and silicon carbide. The resulting values of M_C are summarized in Table 5.2.

To pass the cost criteria of $1/cm^2, the material price per component must be low. Table 5.1 includes raw material and fabrication costs. On the basis of material cost alone, as given in Table 5.2, five candidates remain viable—aluminum, brass, copper, graphite, and silicon carbide. While the initial universe of candidate materials started with thousands of options, the criteria of thermal conductivity, hazard-toxicity, and material cost considerably restricted the candidates.

The next criteria is for a non-magnetic structure and all of these candidates pass that hurdle. This leaves consideration of thermal expansion coefficient, which is a difficulty since silicon semiconductors are about 4×10^{-6} 1/°C. The desire is to approximately match this value. The room temperature thermal expansion coefficients are as follows:

aluminum $= 23 \times 10^{-6}$ 1/°C

Table 5.2 Material cost index for thermal management materials ($/cm^2 at 1 mm thickness)

Material	Material cost index M_C, $/cm^2 at 1 mm
Aluminum	0.05
Aluminum nitride	3
Brass	0.6
Copper	0.6
Diamond	90
Gold	8000
Graphite	0.3
Silver	70
Silicon carbide	0.6
Sterling silver	70

brass $= 17 \times 10^{-6} \, 1/^\circ$C
copper $= 17 \times 10^{-6} \, 1/^\circ$C
graphite $= 1 \times 10^{-6} \, 1/^\circ$C
silicon carbide $= 3 \times 10^{-6} \, 1/^\circ$C.

Thermal expansion coefficient varies with purity. For example, silicon carbide ranges from $3 \times 10^{-6} \, 1/^\circ$C to $5 \times 10^{-6} \, 1/^\circ$C, depending on fabrication route and purity. The listed value corresponds to reactive bonded silicon carbide that has the high thermal conductivity. Likewise anisotropic graphite ranges from 0.5×10^{-6} $1/^\circ$C to $8 \times 10^{-6} \, 1/^\circ$C.

Fabrication is a significant barrier. Silicon carbide is hard, brittle, and difficult to form. The estimated cost for compaction, sintering, and surface grinding is near \$2 per device. Coupled to a material cost of \$0.60 each gives a net cost of \$2.60 per heat dissipation piece, well over the target cost of \$1/cm^2.

Graphite has several complications, one being the anisotropic behavior. Further, it is difficult to form since it is brittle. The compressive strength is good, but graphite only reaches 35 MPa tensile strength, well below the 100 MPa criterion in Table 5.1. Indeed, there is no single material that matches all of the criteria. The metals aluminum, brass, and copper are too high in thermal expansion coefficient. Silicon carbide is attractive if cost is ignored. Indeed, this is the choice for heat dissipation devices in hybrid gasoline-electric automobiles.

Consider now the possible composites. Silicon carbide is low in thermal expansion coefficient, relatively expensive, brittle, and difficult to shape. On the other hand, aluminum and copper are high in thermal expansion, relatively low in cost, ductile, and easy to shape. A combination of ductile metal and hard silicon carbide conceptually provides a composite with property match to the requirements. It is anticipated composites with custom property combinations come from the following six (3 metals, 2 nonmetals) combinations:

Ductile, easily fabricated metal	Hard, difficult to fabricate nonmetal
Aluminum	Graphite
Brass	Silicon carbide
Copper	

The graphite composites are questionable because of anisotropic properties; a preferred orientation must be created during fabrication. Brass contains zinc and is a lower performance option versus copper, so copper is a better choice. Thus, the definition document and evaluation of candidate properties leads to composites such as copper-silicon carbide or aluminum-silicon carbide.

In situations where density is not a concern, solutions included tungsten-copper. In very high performance systems where cost is not a concern, solutions include diamond-copper and diamond-silicon carbide.

This evaluation approach identifies aluminum with silicon carbide as a composite suitable for portable computers, but copper with silicon carbide for stationary devices. If density is ignored, then copper-tungsten and copper-molybdenum are lower cost options. While if cost is removed as a limitation, then composites of copper-diamond are most successful.

Example Definition Documents

The following definition documents are for situations where particulate composites prove successful. The accompanying tables list the application requirements in order of importance. The reader can independently consider the challenge to assess candidates that meet the goals. Later, some of these are used to illustrate applications.

Materials for Wire Drawing Dies

This example relates hard, wear resistant composites used in wire drawing dies. Wire drawing uses a cascade of 20–50 progressively smaller diameter conical dies. The wire is pulled through each die until it reaches the target diameter. Each die is tapered, about 12°, to induce a 15 % area reduction in the wire. Two dies are shown in Fig. 5.3. As the wire diameter decreases the wire length and drawing speed increase. Frictional forces heat the dies so it is imperative the die remain strong to elevated temperatures. Thus, in Table 5.3 the first requirement is hardness [33]. Next is the cost per unit volume. Because of frictional heating, retention of strength to elevated temperature and rapid heat dissipation are desired. Usually the dies are formed by press-sinter technology, with final contouring by laser machining. Due to the operating compression stress, only modest fracture toughness is required.

Fig. 5.3 Two wire drawing ideas where a hard material occupies the conical inner configuration

Table 5.3 Definition for steel wire drawing die materials

Attribute	Description	Measure
Hardness	Wear resistance from high hardness	Minimum 1400 HV
Cost	Material cost per die on volume basis	Maximum $60/cm^3
Strength	Room temperature strength in bending test	Minimum 1200 MPa
Thermal	Withstand prolonged frictional heating	Minimum 400 MPa at 800 °C
Thermal	Dissipation of local frictional heat	Conductivity >90 W/(m °C)
Fabrication	Bulk unfinished die blank production cost	Maximum cost $2 each
Finishing	Surface finish by laser ablation	Roughness Ra below 0.2 µm
Toughness	Fracture toughness	Minimum 15 MPa√m

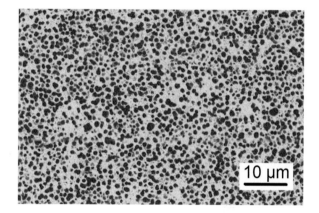

Fig. 5.4 Scanning electron micrograph of a multiple-level particulate composite. The *dark phase* is Ti(C,N) dispersed in a WC-10Co matrix (*white phase*). This composite provides exceptional life in wire drawing dies, due to the combination of hardness, toughness, heat dissipation, and wear resistance [courtesy of J. Keane]

Wire drawing dies have largely migrated to WC-Co composites, termed cemented carbides. Performance gains are possible using a multiple level composite scheme. In this approach, the cemented carbide is reinforced with a dispersed hard titanium carbo-nitride phase. The composite is imaged in Fig. 5.4; the matrix consists of WC-10Co containing about 50 vol% of 2 µm hard Ti(C,N) particles dispersed in that matrix.

For drawing steel wire chord, used to make radial tires, the composite exhibits a life up to 300 h at drawing velocities up to 6 m/s. Long life is valued, since every time any die fails the whole drawing operation is halted to replace that failed die. To avoid frequent interruptions, the protocol is to replace all dies when any die fails.

Self-Winding Watch Weights

In spite of a significant shift to electronic watches, many luxury brands rely on mechanical mechanisms. Watch weights convert random motion into stored energy via springs or batteries. As outlined in Table 5.4, the watch weight requirements

Table 5.4 Definition for watch self-winding weights

Attribute	Description	Measure
Density	High inertial density for winding kinetics	More than $10/cm^3$
Cost	Raw material cost limited	Maximum $40/kg
Strength	Sufficient for assembly	Over 30 MPa
Color/finish	Mechanism is visible, reflective metallic color	Polished silver color
Porosity	No pores apparent on polished surface	Pore size below 0.5 μm
Geometry	Squat asymmetrical shapes	Press-sinter compatible
Fabrication	Limited maximum temperature in sintering	1400 °C
Patentability	Desire to protect formulation	Patentable composition
Nontoxic	No toxic or hazardous materials	Compatible with skin

include a high density to accentuate energy storage [34]. Nominally the denser the weight the smaller the design, so 10 g/cm^3 (corresponding to molybdenum) is a lower limit. Other application requirements are listed in the table and include cost, strength, and aesthetic attributes. The usual geometry is a flat eccentric half-disk, so the main concerns are with ease of fabrication and long service life.

The most popular solutions are tungsten or molybdenum composites with transition metals such as cobalt, iron, copper, nickel, or palladium. Magnetic metals (Fe, Ni, Co) are minimized. Tungsten or molybdenum provides the density and in most cases copper-nickel provides for a low fabrication temperature. The relative cost of copper is small compared to molybdenum or tungsten, so adding copper significantly decreases material cost. Some of the solutions are composite relying on molybdenum and copper with small quantities of nickel to form a high-density, strong, and ductile composite that easily exceeds the mechanical property requirements.

Frangible Ammunition

Billions of bullets are fired yearly, but most are for target practice. Formerly low cost bullets were fabricated from an alloy of lead. Environmental contamination makes lead a problem because of soil and groundwater contamination. The idea of a frangible bullet takes several forms. In one variant the bullet density is matched to the traditional lead bullet, but the projectile contains no lead. One application is for ship-board target practice by amphibious forces. Practice is possible, but no lead contamination is encountered on board the ship.

As outlined in Table 5.5, to avoid lead contamination, the bullet is fabricated from nontoxic materials [35]. However, the density must reach about 11 g/cm^3. One solution is a composite of copper and tin, mixed together as powders, and compacted at 280 MPa. When heated above the melting temperature of tin

Table 5.5 Definition for frangible ammunition

Attribute	Description	Measure
Nontoxic	No toxic or hazardous materials	No environmental issues
Density	Close to lead alloys, higher than steel	Between 9 and 12 g/cm^3
Cost	Low cost for practice ammunition	Maximum $4/kg
Strength	Sufficient for firing from firearm	Minimum of 15 MPa
Ductility	Low to avoid flattening against target	Maximum 1 % elongation
Fracture	Fragmentation on impact with hard target	Impact velocity 300 m/s
Geometry	Elongated and tapered projectile	Press-sinter compatible
Production	Compatible with high volume fabrication	more than 5000/min

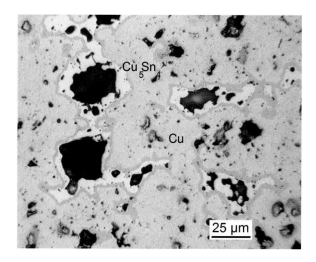

Fig. 5.5 Frangible ammunition formed by heating a mixture of copper and tin to a temperature where Cu_5Sn_4 forms. This brittle intermetallic provides an easy fracture path in the structure important to frangible ammunition; it fractures easily on impact

(232 °C or 505 K) the tin flows and reacts with copper to form a brittle intermetallic. However, when heated above 415 °C (688 K) that intermetallic decomposes. The resulting composite is about 8.9 g/cm^3 in density with a microstructure consisting of ductile copper bonded with Cu_5Sn_4, as illustrated in Fig. 5.5. It is sufficiently strong to survive firing while pulverizing on striking a hard target. Another solution relies on bullets made from copper and steel, giving a density near 8.5 g/cm^3.

As a direct substitute for lead, the density of the Cu-10Sn structure is low. This density shortfall is corrected by adding tungsten. One alternative is 55 vol% tungsten bonded with nylon. This latter mixture meets all of the technical requirements, but proves costly. The difficulty comes from the tungsten. High density tungsten-polymer composites were first employed in birdshot in countries that banned toxic lead.

Radiation Containment

As radioactive materials emerged, scientists such as Marie Curie, who won Nobel
Prizes 1903 in Physics and 1911 in Chemistry, initially failed to recognize the
danger from radiation exposure. By the time of her death in 1934, care in handling
radioisotopes had dramatically improved. The effort continues today, especially as
replacements for lead shielding are sought. Table 5.6 outlines the hierarchical
properties [36]. Applications are in dental offices, radiological laboratories, shipping containers, radiation therapy, and nuclear reactors used on spacecraft and
submarines.

For radio-opacity, a first requirement is a high atomic number that comes from
Hf, Ta, W, Re, Os, Ir, Pt, Au, Hg, Tl, and Pb, and other elements. This material must
be stable, so radioactive elements, such as Th and U, are not allowed. Likewise,
toxicity and cost quickly limit the search. Most of the shapes are relatively simple,
so a mixture of tungsten with a polymer is one solution, available as sheet, tube, or
rod. Early efforts relied on tungsten composites with nickel-copper as the matrix,
and this is still used for higher performance shielding. For nonstructural shielding,
tungsten-polymer composites are most economical. Due to the limitation on packing of tungsten powder, a mixture density about 11 g/cm^3 is common to match lead.
A picture of a rubber-tungsten mixture is shown in Fig. 5.6; this putty mixture fills
seams in radiation containment structures. It has the advantage of easy flow under
pressure, so is simple to shape. If desired, an external container provides handling
strength and resistance to scratching as needed for shipping containers. Thermoplastic variants are often cast into simple brick shapes, and stacked around radiation
sources. The putty is an analogous to mortar.

Table 5.6 Definition for radioisotope containment vessels

Attribute	Description	Measure
X-ray opacity	High atomic number	Atomic number over 71
Radioactivity	Free of radioactivity	Stable isotopes only
Toxicity	Material must be benign	Less toxic than lead
Cost	Raw material cost limited by lead	Maximum of $30/kg
Shape	Containers are cylinders	Up to 200 mm diameter
Fabrication	Pliable, easily formed at low stress	Viscosity of 10 GPa s
Mass	Radiation containers weigh up to 8 kg	Must scale to 8 kg sizes
Strength	Sufficient for handling, transport	Over 30 MPa strength
Surface	Burr free surface	No sharp edges
Hardness	Sufficient to resist scuffing	Minimum 100 HV

Fig. 5.6 An example of a
flexible tungsten-polymer
composite used for
radiation shielding. The
pliable "tungsten putty" is
pressed into cracks in a
radiation shielding structure

Table 5.7 Definition for cement core saw segments

Attribute	Description	Measure
Hardness	High hardness for minimum wear	Minimum of 2000 HV
Material cost	Rely on off-shelf hard materials	Maximum $300/kg
Strength	High compressive yield strength	Over 1000 MPa
Homogeneity	Properties repeatable in all segments	Strength variation ±5 %
Geometry	Flat or simple curved shape	Press-sinter compatible
Size	Typical 8–20 mm length, 3 mm thick	3 mm thick, 20 mm long
Temperature	Limited temperature tolerance	1000 °C for under 5 min
Fabrication	Net shape to minimize fabrication costs	Fabrication cost $0.2 each
Assembly	Compatible with laser welding to steel	Melt below 1800 °C

Concrete Saw Segments

It is fairly common to modify a concrete structure after it is constructed; for
example, to add wiring conduits, windows or doors, or to penetrate the structure
with electrical, water, or sewer lines. Sawing and coring segments are designed to
cut through concrete using structures outlined in Table 5.7 [15]. Concrete structures
have compressive strengths from 20 to 150 MPa. For drilling, a high hardness tool
is required to avoid premature drill failure. Diamond is a favorite as the hard phase,
in spite of the high cost. Wear resistance goes hand in hand with hardness. To keep
tool cost low the diamond content is in the 5–15 vol% range. Drilling involves both
shear and compressive stresses, so the tool needs to resist breakage, a feature
provided by a metallic bond, such as cobalt. The diamond and matrix are consolidated into small, flat segments. These segments are assembled on steel substrates
using laser welding. Core saws then consist of these segments assembled onto the

Table 5.8 Definition for stereo headset speaker magnets

Attribute	Description	Measure
Magnetism	High permanent magnet remanence	900 kA/m
Resistance	Demagnetization resistance from fields	1.3 T coercive force
Corrosion	Resistant to attack by moist air	Ni plate
Cost	Fabricated cost as low as possible	Below $100/kg
Size	Ring shaped hollow cylinder	3.5 mm diameter
Temperature	Able to operate after sitting in the sunlight	Magnetic to 100 °C
Energy	High level of stored magnetic energy	Minimum 300 kJ/m^3

end of a steel tube. For long service life each segment is uniform in properties to avoid premature failure. Because diamond is unstable at elevated temperatures, much attention is devoted to keeping the peak temperature below 1200 °C (1473 K). Otherwise, the diamond will decompose during heating, reverting to graphite.

Stereo Headset Speaker Magnets

Portable music devices rely on a permanent magnet and a soft magnetic core to transform digital electrical pulses into sound waves. The magnets are embedded in speakers, earphones, and other wearable devices. Table 5.8 provides a definition for a typical set of earphone miniature speakers [37]. The selection criteria are dominated by magnetic properties, but to improve efficiency the individual grains are electrically isolated. This combination of properties is satisfied using a mixture of magnetic $Fe_{14}Nd_2B$ as with a rare earth metal, such as Nd, Gd, Sm, Tb, or Er in various combinations. The added rare earth is intentionally oxidized during processing to form an oxide. The particles are compacted into the desired shape and liquid phase sintered to near full density. Figure 5.7 is a microstructure, where about 95 % of the structure is magnetic $Fe_{14}Nd_2B$ and 5 % is nonmagnetic and nonconductive oxide. The oxide is segregated to the interfaces between the magnetic grains. Final shapes are small, simple cylinders that respond to pulsed magnetic fields to move a diaphragm as a means to generate sound waves.

Electric Bicycle Motor Sectors

A bicycle supplemented with a battery operated electric motor is a very efficient transportation device. Table 5.9 provides a definition of a soft magnetic material as required for the electric motor sectors [38]. The design seeks to maximize efficiency with a low mass to avoid weight during peddling and a low cost. The magnetic properties, including electrical resistivity, are summarized in this table. The properties are satisfied at low cost using iron particles bonded with an insulating polymer coating.

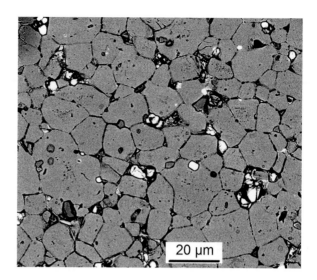

Fig. 5.7 Microstructure for a rare earth permanent magnet. About 95 % of the structure is the $Fe_{14}Nd_2B$ intermetallic phase. The grain boundary oxide is an insulator to interrupt electric conduction. This composition is ideal for situations desiring low energy losses with a strong magnetic response

20 μm

Table 5.9 Definition for electric bicycle motor segments

Attribute	Description	Measure
Magnetism	Strong response to magnetic field	1.2 T at 4000 A/m
Frequency	Able to respond quickly to variable field	Up to 1500 Hz
Conduction	Resistive to minimize eddy current losses	Resistivity of 200 μΩ-cm
Density	Low density to reduce weight	Under 8 g/cm^3
Strength	Resistance to impact, handling	Strength over 75 MPa
Hysteresis	Low energy loss per unit mass in cyclic loads	<70 W/kg, 1 T at 400 Hz
Permeability	Rapid magnetization	450 minimum
Fabricability	Able to form multiple pole geometries	Net-shape die compaction
Shape	Cylindrical disk with projecting poles	24 poles 10 mm diameter
Size	Thickness	20 mm

The coated iron particles are die pressed into the desired shape using sufficient pressure to eliminate most of the pores. The coated particle composite sectors exhibit desirable magnetic properties with low eddy current losses. Polymers suitable for this situation include most of the common plastics, but for cost and ease of processing a typical choice is cellulose dispersed using acetone or an acrylic polymer dispersed using water.

Hot Sheet-Steel Descaling Abrasive

Steel sheets are formed to make automotive bodies. The steel is cast and hot rolled into strip of the desired size. As a last step in sheet steel production, rust, oxides, or scale on the surface are removed prior to coiling. Surface cleaning is achieved using

Table 5.10 Definition for hot steel sheet descaling abrasives

Attribute	Description	Measure
Hardness	Resist wear in descaling steel	Over 1500 HV
Strength	Bending transverse rupture at 600 °C	Minimum 350 MPa
Environment	Resistant to hot steel, iron oxides, air	No degradation to 600 °C
Wear	Resistant to erosion at 1 m/s surface velocity	Under 200 MPa stress, maximum loss of 1 mm/h
Cost	Material cost limited	$15/kg maximum
Fabrication	Able to produce solid wheels	1 m diameter by 0.1 m wide
Reliability	Avoid defects in manufacturing	100 % inspection

high speed, rotating abrasive wheels to skim the steel surface. This enables stamping, bending, galvanization, and painting at the automotive plant without a pretreatment. In addition to abrasive wheels, chemical pickling treatments are used to clean the sheet steel. Accordingly, the requirements for the descaling abrasive are defined in Table 5.10. The steel is hot from heating and working, nominally temperatures up to 600 °C (873 K) are possible [39].

The abrasive descaling occurs at high speeds, with surface velocities more than 1 m/s, where the hot steel moves past the spinning abrasive wheel. Sufficient normal force is applied to abrade the surface layer. For this application, an abrasive is required that is effective, affordable, and long lasting. Typically cost is very low, and this results in a glass bonded abrasive. Some very expensive options are possible, but due to cost the most common abrasive is glass bonded alumina, or a similar hard ceramic composite.

Jet Engine Rotating Bearing

Rotation is key to jet engine compression and thrust. Incoming air is compressed and mixed with fuel. When combusted the expanding gas propels the jet forward. Rotating bearings experience taxing combinations of stress and temperature in the engine. Table 5.11 is a definition for an intermediate temperature bearing designed to operate up to temperatures of 400 °C (673 K) [40]. Resistance is required to oxidation and possible exposure to fuel during prolonged operation, exceeding 5000 h. By design, the loading on the bearing produces a high compressive stress with significant temperature transients. For this application, wear resistance is a key to success. Wear testing is performed under a load of 350 MPa at 1 Hz and 400 °C (673 K) using a pin on disk test applied for 5000 cycles. The opposing material is IN 718 superalloy. For this combination, the maximum allowed friction coefficient is 0.5. Various particulate composites are candidates for this application, including titanium carbide in a tool steel matrix and titanium carbo-nitride bonded with cemented carbide. Proper inspection after fabrication is used to ensure no defects in the final device.

Table 5.11 Definition for jet engine rotating bearing

Attribute	Description	Measure
Temperature	Able to withstand prolonged heating in air	To more than 400 °C
Environment	Resistant to fuel combustion products	To at least 400 °C
Strength	Compressive hot strength	35 MPa at 1200 °C
Wear	Test at 1 Hz, 400 °C, 350 MPa, 5000 cycles	Maximum 0.01 % mass loss
Friction	Coefficient of friction against IN 718	0.5 maximum
Toughness	Fracture toughness	Minimum 12 MPa√m
Fabrication	Able to form thin 10 mm thick ring structure	200 mm diameter
Inspection	Must ensure no critical defects	Test to 1.5 mm

Table 5.12 Definition for chromatography filters

Attribute	Description	Measure
Pore size	No large pores to prevent passage of debris	Below 0.5 μm
Corrosion	Resistant to corrosion in variety of fluids	Minimum stainless steel
Uniformity	Size and properties homogeneous	1 % porosity variation
Permeability	Low flow resistance to fluid passage	Minimum of 10^{-15} m^2
Cost	Material cost limited	$150/kg maximum
Fabrication	Scalable production process	Minimum 10^3/day

Chromatography Filters

Inert porous materials are required to filter samples in chemical analysis by liquid chromatography. The fluid cannot be tainted by the filter. One solution is a composite of an inert solid with controlled pores. The challenge is outlined in Table 5.12, where a pore size below 0.5 μm is required [41]. The corresponding particle size is about 3.5 μm. Likewise, to avoid contamination the filter is inert, and stainless steel is an option for many solutions. The filters are 4–10 mm in diameter, fabricated to deliver uniform flow. Permeability refers to the pressure loss as the fluid flows through the filter. To keep pressure losses low requires high permeabilities, so on the one hand small pores are desired for filtration, but large porosity is desired for permeability. The properly formulated filter is of high value, so material cost is not a large factor. Usually the filters are used in quality control and analytical laboratories. Porous nickel, stainless steel, or polyetheretherketone are the usual choices. The filter disks are compacted using a sacrificial pore forming polymer (stearic acid) that is burned out when the particles are sintered. Figure 5.8 is a scanning electron micrograph of the surface pore structure for a nickel variant.

Fig. 5.8 Scanning electron micrograph of the front surface on a porous nickel filter used in chromatography applications. The pores are tortuous, leading to significant removal of debris prior as a test fluid passed through the filter prior to testing [courtesy T. Young]

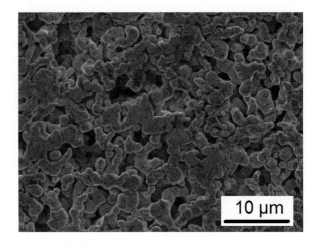

Approach to Constituent Selection

To summarize the constituent selection approach, several factors are pondered in sequence. First, the broad universe of engineering materials is considered. This includes many options, but with a few criteria the options are condensed into a list of leading candidates. Application needs are translated into quantitative criteria presented a hierarchical listing in what is termed the definition document. The most important features are listed in decreasing order of importance. Some properties are given additional emphasis; for example a low elastic modulus might be twice as important versus a high strength.

Evaluation criteria are used to screen candidate constituents. The goal is to reduce the number of candidates at each evaluation step. This is similar to how the personnel or human resources office screens out job applicants that do not pass the minimum criteria for education or relevant experience.

The quality of the selection process depends on the initial definition. The evaluation eventually anticipates how different constituents might be combined to form a composite. Naturally, the selection hones in on "leading" materials—diamond for hardness, polymers for flexibility, chromium-alloys for corrosion resistance, titanium for biocompatibility, aluminum for ease of fabrication, and so on.

Tabulations of material properties exist in many forms, and are the mainstay of online databases, industry promotional literature, engineering handbooks, subscription collections, and government sponsored compilations. There is no need to source those here, especially since these resources are growing rapidly. Trade associations promote their field with data tabulations. These are usually free, but restricted to the interests of the trade association, such as Copper Development Association's online resource. Often the key parameters require extensive investigation. The number of repositories is large, but a few examples are listed below:

American Ceramic Society
ASM International
CRC Materials Science and Engineering Handbook
eFunda
Granta Materials (Cambridge Engineering Selector)
Japan Aerospace Exploration Agency
MatWeb
National Aeronautics and Space Administration
National Institute of Advanced Industrial Science and Technology
National Institute of Standards and Technology
National Physical Laboratory
ProQuest
Smithells' Metals Reference Book
Society of Plastics Engineers.

Porosity Effects

Porosity is an important phase with respect to properties. Voids naturally occur between particles and are not always collapsed during consolidation. In some cases pores are desirable, such as in bearings. Some others include heat insulators, capacitors, biomaterials, and sound attenuation devices. For filters, porosity ranges from 20 to 80 %. In terms of composite properties, pores are considered "zero material"—the pores lack strength, elastic modulus, density, hardness, ductility, thermal conductivity, and so on. Accordingly, a porous material consists of two phases, one with handbook properties and one with zero properties. Accordingly, properties decrease as porosity increases.

The "weak link" character of pores makes it possible to modify properties downward. Figure 5.9 is a demonstration for elastic modulus, showing a plot versus porosity for iron [42]. The full density elastic modulus is 208 GPa. The declination in elastic modulus with increasing porosity allows property customization a wide range. Such a reduction is useful in titanium implants. The elastic modulus of porous titanium is a good match to bone.

Likewise, strength declines with porosity, and routinely falls to half the full-dense value with the introduction of just 15 % porosity. Figure 5.10 plots normalized fracture strength of alumina (Al_2O_3) versus porosity [43]. The plot gives a clear indication of the strength decline with porosity. There is an additional dependence on the pore shape; differences in processing conditions provide a means to modify pore shape. Fracture paths naturally find the weak links. With just a few pores, fracture intentionally jumps from pore to pore, avoiding the strong ligaments. Thus, even a small amount of porosity provides a means to intentionally lower strength.

As a final demonstration of porosity effects, Fig. 5.11 plots the thermal conductivity versus porosity for alumina. The upper end properties reflect full density

Fig. 5.9 Data for porous iron showing the relative elastic modulus (normalized to 208 GPa at 100 % density) versus the porosity level [42]

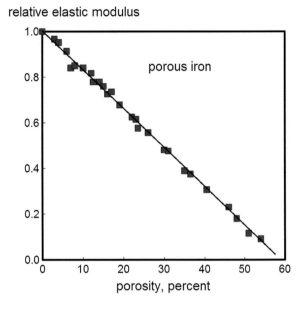

Fig. 5.10 Fracture strength for alumina versus porosity, with a normalized strength of 248 MPa at 100 % density [43]. It is common to see strength decline by 50 % with just 15 % porosity. Other properties are not as sensitive to porosity

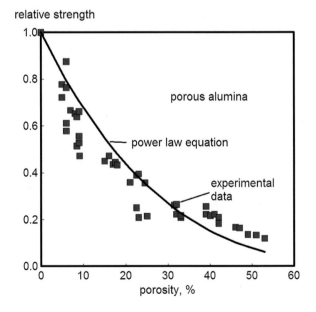

alumina, and as porosity increases the conductivity falls. The ability to tailor properties, such as thermal conductivity, via selected porosity levels is the basis for high temperature insulators. One insulation composite is used to 650 °C (923 K) using foamed calcium silicate with 95 % porosity. That thermal conductivity is low

Fig. 5.11 Relative thermal conductivity for alumina versus porosity. The thermal conductivity for full density alumina is 29 W/(m °C). The data indicates nearly a straight line declination in thermal conductivity with porosity. Air has a conductivity of 0.0257 W/(m °C) at room temperature, so effectively the behavior follows the rule of mixtures for alumina-air

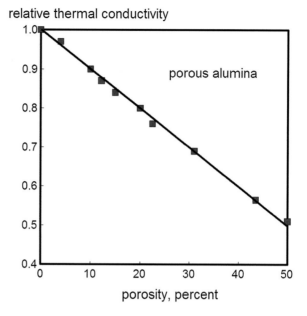

at 0.07 W/(m °C), but the strength is also low at 0.4 MPa. For fracture properties, large pores and irregular pores are more detrimental compared to round or spherical pores.

Common Constituent Materials

Collected here are brief descriptions of several constituents used commonly in particulate composites with a survey of properties and uses [1–31]. These are some of the more favorite constituents in particulate composites, listed alphabetically.

Acrylic (CH_2-$C[CH_3]$-$[COOCH_3])_n$: A colorless, odorless thermoplastic polymer widely used in adhesives and coatings. Several variants exist, with polymethylmethacrylate (PMMA) being widely employed due to toughness and resistance to environmental degradation. It has a density of 1.2 g/cm^3, exhibits good weathering and operates up to 100 °C (373 K). For short-term applications it can withstand a constant stress of 75 MPa, but for long-term applications it is restricted to 10 MPa. The fracture toughness is from 1 to 2 MPa√m.

Alumina (Al_2O_3): A common oxide used in composites because of the combination of hardness, cost, and stiffness. It has a density of 3.98 g/cm^3. The bending or transverse rupture strength varies from 350 to 550 MPa, depending on grain size. The elastic modulus is 405 GPa with a hardness of 1500–1800 HV (depending on purity, density, and grain size), and fracture toughness at 2 MPa√m. It is an insulator used in wear, electrical, thermal, and high temperature structures, for example as the white insulator on spark plugs.

Aluminosilicate Glass (SiO_2-17Al_2O_3-8CaO-7MgO-5B_2O_3): A higher temperature glass that is more expensive versus traditional soda-lime window glass. The density is about 3.5 g/cm^3 with a maximum use temperature of 650 °C (923 K). The room temperature strength is higher than borosilicate glasses, and it is resistant to most chemicals. Uses are in airplane windows, lighting, and electrical systems.

Aluminum (Al): Pure aluminum is a low strength but high ductility metal with a density of 2.70 g/cm^3. It is used in lightweight composites in part due to the easy processing from its 660 °C (933 K) melting temperature. For pure aluminum the thermal conductivity is 237 W/(m °C) with a high thermal expansion coefficient of 23.8×10^{-6} 1/°C. Usually impurities and alloying reduce the thermal conductivity, meaning alloys are significantly lower in conductivity. Aluminum is a good electrical conductor at 4.2×10^6 S/m. The mechanical properties are not unique, since the elastic modulus is 70 GPa and yield strength is 25 MPa; the strength is similar to engineering polymers at half the density and the elastic modulus is considerably below that of steel and titanium. The properties are most attractive on a normalized basis, such as elastic modulus divided by density. In particulate composites, aluminum is combined with low density phases, such as alumina, graphite, aluminum nitride, or silicon carbide, to form hard, stiff, and tough composites.

Aluminum 6061 (Al-1 Mg-0.5Cu-0.5Cr-0.5Si): One of many aluminum alloys, this aerospace alloy is low in density at 2.7 g/cm^3 with adjustable mechanical properties depending on heat treatment. The thermal expansion coefficient is relatively high at 24×10^{-6} 1/°C with a thermal conductivity of 170 W/(m °C). The elastic modulus is near 70 GPa, while the typical yield strength is 250–290 MPa with 17 % fracture elongation; it can be heat treated to promote more strength with less ductility.

Borides: Several high performance ceramic phases arise from boron compounds. Usually they are hard but expensive due to the difficulty in forming the compounds. An important boride is B_4C, which has a 2450 °C (2723 K) melting temperature and exceptional hardness. The density is 2.51 g/cm^3. Other common borides and their densities are as follows: chromium boride CrB (6.14 g/cm^3), hafnium boride HfB_2 (11.19 g/cm^3), tantalum boride TaB_2 (12.54 g/cm^3), titanium boride TiB_2 (4.52 g/cm^3), vanadium boride VB (5.60 g/cm^3), and zirconium boride ZrB_2 (6.10 g/cm^3).

Boron (B): Boron melts at approximately 2300 °C (2573 K). The density is 2.34 g/cm^3. The elastic modulus is 300 GPa, hardness 3300 HV, but fracture strength is low at 200 MPa. Its use is mainly to form boride compounds. For example, stainless steel powder with added boron generates a composite of chromium boride dispersed in stainless steel. The stainless steel provides strength, toughness, and corrosion resistance, while the boride provides hardness and wear resistance.

Boron Carbide (B_4C): This hard material is fabricated in a manner similar to diamond, relying on high pressure-temperature synthesis. Accordingly it is expensive. The estimated melting temperature is 2450 °C (2723 K) but it is unstable on heating unless pressurized. The density is 2.51 g/cm^3, elastic modulus is 440 GPa, and fracture strength ranges from 340 to 440 MPa. The hardness ranks at 3200 HV.

Boron Nitride (BN): Boron nitride is a high melting temperature compound with 2.08 g/cm^3 density. Strength is 310 MPa, hardness is 250 HV, and elastic modulus is 78 GPa. It has good chemical resistance up to high temperatures, making it a good choice for high temperature applications. Hexagonal boron nitride is soft and easily machined and is used for contact with molten metals. The cubic structure is often used in composites for cutting tool applications, especially where diamond fails.

Borosilicate Glass (SiO$_2$-24B$_2$O$_3$-6Na$_2$O-2Al$_2$O$_3$): Borosilicate glasses are used in laboratory and lighting devices. They are noted for thermal shock resistance, low thermal expansion coefficient, and chemical durability. Several compositions variants are possible. It is amorphous, so lacks crystal structure. The glass is transparent and useful to 440 °C (713 K). It is relatively low in strength (30 MPa tensile and 130 MPa bending) and toughness (0.6 MPa√m), so easy forming is the main attribute. Elastic modulus is 70 GPa and density is about 2.4 g/cm^3.

Bronze (Cu-10Sn): This solid solution alloy of copper and tin has exceptional durability in solutions (except strong acids) and to temperatures of 500 °C (773 K). Bronze is widely used in bearings, corrosion resistant hardware, marine hardware, and industrial fittings. The yield strength is 190 MPa with 50 % elongation to fracture and 117 GPa elastic modulus. For bearings it is combined with iron or graphite to improve strength, lower friction, and lower cost.

Carbides: Carbon reacts with several metals to form carbides, such as iron (Fe$_3$C), niobium (NbC), vanadium (VC), chromium (Cr$_3$C$_2$), hafnium (HfC), molybdenum (Mo2C), tungsten (WC), titanium (TiC), tantalum (TaC), silicon (SiC), and zirconium (ZrC). Carbides are hard and generally stable at high temperatures in atmospheres such as nitrogen and vacuum. Refractory metal carbides are used in wear or cutting composites.

Carbo-Nitrides: Carbon and nitrogen compounds are able to form mixtures termed carbo-nitrides. Some of the most useful are iron and titanium compounds, with strengths reaching to 3 GPa. The similar atomic bonding of carbon and nitrogen with various metals allows for substitution one for the other, so the compounds are nominally termed carbo-nitrides to reflect the interchange in ingredients.

Cobalt (Co): Cobalt is a magnetic, polymorphic element used frequently as a bonding agent for carbides and diamonds. It has a density of 8.92 g/cm^3 and melts at 1495 °C (1768 K). It is widely used in composites because of good hardness (160–370 HV), high yield strength of 300–550 MPa, depending on heat treatment, and good wetting of hard phases such as TiC, WC, and NbC. The elastic modulus is 211 GPa. It has a fracture toughness that can reach 150 MPa√m.

Copper (Cu): Copper is a high ductility metallic phase widely used in electrical and thermal devices. The density is relatively high at 8.96 g/cm^3 and the melting temperature is 1083 °C (1356 K), but most attractive is the high thermal conductivity at 403 W/(m °C) and low electrical resistivity 1.6 μΩ-cm. The elastic modulus is 145 GPa. Without alloying it is a relatively soft metal, with yield strength of

Fig. 5.12 Polyhedral
diamond grains, showing
flat faces and sharp corners
that are usually retained in
the final composite for use
in cutting, abrasion, or
polishing applications

120 MPa and 45 % elongation to fracture. For this reason it is alloyed with tin to
form bronze or zinc to form brass.

Diamond (C): Diamond is pure carbon stabilized from graphite by ultrahigh
pressures. Without pressure, diamond converts to graphite on heating over about
1000 °C (1273 K). The higher the peak temperature the shorter the time to
decomposition. Diamond is used in the production of abrasive tools, grinding
wheels, and surface finishing tools. Synthetic diamond is formed from graphite at
high temperature and high pressure using transition metal catalysts to produce
polyhedral grains as seen in Fig. 5.12. The density is 3.51 g/cm^3 with an elastic
modulus of 900 GPa. It is one of the hardest materials at 10 on the Mohs hardness
scale, reaching 4000–5000 HV. In the proper orientation and with high purity,
diamond thermal conductivity reaches 2100 W/(m °C), but often measured values
are in the 400 W/(m °C) range—still feeling "cold as ice". Because diamond
decomposes on heating, composites rely on cobalt, copper, iron, titanium, and
other metals as matrix phases. The diamond-metal composites are consolidated in
hot pressing treatments at peak temperatures from 800 to 1200 °C (1073–1473 K)
with rapid heating and cycles lasting just minutes.

Epoxy: Thermosetting compounds that react to crosslink into a hard, rigid
polymer are known as epoxies. The typical density is 1.2 g/cm^3. Epoxy is lower
in strength (70 MPa) versus some of the high performance thermoplastics, but
epoxy is higher in hardness at 15 HV. The fracture toughness is low at 0.4 MPa√m.
Maximum service temperatures are near 138 °C (411 K). The primary interest in
epoxy formulations arises since they are easy to mix with other phases and the
incorporated filler particles provide electrical, thermal, or wear properties. Epoxy is
used in golf clubs, aerospace assembly, home repairs, electrical circuit encapsulation, and routine assembly.

Fly Ash: A waste product from coal combustion, this pulverized powder must be disposed of in land-fills, but many efforts are made to use it in engineering products. It is available from coal-fired power plants in excess of 10^{11} kg/year in the USA. The composition is variable with the coal source, but usually consists of silica (SiO_2), alumina (Al_2O_3), and calcia (CaO). The exact composition is not controlled. Fly ash is captured by pollution control devices. Typical particle sizes range up to 150 μm and the density variation from 1.6 to 2.6 g/cm^3 reflects the compositional variation. As a filler it is used in concrete (Portland cement), asphalt, and bricks, but it has not found acceptance for high performance particulate composites.

Graphite (C): The hexagonal crystal form of carbon (versus the cubic crystal form of diamond) has significantly anisotropic properties. Graphite is naturally occurring with high impurity levels, but is also manmade from methane or other gas reactions, resulting in higher purity. The manmade form comes from pyrolytic events, such as decomposition of methane or rayon. Further it is possible to form amorphous graphite. Due to a low hardness, typically near 30 HV, graphite is easy to machine. Although brittle, the elastic modulus varies with purity and orientation, ranging from 14 to 500 GPa. This stiffness with a 2.0–2.5 g/cm^3 density gives a very high specific modulus. By itself graphite is used for electronic resistors, since it provides limited conductivity. Also, the retention of modest strength (30–100 MPa) to high temperatures makes graphite a favorite material for applications such as furnace heating elements, electrical contacts, and hot pressing dies. It has a low coefficient of friction against steel, leading to graphite-bronze composites for use in bearings.

Hafnium Carbide (HfC): It is one of the highest melting temperature materials known with an estimated melting point of 4160 °C (4433 K). The hardness is 2000 HV, density is 12.6 g/cm^3, elastic modulus is 460 GPa, but strength is low at 240 MPa. It is used as a reinforcement for high temperature rocket structures formed from tungsten, rhenium, or molybdenum as the matrix phase.

Hydroxyapatite ($Ca_5(PO_4)_3OH$): This calcium phosphate is similar to human bone and teeth. Accordingly hydroxyapatite is widely recognized as being biocompatible. In practice it is often combined with metallic systems such as titanium to fabricate strong and biocompatible composites for implants. Hydroxyapatite has a density of 3.16 g/cm^3, melting temperature near 1200 °C (1473 K), strength near 40–100 MPa, and elastic modulus of 42 GPa.

IN 600 (Ni-16Cr-7Fe-0.2Mn-0.2Si-0.1Cu): Also known as Inconel 600, this is a mainstay high temperature alloy for use in air up to 1100 °C. The density is 8.48 g/cm^3 and elastic modulus is 214 GPa. Although the room temperature yield strength is just 205 MPa, the fracture elongation is high at 50 % and strength is retained to high temperatures. It is an example of a low-grade superalloy with an attractive combination of temperature resistance, strength at high temperature, and oxidation resistance.

Iron (Fe): This polymorphic element is widely used as the basis for steels, stainless steels, and soft magnets. It has a density of 7.87 g/cm^3, elastic modulus of 196 GPa, and melting temperature of 1536 °C (1809 K). The body-centered cubic phase is stable below 910 °C (1183 K). The higher temperature face-centered cubic phase known as austenite has solubility for several species, such as carbon and nitrogen. The dissolution and precipitation of carbides and nitrides provides a means to manipulate properties by heat treatment. In the unalloyed condition iron is soft and ductile, with yield strength of 150 MPa, tensile strength of 262 MPa, and 45 % elongation to fracture. Alloying and heat treatment increases strength tenfold.

Iron-Neodymium-Boron ($Fe_{14}Nd_2B$): This compound is most useful in the amorphous state, where it has a density of 7.7 g/cm^3 and provides outstanding magnetic performance, exhibiting the highest magnetization energy product of any known material. To avoid crystallization, it is produced by rapid solidification into thin foils, pulverized, and compacted to form magnets. The compaction is performed with an aligning magnetic field. An excess of rare earth enables liquid phase sintering with low grain growth. After sintering the elastic modulus is 142 GPa and ultimate tensile strength is 600 MPa, but there is no ductility. Magnets from this compound are widely used in automotive, electronic, and computer devices, including stereo music speakers and headphones. When attritioned to small particles, the compound will spontaneously ignite, so inert handling is required.

Magnesia (MgO): The stoichiometric oxide of magnesium is a favorite for higher temperature structures, such as furnace linings for steel and glass melting furnaces. In former times it was reinforced with asbestos, but now for high temperature refractories the compound is mixed with alumina or silica. Alumina additions largely reduce cost with modest property degradation, but silica is more detrimental. Magnesia has a density of 3.58 g/cm^3, melting temperature of 2825 °C (3098 K), hardness of 890 HV, elastic modulus near 300 GPa. The fracture strength is low at 200 MPa. The thermal expansion coefficient and thermal conductivity match steels, thereby avoiding thermal stresses in high temperature processing equipment.

Maraging Steel (Fe-18Ni-12Co-4Mo-0.8Ti): Highly alloyed martensitic steels that lack carbon, but harden instead by precipitation of small reinforcing phases, are known as maraging steels. Variants range up to 25 wt% nickel, manganese, and various combinations of aluminum, titanium, or niobium for hardening. They generally melt over 1400 °C (1673 K). When heat treated, small precipitates form to induce extraordinary levels of strength and toughness. Yield strengths range from 1400 MPa to over 5000 MPa, depending on alloying and heat treatment. Typical densities are 8.1 g/cm^3, and like other ferrous alloys the elastic modulus is 210 GPa. The fracture toughness is 175 MPa√m with 15 % elongation to fracture.

Mica (SiO_2-Al_2O_3-Fe_2O_3 and related oxides): Mica is a platelet mineral that provides a low-cost reinforcement for polymers. It is a silicate of alumina that contains various other elements in contents from a few percent or less, such as iron, sodium, calcium, potassium, and magnesium. There are 37 recognized variants. Although capable of operating at moderate temperatures, up to about 500 °C (773 K), it is mostly suitable for polymers such as polyethylene and polypropylene that melt at lower temperatures. It is anisotropic, so properties vary by a factor of ten depending on the orientation. Density is over 2.2 g/cm^3 and can range to 3.2 g/cm^3 depending on composition. Hardness is moderate, so mica is considered a soft mineral, but even so it provides a low-cost means for improving properties in plastics.

Molybdenum (Mo): Molybdenum is a high temperature metal with density at 10.22 g/cm^3 significantly lower than other most other refractory metals such as tungsten, tantalum, or rhenium. It melts at 2610 °C (2883 K) and has a 5.4×10^{-6} 1/°C thermal expansion coefficient. At room temperature the elastic modulus is 328 GPa, yield strength is 225 MPa, and fracture elongation is 3 %. Molybdenum with second phase dispersions for creep strengthening, such as lanthanum oxide, is used in furnace hardware. In microelectronics the mixture of molybdenum and copper delivers a low thermal expansion coefficient in a high thermal conductivity composite. Such composites are useful in heat dissipation devices associated with microelectronics using copper additions. Molybdenum retains strength to elevated temperatures, exceeding 100 MPa at 1000 °C (1273 K). The TZM alloy, consisting of molybdenum, titanium, and zirconium is especially resistant to high temperatures. Deformation processing delivers high room temperature strengths, near 620 MPa at room temperature. At 1095 °C (1368 K) molybdenum has a stress rupture life of 100 h at a stress of 85 MPa.

Molybdenum Disilicide (MoSi$_2$): This compound is unique in the combination of temperature and oxidation resistance. The theoretical density is 6.26 g/cm^3, elastic modulus is 160 GPa, and room temperature fracture strength is 450 MPa with no ductility. The major attraction is the ability to operate in air up to 1600 °C (1873 K) or higher. Indeed, the compound requires surface silicon oxides for best properties. The attributes make molybdenum disilicide useful for heating elements. Besides lanthanum oxide, other strengthening reinforcements include silicon carbide.

Nickel (Ni): Nickel is magnetic and corrosion resistant, proving useful for applications in batteries. It is heavy, with a density of 8.9 g/cm^3, and melting temperature of 1453 °C (1726 K). Similar to iron and cobalt, nickel has solubility for many of the refractory metal carbides, making it a common ingredient in cermets (ceramic-metal composites such as VC-Ni, TiC-Ni, and TaC-Ni), as well as cemented carbides and refractory metal composites. Generally it is a soft, tough metal with a yield strength of just 148 MPa, but 47 % elongation to fracture. Nickel alloys are a mainstay of elevated temperature aerospace structures.

Nickel Aluminides (NiAl and Ni$_3$Al): These are intermetallic compounds formed by exothermic reaction of nickel and aluminum at either the 1:1 or 3:1 stoichiometry. The NiAl compound melts at 1640 °C (1913 K), has a density of 6.05 g/cm^3 and room temperature elastic modulus of 290 GPa. Nickel aluminides increase strength on heating; the Ni$_3$Al compound is ductile and reaches peak strength at 800 °C (1073 K). The density is 7.25 g/cm^3 with a melting temperature of 1380 °C (1653 K). The room temperature yield strength is 450 MPa with 50 % elongation to fracture. It is more oxidation and corrosion resistant than stainless steels. It is used for heat treating equipment and thermal barrier coatings.

Niobium (Nb): This refractory metal is used as a high-temperature replacement for nickel alloys in turbines. At a density of 8.57 g/cm^3 and melting temperature of 2468 °C (2741 K) niobium provides several advantages over nickel alloys, most especially a higher operating temperature. The unalloyed yield strength is 255 MPa with 26 % fracture elongation and 113 GPa elastic modulus. Niobium composites contain silicides.

Nitinol (NiTi): This is the leading shape memory alloy. It is an equiatomic intermetallic compound named after the elements Ni and Ti and the site of discovery, Naval Ordnance Laboratory. It has attributes in austenite-martensite phase transformations that allow a device to be trained to two shapes. It can be induced to switch back and forth between those two shapes during thermal cycling or via plastic deformation. The melting temperature is 1310 °C (1583 K) with 75 GPa elastic modulus, 1150 MPa tensile strength, and 10 % fracture elongation. After plastic deformation (usually less than 8 % strain), modest heating causes a return to the original shape. Besides the shape memory feature, it is corrosion resistant with a low elastic modulus, making it useful for orthodontia.

Nitrides: Similar to borides and carbides, many nitrogen compounds form with transition metals, usually resulting in a hard phase. The most stable nitrides occur via reaction with refractory metals, such as chromium to give CrN or tantalum to give TaN, but include aluminum nitride (AlN), boron nitride (BN), hafnium nitride (HfN), niobium nitride (NbN), silicon nitride (Si_3N_4), titanium nitride (TiN), vanadium nitride (VN), and zirconium nitride (ZrN). A popular nitride is TiN, which has a gold color and excellent wear and corrosion resistance, so it is deposited on structures to form a hard and protective coating. One common application is in cutting tools, plumbing fixtures, and decorative jewelry. Aluminum nitride is an electrical insulator with high thermal conductivity. Cubic boron nitride is similar to diamond in hardness. Silicon nitride has excellent wear resistance and is often employed to high temperatures in air, such as in turbochargers.

Polyamide or Nylon ($NH-[CH_2]_{10}-CO)_n$: Nylon is widely used for industrial components. It resists most common solvents and flames. Service is possible to 100 °C (373 K) since it is susceptible to moisture adsorption. Density is 1.14 g/cm^3 with strength near 83 MPa and 60 % fracture elongation. Compared to other common plastics, the mechanical properties are all good, but not outstanding. However, it is widely available and easy to form and often reinforced with mineral or glass phases.

Polycarbonate ($O-[C_6H_4]-C[CH_3]_2-[C_6H_4]-CO)_n$: This polymer is widely employed for high impact applications, including bullet-proof windows, eye protection goggles, and safety helmets. The polymer is best recognized for high impact strength. Density is 1.15 g/cm^3, elastic modulus is 2.4 GPa, and tensile strength is 90 MPa with 110 % fracture elongation. Maximum use temperature is limited to 144 °C (417 K).

Polyetheretherketone ($O-[C_6H_4]-O-[C_6H_4]-C[O]-[C_6H_4])_n$: Commonly called PEEK, this is a chemically durable polymer with 1.32 g/cm^3 density, relatively good hardness of 30 HV, and strength of 92 MPa. It is widely employed for the chemical durability, as it is inert with respect to all environments except strong acids. The melting range is from 322 to 346 °C (595–619 K). It is used in extruded products, but cannot be formed using injection molding. Common additives are graphite and glass to improve durability.

Polyethylene Terephthalate $(CO\text{-}[C_6H_4]\text{-}CO\text{-}O\text{-}[CH_2]_2\text{-}O)_n$: This popular plastic comes in various levels of crystallinity and is recognized by the more common term polyester. The amorphous form of PET is commonly employed in beverage bottles. It is transparent, has a good toughness and strength (75 MPa) with 75 % elongation to fracture, but low thermal conductivity. Processing temperatures are in the 280 °C (553 K) range, but use temperatures are limited to a maximum of 65 °C (338 K). Generally polyethylene terephthalate is considered to be a thermoplastic, but thermosetting variants exist. Besides beverage bottles, it is used for clothing, packaging, and as the support phase in flexible magnetic tape.

Polyoxymethylene $(CH_2\text{-}O)_n$: Also commonly called polyacetal, the polymer has good strength and hardness. However, it is unstable in contact with nitric acid, halogens, and zinc ions. It is commonly used as binder for metal and ceramic powders where the mixtures are injection molded at 160 °C (433 K) to form complicated shapes.

Polytetrafluoroethylene $(CF_2\text{-}CF_2)_n$: The PTFE polymer is widely known for its trade name Teflon®. It is difficult to form so usually it is employed as a simple shape or coating. The mechanical properties are not impressive, but the inability to wet this polymer makes it useful in industrial and consumer products, including dental floss and rainproof clothing and shoes.

Polypropylene $(CH_2\text{-}CH[CH_3])_n$: Polypropylene is the second most commonly used polymer after polyethylene. Accordingly, it is broadly applied in consumer products, ranging from toys to lawn furniture. It combines easy forming by injection molding or extrusion, with excellent durability in a wide variety of environments. The mechanical properties are similar to epoxy compositions, being a low cost thermoplastic in contrast with the thermosetting character of epoxy.

Rhenium (Re): This is a very high density metal at 21.02 g/cm³, often used in high temperature metallic alloys, especially molybdenum and tungsten. The elastic modulus is 466 GPa, yield strength 315 MPa, tensile strength 1034 MPa, with elongation to fracture of 10 %. Rhenium improves ductility of other refractory metals. After deformation processing it helps resist recrystallization up to 1600 °C (1873 K).

SiAON $(Si_3N_4\text{-}Al_2O_3\text{-}AlN)$: A group of covalent ceramics based on silicon nitride, with additives that include alumina, aluminum nitride, silicon oxide, and yttria; the abbreviation stands for Si-Al-O-N compounds. These are strong and low density materials with high properties useful up to temperatures over 1200 °C (1473 K) in air. The compound is used in high temperature oxidizing conditions, such as furnaces, metal cutting tools, diesel engines, gas turbines, and automotive turbochargers. The properties are variable with compositions, where the typical values might be 3.2 g/cm³ density, 290–300 GPa elastic modulus, 450 MPa bending fracture strength, and 3,500 MPa compressive strength. Fracture toughness is 6.5 MPa√m.

Silicon Carbide (SiC): This man-made covalently bonded ceramic compound SiC is very hard and stiff, desirable for use in wear applications. As a semiconductor it operates at much higher temperatures when compared with silicon. As a structural ceramic it retains strength in air to high temperatures, making it useful for

Fig. 5.13 Silicon carbide whiskers as used in metal matrix composites. The whiskers shown in this scanning electron micrograph are not wetted by metals, so it is necessary to add wetting agents such as magnesium to enable a strong interface bond

heaters and combustion components. The compound has several polymorphic forms. Silicon carbide has a density of 3.145 g/cm^3, approximate melting temperature of 2700 °C (2973 K), room temperature elastic modulus of 414 GPa, and thermal conductivity that ranges up to 210 W/(m °C) depending on purity. At room temperature the typical bending fracture strength is 400 MPa with no ductility and a fracture toughness of 4.5 MPa√m. Whiskers or particles of silicon carbide are frequently used in particulate composites, especially with aluminum alloys as the matrix. Figure 5.13 is a scanning electron micrograph of a SiC whisker designed for such composites. The whiskers come in sizes from 0.1 to 1.0 μm diameter and lengths up to 200 μm.

Silicon Nitride (Si$_3$N$_4$): Most silicon nitride is prepared with additives to improve processing, leading to SiAlON compositions with properties mentioned earlier. The additives provide a high temperature viscous phase that assists processing by hot pressing or related consolidation routes. Pure silicon nitride has a density of 3.18 g/cm^3, hardness of 1700–1800 HV, and elastic modulus of 300 GPa. It is attractive because of high hardness, stability to high temperatures, oxidation resistance, and good mechanical properties in a low density compound. At high temperatures it is unstable and requires nitrogen to prevent decomposition. Typical room temperature mechanical properties are tensile strength near 580 MPa, no ductility, and fracture toughness of 6 MPa√m.

Silver (Ag): This precious metal is widely used in electrical and thermal systems because of exceptional conductivity, both electrical (6.3·10^7 S/m) and thermal (431 W/(m °C). In the unalloyed form silver is soft (27 HV), weak (10 MPa yield strength), but ductile (45 % elongation). It tends to tarnish in air, so it requires protection against surface attack during long-term use. Electrical contacts often contain Mo, W, CdO, or WC as a hard and arc erosion resistant phase. Silver has a density of 10.5 g/cm^3, so it is used mostly in stationary structures, such as circuit breakers and power switches. The thermal expansion coefficient is 19.2×10^{-6} 1/° C. For pure silver the melting temperature of 961 °C (1234 K) enables infiltration into refractory metals to form composites.

Soda-Lime Glass (SiO_2-17Na_2O-5CaO-4MgO-1Al_2O_3): This transparent glass has a density near 2.45 g/cm^3, and it is slightly stronger than borosilicate glass. It is used in lamp applications and safely operates in air to 460 °C (733 K).

Spinel ($MgAl_2O_4$): A compound of alumina and magnesia, spinel is generally sought for optical properties. The density is 3.55 g/cm^3 with a high elastic modulus at 277–284 GPa. It is close to alumina in hardness (1200 HV) with a melting temperature of 2135 °C (2408 K). Uses include watch covers and radar covers. Since the compound remains strong to high temperatures, it also proves difficult to consolidate except at very high temperatures.

Stainless Steel 17-4 PH (Fe-17Cr-4Ni-4Cu-0.3Nb): A precipitation hardenable martensitic stainless steel with excellent strength and hardness, and good corrosion resistance. It is also known as AISI 630 stainless steel or surgical stainless steel. After a solutionization heat treatment, best properties are achieved by subjecting the material to an aging heat treatment. Depending on the aging temperature, the strength and hardness increase while ductility and toughness decrease. For example, when heat treated to 285 HV, the yield strength is 900 MPa and the elongation to fracture is 17 %. When heat treated to a higher hardness of 460 HV, the yield strength is 1150 MPa with a ductility of 14 % elongation and fracture toughness of 142 MPa√m.

Stainless Steel 304 (Fe-18Cr-10Ni-1Mn): This is sometimes designated at 18-8 or 18-10 stainless steel, denoting the two main alloying ingredients of chromium and nickel. It is austenitic and nonmagnetic. The variant known as 304 L requires a carbon level below 300 ppm (0.03 wt%) to avoid embrittlement and corrosion difficulties from chromium carbide formation during welding, sintering, heat treatment, or thermal cycle. The composition is allowed to vary by ±2 wt% for Ni and Cr, meaning the density likewise ranges from 7.85 to 8.05 g/cm^3. The annealed room temperature elastic modulus is 195 GPa, yield strength is 205 MPa, and tensile strength is 510 MPa, with at least 45 % fracture elongation. The fatigue strength is 230 MPa and the fracture toughness is 55 MPa√m. It has good corrosion resistance in most common situations.

Stainless Steel 316 (Fe-17Cr-12Ni-2Mo-2Mn): This stainless steel grade is similar to 304 (except for molybdenum) with slightly better corrosion resistance, especially pitting corrosion resistance. It is the first choice for general purpose corrosion applications. The density is 8.05 g/cm^3, elastic modulus of 193 GPa, yield strength of 255 MPa, tensile strength of 530 MPa, with 55 % fracture elongation, and fracture toughness of 72 MPa√m. The thermal conductivity is relatively low at 16.2 W/(m °C) with a thermal expansion coefficient of 15.9×10^{-6} 1/°C. In the annealed condition it is nonmagnetic, but deformation can induce magnetization. Because the alloy can vary in composition by a few percent, the melting temperature ranges from 1375 to 1440 °C (1648–1713). It is available in the low carbon composition designated as 316 L.

Stainless Steel 410 (Fe-13Cr-1Mn-1Si-0.3C): A hard but less corrosion resistant stainless steel. When heat treated to give a 260 HV hardness the tensile strength is 800 MPa with 27 % elongation to fracture. It is generally used for applications requiring wear resistance.

Steel 4140 (Fe-1Mn-0.2Mo-0.2Si-0.4C): Approximately 10,000 steels are in common use, so this is only representative of the broad array of compositions. The 4140 grade steel is a common, lower-cost, heat treatable composition with excellent properties. The density is 7.86 g/cm^3 with 211 GPa elastic modulus. When heat treated to a hardness of 400 HV the yield strength is 1200 MPa with 15 % fracture elongation and 50 MPa√m fracture toughness. These properties are characteristic of many inexpensive low alloy steels.

Steel 4640 (Fe-2Ni-0.3Mn-0.2Mo-0.4C): This steel relies on nickel and molybdenum for strengthening and is heat treatable to provide excellent property combinations. The melting temperature is 1410 °C (1683 K), density is 7.75 g/cm^3, and elastic modulus is 209 GPa. When annealed the yield strength is 855 MPa with 16 % fracture elongation and 50 MPa√m fracture toughness. When heat treated, the yield strength reaches 1100 MPa, tensile strength reaches 1275 MPa, and is elongation 14 %. These properties are characteristic of many medium carbon steels. High alloy steels, with more than 8 wt% nonferrous content, include ferrous superalloys, tool steels, and maraging steels.

Superalloy IN 718 (Ni-19Cr-18Fe-5Nb-3Mo-1Ti-0.5Al): This is one of many nickel-base superalloy compositions developed for demanding applications such as in aircraft engines. The retention of strength to high temperatures in demanding combustion atmospheres without creep, oxidation, or erosion makes this one of the outstanding candidates for high temperature composites. It has a relatively high density of 8.19 g/cm^3 with 1190 MPa yield strength and 21 % fracture elongation.

Tantalum (Ta): Tantalum is used as a porous structure in capacitors and biomedical implants. The production of porous electrical capacitors is the dominant application. Small particles with controlled pores provide the surface area, anodization, and electrical conductivity combination needed for high reliability capacitors. It has a density of 16.6 g/cm^3 and melting temperature of 2996 °C (3269 K). Tantalum is biocompatible, electrically conductive, and ductile (40 % fracture elongation). It retains strength to high temperatures; for pure tantalum at 500 °C (773 K) the strength is 210 MPa and at 1000 °C (1273 K) the strength is 100 MPa.

Tantalum Carbide (TaC): This equiatomic compound of carbon and tantalum is used in metal cutting tools. The estimated melting temperature is near 4820 °C and the density is 14.53 g/cm^3. The Mohs hardness is in excess of 9 with a Vickers hardness of 1952 HV, approaching the hardness of diamond. The fracture strength is 310 MPa. The high hardness is a key attribute for the use of TaC in wear components.

Titanium (Ti): Titanium is a polymorph that is usually alloyed (see Ti-6-4) to improve strength, but CP or commercially pure titanium is employed in implants. The metal is recognized for high strength (500–900 MPa depending on alloying, impurities, and heat treatment), ductility (12–23 % fracture elongation), low density (4.5 g/cm^3), biocompatibility, and corrosion resistance. It has a high melting temperature of 1668 °C (1941 K). It is used in applications as diverse as jet engines, petrochemical plants, and biomedical devices. Because titanium is a polymorph, different properties are possible depending on the crystal structure as stabilized by different combinations of deformation and heat treatment.

Titanium Aluminide (TiAl, Ti$_3$Al): Both intermetallic compounds are used in aerospace applications due to the combination of low density (3.8 and 4.7 g/cm^3), oxidation resistance, and high yield strength (600 and 990 MPa at room temperature), with some modest ductility (4 and 10 % elongation to fracture). Often they are used in coatings, possibly formed by thermal spray or plasma spray routes. A composite with a fracture toughness more than 100 MPa√m is used in lightweight armor systems.

Titanium Carbide (TiC): A very hard, refractory, stoichiometric compound used extensively in hard metals, cutting tools, and wear resistant cermets. This compound has a broad stoichiometry range, so the properties are highly variable with composition or Ti:C ratio. The density is typically between 4.9 and 5.0 g/cm^3 and the melting temperature peaks near 3070 °C (3343 K) at 18 wt% carbon (45 at. % C). The hardness ranges up to 3200 HV but can be as low as 2500 HV, with elastic modulus from 315 to 440 GPa. Because of its hardness and stiffness, the brittle compound is used in composites involving a tough matrix such as Ni, Co, or Fe. It is attacked by oxidizing chemicals.

Titanium Diboride (TiB$_2$): This compound with 2700 HV hardness is electrically conductive, making it useful in elevated temperature refining devices. It is oxidation resistance to 1500 °C. The low density of 4.5 g/cm^3 coupled to a high elastic modulus of 420 GPa makes and the related TiB compound useful for steel matrix composites. The melting temperature is above 2900 °C (3173 K). Titanium diboride resists wetting and attack by molten aluminum, leading to uses in aluminum refining in drained cell designs intended to lower electricity consumption.

Titanium Dioxide (TiO$_2$): Also known as titania, this white powder is the most common man-made ceramic. For titania, the density is 4.17 g/cm^3, melting temperature is 1830 °C (2103 K), hardness is 1000 HV, and elastic modulus is 282 GPa. Because of low cost and high thermal stability it is used in a wide variety of applications, ranging from toothpaste to thermal insulation. It also has solar absorption attributes that are employed in solar cells and sunblock skin cream.

Titanium Nitride (TiN): This gold colored equiatomic compound is used to protect carbide cutting tools used in metal cutting and recently it is employed in jewelry. The room temperature hardness is 1800 HV, density is 5.4 g/cm^3 and it melts at 2930 °C (3203 K). Whiskers are used to reinforce hard composites for cutting ferrous metals, since TiN is inert with respect to iron.

Titanium Ti-6-4 (Ti-6Al-4 V): Over half of all titanium production is used to form this alloy. The density of Ti-6-4 is 4.46 g/cm^3. This couples with a high yield strength of 835 MPa and 15 % elongation to fracture with 9 ppm/°C thermal expansion coefficient. The combination of strength and thermal expansion coefficient is attractive for aerospace applications, especially graphite composite aircraft. Fracture toughness is another attraction, with typical values of 55 MPa√m. A wide variety of strength-ductility combinations are possible depending on the heat treatment.

Tool Steel (Fe-9 W-3Cr-0.5 V-0.3Ni-0.3Mn-0.3Si-0.3C): Tool steels are highly alloyed ferrous materials. They rely on carbide formers, such as tungsten, chromium, or molybdenum, and the heating/cooling cycles to produce high hardness. Listed here is the composition for H21, a tungsten tool steel with good strength retention to high temperatures. It can be used as the matrix for titanium carbide. By itself, this alloy delivers a yield strength of 1600 MPa, 7–17 % elongation to fracture, and fracture toughness around 30 MPa√m. The heat treated hardness reaches 600 HV at room temperature. The largest use in particulate composites is for hot working using about 50 vol% TiC reinforcement. Machining the composite is at the limit possible using metal cutting tools, so tool steel with TiC is considered the ultimate in traditionally machined materials.

Tungsten (W): This pure metal is one the highest melting temperature metals with a melting temperature of 3410 °C (3683 K) and density of 19.26 g/cm^3. The former attribute is useful in high temperature systems, including lighting filaments and furnaces heating elements, and the high density is used in inertial weights and munitions. At room temperature tungsten has a hardness of 430 HV and yield strength of 550 MPa. Tungsten can be significantly strengthened by deformation processing. Although brittle at room temperature, tungsten undergoes a brittle to ductile transition on heating. Impure tungsten is brittle up to 450 °C (723 K), but pure tungsten is ductile at 150 °C (423 K). As a fiber the tensile strength increases to 3900 MPa. Alloying tungsten with various combinations of nickel, iron, copper, or cobalt provides a high density composite with impressive ductility and toughness.

Tungsten Carbide (WC): This carbide is extremely hard and relatively inexpensive. Accordingly, it is the basis for many metal cutting tools, as well as oil and gas drilling bits, and highway construction or mining wear components. Depending on the stoichiometry, the compound has a density of 15.5–15.8 g/cm^3. It decomposes on heating, so only an estimated melting temperature is given, namely 2800 °C (3073 K). The combination of high hardness (various reports give 2200–2500 HV), good fracture strength (540 MPa), and high elastic modulus (540 GPa) proves useful in precise, long-life metal cutting tools.

Vanadium Carbide (VC): The equiatomic compound has a density of 5.81 g/cm^3 and because of a high hardness in the 2100–2500 HV range, it is used in wear materials. As a dispersed phase, VC provides desirable wear resistance in premium tool steels.

Yttria (Y_2O_3): The density of 5.03 g/cm^3 and melting temperature between 2460 and 2700 °C (2733–2973 K) make yttria useful in high temperature structures. The elastic modulus is 280 GPa and fracture strength is 300 MPa. It is often combined with other materials such as nickel-base superalloys, stainless steels, and ceramics such as aluminum nitride, silicon nitride, and zirconia.

Zirconia (ZrO_2): The zirconium oxide compound has a melting temperature near 2700 °C (2973 K) with a density more than 6 g/cm^3. The strength in tension is low, but due to toughening the bending strength can range from 500 to 1200 MPa. Hardness varies from 700 to 1200 HV, depending on additives, density, and grain size. When zirconia is stabilized, the fracture toughness approaches 10 MPa√m. It is a good electrical insulator with a low thermal conductivity, near 2 W/(m °C).

Study Questions

5.1. In a novel composite knife blade, the desire is to combine a hard particle with stainless steel, where the former provides resistance to wear and the latter provides corrosion resistance. Outline an approach to identification of the most suitable hard phase.

5.2. For the knife blade composite described in Study Question 5.1, what might be a starting target composition?

5.3. For the knife blade composite described in Study Questions 5.1 and 5.2, estimate the hardness possible with some of the candidate materials. What is the most compatible composite in terms of the phases, phase properties, hardness gain, and cost?

5.4. In each major material category (metals, ceramics, polymers) there are thousands of options. To sense the uniformity of properties in each material category, determine the tensile strength and fracture elongation for three common plastics—polyethylene, polypropylene, and polyethylene terephthalate. Plot strength versus elongation to determine how closely they cluster. Compare these values with a common steel, include specific strength (tensile strength divided by density).

5.5. One option for frangible ammunition is a composite of tungsten and tin. Calculate the weight fraction of tungsten required to form 11 g/cm^3 bullets (assuming no porosity). What is the cost ratio for this composite compared to lead, assuming tin costs \$20/kg, lead costs \$2.20/kg, and tungsten costs \$45/kg?

5.6. Your employer asks for a definition document for a new art medium to compete with "bronze clay" and "precious metal clay". These products rely on bronze or silver powders mixed with a flexible, rubbery polymer. The composite allows shaping by artists similar to how clay is formed. After the object is formed, the polymer is burned out while the powder sinters to produce the desired object. What are the advantages behind such a product?

5.7. Usually material cost is on a weight basis. In production, often component volume is constrained, so the number of components produced per unit mass depends on density. Find the current cost for molybdenum and convert this to a cost per unit volume.

5.8. Compare the cost of common plastic such as polyethylene, steel such as 1060, and concrete on a per unit volume basis.

5.9. An inventor has synthesized the B$_{14}$AlMg compound (designated as BAM) with a hardness higher than 3200 HV. He proposes combining the compound with titanium diboride to form metal cutting tools. What laboratory properties should be measured to evaluate the composite for this application? Besides cost, what are some other practical limitations to such a composite?

5.10. A composite of bronze and graphite is frequently used to make a sliding bushing in metal stamping machines to ensure precise motion and low friction. Besides low wear and good stiffness, what other merits are evident in this composite?

5.11. Electrical power switches and circuit breakers rely on tungsten-silver composites to obtain arc erosion resistance and electrical conductivity. It is proposed to switch the silver to aluminum to lower material cost. What problems you might anticipate from this change?

5.12. Porous insulators are created from ceramic slurries with up to 95 % porosity. If a borosilicate glass is selected with a full density thermal conductivity of 1.3 W/(m °C), what is the estimated thermal conductivity expected with 95 % porosity?

5.13. Filters with 25 % porosity are fabricated from 316 L stainless steel. What is the expected strength?

References

1. Anonymous, *Cambridge Engineering Selector* (Granta Design, Cambridge, updated annually)
2. Anonymous, *NIST Standard Reference Data—Materials Properties*, Gaithersburg, www.nist.gov/srd/materials.cfm and www.ceramics.nist.gov/srd/summary
3. G.S. Brady, H.R. Clauser, J. Vaccari, *Materials Handbook*, 15th edn. (McGraw-Hill, New York, 2002)
4. K.J.A. Brookes, *Hardmetals and Other Hard Materials*, 3rd edn. (International Carbide Data, East Barnet, 1998)
5. H.E. Boyer, T.L. Gall (eds.), *Metals Handbook Desk Edition* (ASM International, Materials Park, 1998)
6. Q. Chen, G.A. Thouas, Metallic implant biomaterials. Mater. Sci. Eng. **R87**, 1–57 (2015)
7. Y.M. Chiang, D. Birnie, W.D. Kingery, *Physical Ceramics* (Wiley, New York, 1997)
8. J.R. Davis (ed.), *Handbook of Materials for Medical Devices* (ASM International, Materials Park, 2003)
9. J. Emsley, *The Elements*, 3rd edn. (Oxford University Press, Oxford, 2000)
10. P. Ettmayer, Hardmetals and cermets. Annu. Rev. Mater. Sci. **19**, 145–164 (1989)
11. W.F. Gale, T.C. Totemeier, *Smithells Metals Reference Book*, 8th edn. (Elsevier, Oxford, 2004)
12. R.M. German, *A-Z of Powder Metallurgy* (Elsevier Scientific, Oxford, 2005)
13. Y. Hirata, Representation of thermal expansion coefficient of solid material with particulate inclusion. Ceram. Int. **41**, 2706–2713 (2015)
14. K. Kondoh, in *Titanium Metal Matrix Composites by Powder Metallurgy Routes*, ed. by M.A. Qian, F.H. Froes. Titanium Powder Metallurgy (Elsevier, Oxford, 2015), pp. 277–297
15. J. Konstanty, *Powder Metallurgy Diamond Tools* (Elsevier, Amsterdam, 2005)
16. H. Lange, G. Wotting, G. Winter, Silicon nitride—from powder synthesis to ceramic materials. Angew. Chem. Int. **30**, 1579–1597 (2003)
17. W. Lassner, W.D. Schubert, *Tungsten: Properties, Chemistry Technology of the Element, Alloys, and Chemical Compounds* (Kluwer, New York, 1999)
18. P.W. Lee, Y. Trudel, R. Iacocca, R.M. German, B.L. Ferguson, W.B. Eisen, K. Moyer, D. Madan, H. Sanderow (eds.), *Powder Metal Technologies and Applications*, vol. 7 (ASM Handbook, ASM International, Materials Park, 1998)
19. J.V. Milewski, H.S. Katz, *Handbook of Reinforcements for Plastics* (Van Nostrand Reinhold, New York, 1987)

20. R. Morrell, *Handbook of Properties of Technical and Engineering Ceramics* (Her Majesty's Stationery Office, London, 1987)
21. R.G. Munro, Material properties of titanium diboride. J. Res. Natl. Inst. Stand. Technol. **105**, 709–720 (2000)
22. J. Ormerod, The physical metallurgy and processing of sintered rare earth permanent magnets. J. Less Common Met. **111**, 49–69 (1985)
23. H.O. Pierson, *Handbook of Refractory Carbides and Nitrides* (Noyes, Westwood, 1996)
24. G.D. Rieck, *Tungsten and Its Compounds* (Pergamon Press, Oxford, 1967)
25. R. Riedel (ed.), *Handbook of Ceramic Hard Materials* (Wiley-VCH, Weinheim, 2000)
26. S.J. Schneider, in *Ceramics and Glasses*, Engineered Materials Handbook, vol. 4 (ASM International, Materials Park, 1991)
27. P. Schwarzkopf, R. Kieffer, W. Leszynski, F. Benesovsky, *Refractory Hard Metals: Borides, Carbides, Nitrides, and Silicides* (MacMillan, New York, 1953)
28. A.B. Strong, *Plastics—Materials and Processing*, 3rd edn. (Prentice Hall, Upper Saddle River, 2006)
29. G.S. Upadhyaya, *Nature and Properties of Refractory Carbides* (Nova Science, Commack, 1996)
30. J.B. Watchman, *Mechanical Properties of Ceramics* (Wiley, New York, 1996)
31. A.W. Weimer, *Carbide, Nitride, and Boride Materials Synthesis and Processing* (Chapman and Hall, London, 1997)
32. C. Zweben, Advances in high performance thermal management materials: a review. J. Adv. Mater. **39**, 3–10 (2007)
33. K.J.A. Brookes, *Hardmetals and Other Hard Materials*, 3rd edn. (International Carbide Data, Hertsfordshire, 1998)
34. J.P. Schluep, Component parts for watch movements, U.S. Patent 3,942,317 A, 1976
35. A.V. Nadkarni, J.T. Abrams, Lead-free frangible bullets and process for making same, U.S. Patent 6,536,352 B1, 2003
36. S.G. Caldwell, A.L. Madison, A review of tungsten heavy alloy utilization in isotope transport containers. Proceedings WM Symposia, Tempe, AZ, 2013, paper 13380
37. J.S. Cook, P.L. Rossiter, Rare-earth iron boron supermagnets. Crit. Rev. Solid State Mater. Sci. **15**, 509–550 (1989)
38. S. Tiller, Soft magnetic composites in the development of a new compact transversal flux electric motor. Powder Metall. Rev. **3**, 75–77 (2013)
39. R.B. Cauffiel, Apparatus for descaling metal strips, U.S. Patent 4,019,282 A, 1977
40. P. Gloeckner, K. Dullenkopf, M. Flouros, Direct outer ring cooling of a high speed jet engine mainshaft ball bearing: experimental investigation results. J. Eng. Gas Turbines Power **133**, (2011). paper 062603
41. W.J. Cheong, Fritting techniques in chromatography. J. Sep. Sci. **37**, 603–617 (2014)
42. R. Haynes, *The Mechanical Behaviour of Sintered Metals* (Freund Publishing House, London, 1981)
43. R.L. Coble, W.D. Kingery, Effect of porosity on physical properties of sintered alumina. J. Am. Ceram. Soc. **39**, 377–385 (1956)

Chapter 6
Powder Selection

Once the chemical composition of each phase is decided, attention then turns to selection of the constituent powders. Focus here is given to the powder options in terms of particle fabrication route, particle characteristics such as size, shape, packing, and the important morphological relations between phases controlled by the particle characteristics.

General Concerns

Once a composition is determined, the next important step in building a particulate composite focuses on identification of the powders to deliver the desired performance. Sometimes the particles are custom produced to suit the application requirements, but cost usually dictates selection of existing powders whenever possible. Available powders may be modified to match the consolidation process or envisioned final microstructure. Proper particle selection is crucial to attaining the desired phase morphology to ensure proper performance.

On the one hand, larger particles reduce handling problems, making them desirable. Yet on the other hand, small particles, especially those smaller than about 10 μm, are difficult to handle and are easily suspended in air. Special care to needed with small particles to avoid inhalation, skin irritation, ingestion, and possible fires or even explosions. A powder that burns in air is termed pyrophoric. These powders are inherently unstable and will spontaneously burn to form an oxide or nitride once ignited [1]. It is for this reason small powders are routinely used in pyrotechnics, explosives, and rocket fuels. A most dramatic demonstration for the author was titanium powder immersed in liquid nitrogen at 77 K ($-196\,°C$). When an electrical spark initiated a reaction, titanium nitride rapidly formed in a sustained exothermic reaction in spite of the very low starting temperature. Fortunately titanium and related powders require an ignition source. For titanium, zirconium, and tantalum, reactions with air occurs with a heat source of just 400 °C (673 K), a temperature easily reached by a spark. Once ignited, a fire or

© Springer International Publishing Switzerland 2016
R.M. German, *Particulate Composites*, DOI 10.1007/978-3-319-29917-4_6

Table 6.1 Common polymers used as process control agents

Polymer name	Structure	Melting temperature, °C	Decomposition temperature, °C
Stearic acid	CH_3-$(CH_2)_{16}$-CO_2H	70	383
Paraffin wax	C_NH_{2N+2}; N = 18–45	45	370
Mineral oil/kerosene	C_NH_{2N+2}; N = 6–40	−45	40
Ethylene-bis-stearmid	2 stearic acids coupled by N-O-C bond	144	350
Polyoxymethylene (polyacetal)	-$[CH_2$-O$]_N$-	162	160 (in N_2+HNO_3)
Polyethylene	-$[CH_2$-$CH_2]_N$-	110	400
Polyethylene oxide (poly-ethylene glycol)	OH-$[O$-CH_2-$CH_2]_N$-H	60	182
Polypropylene	-$[CH_2$-$C_2H_4]_N$-	130	450

explosion is possible, generating pressure waves of 20 MPa in a split second. Besides the recognized reactive species, several other metals are easily oxidized, including aluminum, copper, iron, molybdenum, niobium, silicon, and chromium. Students working in the author's laboratory have experienced fires with these powders, fortunately with no injuries, but fire trucks were required on one occasion.

The worst explosions occur when powders are dispersed in air at concentrations greater than about 40 g/m^3. One danger is from static discharges or inadvertent sparks during powder loading or unloading, such as with emptying mixers, storage containers, or attritioning mills. The problem is acute with small powders and is formally known as dusting. When poured, smaller particles tend to undergo turbulent flow, become levitated in air turbulence, and scatter to create a dust cloud. For this reason, wetting agents are added to agglomerate the particles to avoid dusting. Some lubricant and agglomeration additives are listed in Table 6.1. The common and inexpensive agents are paraffin wax or hydrocarbon fluids such as kerosene, mixed at concentrations of 0.5–1 wt% with the powder. Fundamentally the common agents are short chain polymers with a hydrocarbon makeup of C_NH_{2N+2}, where N is an integer between 18 and 45. For example, paraffin wax, mineral oil (N is an integer from 15 to 40), or kerosene (N is an integer from 6 to 16) are low molecular weight versions of the same repeating -CH_2- hydrocarbon structure. All are used to control dusting. Other polymers in widespread use include ethylene vinyl acetate (the basis for hot glue) and water soluble polyethylene glycol or methyl cellulose. These might be dispersed as tiny droplet emulsions, or dissolved in water. Once the powder is mixed with the liquid dispersion, the water is evaporated to leave agglomerated clusters consisting of hundreds of individual particles.

The scanning electron micrograph in Fig. 6.1 is an example of intentionally agglomerated particles held together by 0.5 wt% paraffin wax. Agglomeration was achieved by mixing the two powders (tungsten carbide and cobalt) with paraffin wax and heptane to form a slurry with the consistency of house paint. This slurry

Fig. 6.1 Scanning electron micrograph of an agglomerated powder prepared by spray drying a slurry of powder, paraffin wax, and heptane. The slurry was sprayed into a heated chamber to evaporate the heptane solvent, leading to these spherical balls

was sprayed to form droplets. During freefall in a warm chamber the droplets spheroidized while the heptane evaporated. This process is known as spray drying. The early concept was developed as a means to form powdered milk by removing water from sprayed milk, leaving only solids as dry milk. During particle handling, such as loading into compaction equipment, an agglomerated system avoids the tendency for the tungsten carbide and cobalt to separate; separation is a concern whenever powders differ in density, size, or shape.

Beyond handling concerns, attention is needed to properly select the starting powders to attain the desired microstructure. Powders have artifacts from the process used in their formation, and those artifacts carry over into the consolidated product. For example, hollow particles retain the void space in the consolidated structure. Thus, the powder may need to be tailored to present the desired size, shape, or surface condition needed for the consolidation process. Both a language and mathematical approach exist for defining powders and particles to ensure proper specifications as detailed in this chapter [2, 3].

Cost and Availability

Powder production is a substantial industry, resulting in a wide array of compositions, purities, particle sizes, particle shapes, and surface chemistries. Globally, the powder production geared to engineering structures exceeds $20 billion in value each year. That ignores the large tonnages of everyday powders, such as flour, salt, sugar, paint pigment, inks, fertilizers, and abrasives. Sorting through the array of offerings to find the right powders requires discriminants to select the best match to the application requirements.

Powders follow a classic use-price curve, where the highest prices are associated with lowest consumption. Most likely a high cost deflects users to lower cost powders. Powder price also varies with requirements. An example would be a surgical stainless steel powder in 1000 kg quantities. The price increases as the particle size decreases; the mainstay is 20 μm powder at approximately $15/kg; 15 μm powder is 10 % more expensive, 10 μm powder is 25 % more expensive, and 5 μm powder is 100 % more expensive. Relative prices for some of the more common particulate composite constituents are listed in Table 6.2 [4]. These prices are normalized to the lowest cost, nominally corresponding to pure 120 μm irregular iron and 10 μm calcite. Additional factors are order quantity and transportation costs. For some of the low cost powders, such as iron, transportation costs add 10 % to the purchase price.

The powder cost is a significant consideration for many applications. As an illustration, a modern automobile is selling for about $20/kg. In such products the manufacturing cost is about half the selling price, or about $10/kg, the remainder goes to advertising, management, stock dividends, interest on loans, and other indirect burdens. Raw materials typically account for just 8–15 % of the manufacturing cost. For particulate composites to penetrate into automotive applications requires judicious attention to cost. Thus, polymer reinforced with mineral phases, such as calcite, are easy to qualify. On the other hand there are several high performance applications tolerant of higher costs, such as oil well drilling, medical implants, wire drawing, metal cutting, high performance computing, industrial sensors, and aerospace devices. Some of these allow for powder costs up to $1000/kg. The normal expectation is for cost to decrease as consumption increases. The opposite situation arises for materials with restricted production capacity—cost increases with consumption. This is a difficulty for titanium, since limited powder supplies imply cost would climb with any large scale application. Automotive firms feared increasing costs and refused to be early adopters, since they require cost decreases as use expands.

Materials Selection Protocol

The constituent selection process involves concerns as follows:

1. What are the needed powder characteristics for the minor phase (the phase constituting below 50 vol% of the composite)?
2. What characteristics (powder or liquid) are required for the major phase (the phase constituting more than 50 vol% of the composite)?
3. What level of connectivity is desired for each of the phases?
4. Is porosity accounted for in the design?

These factors are interrelated as treated in this chapter. Porosity was treated previously and generally is a means to discount properties to lower strength, toughness, conductivity, or other attributes. There are occasions where porosity is beneficial, such as in materials designed for easy fracture or for drainage. Porosity is intentionally specified in structures used for filtration, sound absorption, fluid

Table 6.2 Comparative prices for powders (Normalized to 120 μm Iron Powder as Unity)

Material	Particle shape and size	Relative price
Alumina (Al_2O_3)	Platelet, 0.4 μm	40
Alumina (Al_2O_3)	Angular, 4 μm	2
Alumina (Al_2O_3)	Angular, crushed, 50 μm	5
Aluminum alloy (Al)	Ligamental, 20 μm	12
Aluminum nitride (AlN)	Rounded, 0.5 μm	150
Boron nitride, cubic (cBN)	Angular, 40 μm	2500
Bronze (Cu-10Sn)	Spherical, 24 μm	14
Calcite ($CaCO_3$)	Angular, 10 μm	1
Carbon black (C)	Agglomerated, 0.025 μm	2
Cobalt (Co)	Rounded, agglomerated, 1.5 μm	55
Cobalt-chromium (66Co-28Cr-6Mo)	Spherical, 20 μm	160
Copper (Cu)	Rounded, 4 μm	50
Copper (Cu)	Rounded, 10 μm	11
Copper (Cu)	Irregular, 100 μm	8
Diamond (C)	Polygonal, 6 μm	2200
Glass (alumino-silicate)	Angular, irregular, 7 μm	10
Invar (Fe-36Ni)	Irregular, rounded, 30 μm	18
Iron (Fe)	Spherical, 5 μm	12
Iron (Fe)	Irregular, 120 μm	1
Iron-neodymium-boron ($Fe_{14}Nd_2B$)	Flake, angular, 3 μm	48
Magnesia (MgO)	Irregular, 3 μm	12
Manganese (Mn)	Rounded, 10 μm	20
Molybdenum (Mo)	Polygonal, 3 μm	55
Nickel (Ni)	Spherical, 4 μm	40
Nickel aluminide (Ni_3Al)	Irregular, rounded, 22 μm	25
Silica (SiO_2)	Spherical, agglomerated, 0.15 μm	2
Silicon carbide (SiC)	Angular, 0.2 μm	88
Silicon carbide (SiC)	Whisker, 2 μm diameter	150
Silicon nitride (Si_3N_4)	Angular, 0.1 μm	101
Stainless steel (Fe-19Cr-9Ni)	Spherical, 22 μm	15
Stainless steel (Fe-19Cr-9Ni)	Irregular, 100 μm	8
Stainless steel (Fe-17Cr-4Cu-4Ni)	Rounded, 16 μm	16
Steel (Fe-2Ni-0.5Mo-0.4C)	Irregular, 120 μm	2
Titania (TiO_2)	Spherical, 0.25 μm	4
Titanium (Ti)	Irregular sponge, 30 μm	35
Titanium (Ti)	Spherical, 45 μm	150
Titanium alloy (Ti-6Al-4V)	Irregular, angular, 150 μm	35
Titanium alloy (Ti-6Al-4V)	Rounded, 40 μm	85
Titanium carbide (TiC)	Rounded, 1 μm	68
Titanium nitride (TiN)	Angular, 25 μm	1000
Tool steel (88Fe-5Mo-4Cr-2V-1C)	Spherical, 20 μm	23
Tungsten (W)	Polygonal, agglomerated, 1 μm	45
Tungsten carbide (WC)	Angular, 6 μm	98
Zirconia (ZrO_2)	Rounded, 0.2 μm	80

permeation, or fluid storage. Controlled porosity structures are used in water purification, insulation, magic marker pens, fuel filters, and biomedical implants.

Melting Temperature Difference

The two phases in a composite have different melting ranges. If the melting temperatures are very different, then little interaction is expected during processing. The higher melting temperature phase remains inert during consolidation. Without atomic diffusion and dissolution, the interface remains weak unless a special adhesion agent is used. For example, adding a high temperature mineral or ceramic (say calcite $CaCO_3$ with a melting temperature of 1339 °C or 1612 K) to polypropylene (melting temperature near 160 °C or 433 K) results in little interaction [5]. Calcite improves wear resistance, hardness, and lowers cost. However, since the phases do not interact the composite strength is degraded by the calcite addition. In such situations, strength loss is mitigated by first adding an active polymer coating on the calcite prior to consolidation, but this proves expensive so it is not always used.

On the other hand, some particulate composites develop strong interfacial bonds during heating as part of the fabrication cycle. For example, WC-Co, SiC-ZrO_2, Al_2O_3-SiC, and Ti-TiC are consolidated at high temperatures where interdiffusion occurs between phases. Densification, with some microstructure coarsening, is achieved using sintering, hot pressing, liquid phase sintering, spark sintering, or liquid infiltration. Indeed, interdiffusion at high relative temperatures provides for excellent bonding, as long as reactions are avoided [6]. Although there is no fixed rule, practically the consolidation temperature needs to be over half the absolute melting temperature of the two phases. For example, alumina-zirconia composites (Al_2O_3-ZrO_2) are consolidated at 1500 °C (1773 K) corresponds to consolidation at 60 % of the zirconia melting point and 76 % of the alumina melting point. The relatively high temperature induces atomic bonding. Likewise, for Ti-TiC composites, consolidated at 1400 °C (1673 K) corresponds to 52 and 86 % of the respective melting temperatures.

Consequently, one important consideration is how the two phases will bond to each another. Consultation of appropriate phase diagrams is one means to assess interactions. Unfortunately, many of the composites are based on complex systems where phase diagrams are absent. However, in cases where phase diagrams provide solubility information there is evidence that increased composite strength comes from some phase intersolubility during fabrication, as evident by systems such as Al-Si, Al-Al_2O_3, Be-BeO, Fe-TiB_2, W-(Ni, Fe), Co-WC, WC-Co, and TiC-Ni. Take for example the Fe-TiB_2 system. At 1450 °C (1723 K) iron dissolves TiB_2 to a solubility limit of 8 at.% (atomic percent) Ti and 16 at.% B. On cooling to 1000 °C (1273 K) the solubility declines to less than 1 at.%. Thus, fabrication of the Fe-TiB_2 composite at high temperatures induces solvation (dissolution of the boride into iron) with concomitant interface wetting and bonding. As a rule of thumb wetting and solvation go hand in hand. The wetting improves bonding, evident by higher composite strength. Low processing temperatures thwart chemical interactions, meaning the high temperature phase is inert with little bonding. In such cases surface treatment of the powder, possibly with a lower temperature phase, is required to induce the bonding. A good example is SiC (melting temperature generally estimated at 2700 °C or 2973 K) which is coated with Mg (melting temperature of 650 °C or 923 K) to induce wetting by Al (melting temperature 660 °C or 933 K) to form Al-SiC composites.

Powder Considerations

A variety of powders are available, ranging from the size of sand (1 to 0.1 mm) to nanoscale particles with dimensions smaller than 100 nm (0.1 μm). Larger particles are usually less expensive, more prevalent, and less difficult to handle. For example, 100 μm particles are free flowing, so they easily pass though hoses, pipes, hoppers, and discharge valves in automatic forming equipment. Such large particles are a favorite for hot consolidation, such as via hot isostatic pressing [7]. On the other hand, nanoscale particles agglomerate, pack poorly, and are difficult to handle, adding to the material and the handling expenses [8]. They tend to stick to walls and containers, refusing to flow even under vibration or mechanical agitation.

Contrast the powders shown in Figs. 6.2 and 6.3; the former is a large spherical particle of titanium with excellent flow, mixing, and packing attributes. The second powder is highly agglomerated, nanoscale titanium that resists flow, mixing, and packing. It is unstable in air, adding to the handling difficulty. Nanoscale particles naturally bond together during synthesis, forming long chain-like structures, making it difficult to mix them with other powders. On the other hand, when properly dispersed the smaller particles promote strength and hardness. Clearly, simple synthesis of a small particle size is not sufficient, since dispersion is required to enable any practical forming step.

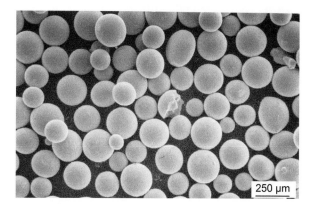

Fig. 6.2 Large spherical titanium particles formed by centrifugal atomization. The powder has excellent flow and packing due to the large spherical particle characteristics

Fig. 6.3 Sponge titanium powder consisting of nanoscale grains formed by hydrogen plasma reduction of titanium tetrachloride (TiCl$_4$). This powder has extremely low packing density, poor flow, and a high impurity level. In addition it tends to easily react and burn when exposed to air

Surface bonding involves some form of chemical interaction, but mechanical bonding is also possible [9]. Gains in adhesion are possible by selecting larger particles with rough surfaces, but an offsetting factor is that sharp corners act to nucleate cracks and should be avoided. Thus, for composites that will experience high tensile stresses, the preference is for a rounded particle shape. Efforts to identify an optimal particle shape, using computer simulations, generally prove inconclusive. Probably ellipsoidal particles are best, since they provide more surface area for bonding without the introduction of stress concentrations. Solvation and diffusion, as associated with high temperature fabrication, are generally beneficial, as long as the consolidation cycle avoids forming a brittle surface compound. Also, high temperature diffusion naturally tends to remove asperities, making the grains rounded after consolidation. So in those cases, the starting particle shape is modified by the fabrication process, resulting in a desirable rounded grain shape.

Connectivity Considerations

Two factors dominate the phase connections in a particulate composite—composition and particle size ratio. At least one phase is always connected throughout the structure, but typically both phases are connected. The idea of phase connection, or percolation, arose to explain why breathing masks for miners suddenly became plugged. As an illustration, Fig. 6.4a sketches a two-dimensional ordered packing of conductive disks. Top and bottom electrodes provide voltage, so current flows through the contacting disks and returns through the external circuit. Adding a light bulb in series is a means to indicate conduction. With all sites occupied there are many pathways for electric current to flow through the disks and the light bulb glows.

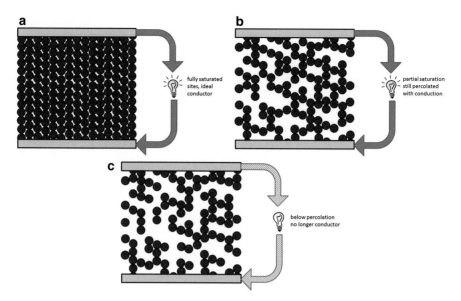

Fig. 6.4 A two-dimensional array of conductive disks with an external circuit to apply voltage and assess conductivity; (**a**) all sites in the array are filled to give full conduction, (**b**) some of the sites are removed, but there remain conductive pathways through the array, (**c**) sufficient conductor has been removed to where there is no conduction through the array. The transition from conductor to nonconductor occurs at what is called the percolation limit

Ponder the situation as a few disks are randomly removed, while the top and bottom electrodes remain. The structure is still conductive and the light bulb remains lit, as sketched in Fig. 6.4b. However, when sufficient disks are removed there emerges a case where the removal of one more disk switches the circuit off. This point of a small change producing a large response corresponds to the percolation limit. It is known as a critical event, similar to how avalanches start with little agitation. As sketched in Fig. 6.4c a continuous path no longer exists so the light bulb is not illuminated. The same behavior would be observed if the disks were randomly mixed conductor and nonconductor, instead of removing conductor disks a simple switch of a conductor disk to nonconductor status would produce a similar change.

Unlike the illustration with two-dimensional disks, the connectivity in a particulate composites is a three-dimensional problem. For illustration purposes, we ponder a mixture of conductor and nonconductor grains; but the same concept would be true for cases of ductile and brittle phases, high and low thermal conductivity phases, or corrosive and noncorrosive phases. For now stay with the case of a conductor and nonconductor. If the conductor grains are connected throughout the body, then the structure is conductive. On the other hand, if the conductor is isolated or only forms small clusters, then the body is nonconductive. Very close to the percolation limit, the point of critical change, the properties

Fig. 6.5 Data reported by
Gurland [11] for the
electrical conductivity of
silver-epoxy composites
versus the silver content.
The conductivity increases
by a factor of 10^{13} between
20 and 40 vol% silver

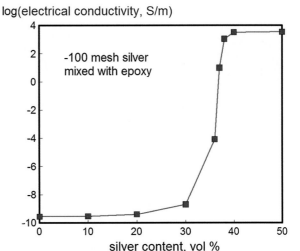

undergo substantial change with small composition change. Near this critical
composition behavior is very sensitive to composition [10].

Early three-dimensional experiments were performed using a mixture of silver
powder and epoxy [11]. Data from these experiments are plotted in Fig. 6.5. In this
plot, the conductivity is on a logarithmic scale. For these tests, silver powder of
particle size ranging from 150 μm and smaller, was mixed with liquid epoxy to form
test samples. After hardening the mixture was tested for conductivity. As the silver
content increased a dramatic conductivity change occurred at about 36 vol% (near
80 wt% silver). The percolation limit corresponds to the composition where the
system switches behavior, defining how much conductor is needed to invoke
conductive behavior. The transition is over a narrow composition range and shifts
slightly in repeat tests. Part of the variation is because the heavy silver particles tend
to settle and the particles were not all the same size. Subsequent study with single
size spherical particles puts the percolation limit closer to 19–20 vol% conductor. A
host of models are available to treat the problem [12].

In three-dimensions it is possible for one phase or both phases to be connected.
The composition over which the conductor is connected overlaps the composition
where the nonconductor is connected. At the intermediate compositions, say
50 vol% silver, both phases are connected, forming three-dimensional interlaced
networks. In such cases it is unclear which phase is best termed the matrix, since
both phases form long-range skeletons.

Percolation conditions are important to the design of particulate composites
[13]. Further, percolation can be manipulated by powder selection. It is possible
to estimate if a phase will be connected as part of the powder selection process. A
few examples where this occurs are listed below:

- porous filters require the pores to be connected to allow fluid passage while debris passage is blocked,
- a nonconductor is made conductive with a metal addition, yet cost mandates the lowest concentration of conductor,
- wear resistance increases with the addition of a hard phase yet thermal conductivity might degrade to cause the composite to over-heat during use, but this can be offset by ensuring the higher thermal conductivity phase is connected in the structure,
- if one phase has a low fracture toughness, then a low level of connectivity for that phase improves toughness,
- a brittle phase enmeshed in a ductile matrix is less susceptible to fracture if the brittle phase is not connected,
- to avoid creep at high temperatures the highest melting high temperature phase needs to be connected to provide a rigid skeleton,
- spark sintering of powders requires electric current passage through the structure, so this rapid consolidation technique requires a connected conductive phase.

Not all problems in percolation are the same; several variants are known. Unfortunately, the several variants to the percolation problem result in calculation of a different percolation limit. Part of this variation comes from assumptions applied to the problem—random loose powders, ordered powders, dense granular materials, or particles dispersed in a fluid as examples. It is common to rely on computer simulations to estimate critical conditions for different particle sizes and porosity levels. Assumptions are required on phase dispersion or homogeneity. Solutions are known for the following cases (for illustration purposes the two phases are identified as conductor and nonconductor) [13–17]:

1. Uncompacted spherical conductor and nonconductor powders of the same size are randomly placed in a container at 60 % density. Both phases are percolated between 28 and 72 vol% solid, but only the majority phase is percolated outside this range.
2. Uncompacted spherical conductor and nonconductor powders of the same size are placed in an ordered structure (simple cubic, body-centered cubic, so on). The packing density and percolation limit are given in Table 6.3. For example, the percolation limit is 25 vol% for the cubic arrangement. That means that between 25 and 75 vol% both the conductor and the nonconductor phases are connected.

Table 6.3 Percolation ranges for different ordered particle structures

Structure	Packing density, %	Coordination, N_C	Percolation limit, vol%
Tetrahedral	34	4	39
Cubic	52	6	25
Body-centered cubic	68	8	18
Face-centered cubic	74	12	12

Fig. 6.6 Data on the conductor content at the percolation threshold versus particle size ratio for the nonconductor to conductor for two situations; loose powders at 60 % density and fully dense structures. When the grains are the same size the loose structure percolation limit nominally occurs at 28 vol% while the dense structure percolation limit is near 19 vol%

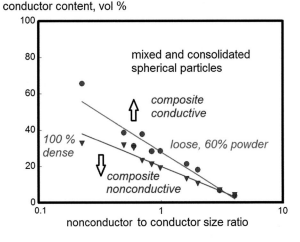

Fig. 6.7 Coordination distribution for a full density structure consisting of two phases in equal proportion (50 vol%). Each grain on average has 13 contact faces, and about half the grains have 4–6 contacts with the same composition

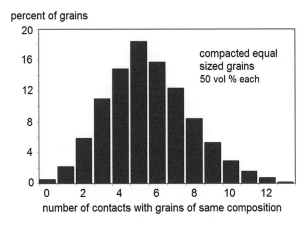

3. Uncompacted randomly mixed conductor and nonconductor spheres of differing sizes are filled into a container. The percolation threshold depends on the particle size ratio. The results are plotted in Fig. 6.6 using the logarithm of the nonconductor to conductor sizes for the lower axis. At conductor contents above the line labeled "loose" the structure is conductive. At small size ratios the nonconductor is effectively acting as a coating to interfere with conductor contact even at a high concentration.

4. Conductor and nonconductor powders are randomly mixed and consolidated to a pore-free condition giving equal sized grains of both phases. The corresponding percolation occurs for any phase over 19 vol%. Both phases are percolated for compositions between 19 and 81 vol%. For a composition of 50 vol% the typical grain has 13 contacting neighbors and typically with 5–7 of those neighbors are of the same composition. Figure 6.7 plots the distribution in contacts, showing only 3 % of the grains have 0 or 1 neighbors with the same

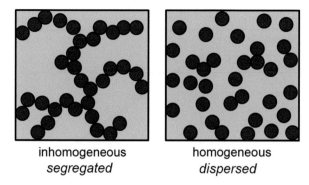

inhomogeneous
segregated

homogeneous
dispersed

Fig. 6.8 This two dimensional illustration shows how an inhomogeneous particle distribution, due to clustering, gives percolation. At the same composition the homogeneous structure is not percolated. Thus, experimental measures of the percolation limit tend to be scattered due to differences in homogeneity

composition and only 3 % have more than 10 neighbors of the same composition.

5. Mixed conductor and nonconductor grains are consolidated to a pore free condition but the grains are different in size. The percolation limit versus the logarithm of the size ratio is included in Fig. 6.6, corresponding to the "dense" line. Since the coordination number is higher for dense systems, versus loose systems, the critical percolation limit occurs at a lower concentration.

6. The last case corresponds to particles dispersed in a continuous matrix phase, such as the silver particles in epoxy as illustrated earlier. Since there is no particle size to the nonconductor phase, randomly placed spheres form a conductive network at 20 vol% when the spheres are the same size. Some calculations give 17–18 vol% as the percolation limit, but to be on the safe side it is best to use 20 vol%. Part of the variation in percolation limit comes from the structure homogeneity, as illustrated in Fig. 6.8. The inhomogeneous, clustered structure is percolated at a lower volume of conductive phase.

As mentioned, the percolation limit is not crisp and varies due to random events. For example, since particle placement is random, then different clusters form prior to actual conduction. In reality, particle contacts are imperfect. Also, there are practical problems with obtaining ideal spherical powders and making homogeneous samples, so exact experimental trials are difficult. The values cited above largely come from computer simulations, run repeatedly to give average values to the percolation limit.

Advanced models are available to predict percolation based on particle size, particle shape, composition, and mixture homogeneity. For example, if the conductor particles are very small compared to the nonconductor, then only a small content of conductor is required to induce bulk conductivity. This is similar to an inhomogeneous distribution of the conductor. Consider that if each conductor particle is coated with small nonconductor particles, then the bulk structure is

Fig. 6.9 A log-log plot of the critical volume percentage of whiskers needed to induce percolation in a matrix phase, versus the whisker length to diameter ratio

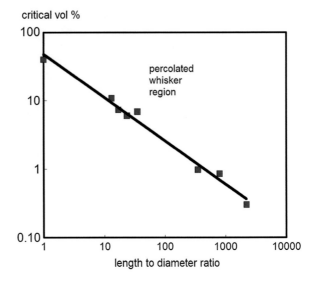

nonconductive, even with 95 % conductor. On the other hand, solid particles infiltrated with liquid are connected in both phases independent of the composition, simply because a connected solid structure is required for rigidity and melt flow requires connectivity to fill the voids.

The role of particle shape is a further consideration. Long filament particles mixed in a fluid form a percolated structure at just 1 vol%. This is the basis for electrically conductive polymers filled with small carbon wires or tubes. Filament particles are percolated at lower concentrations as their length to diameter ratio increases. Figure 6.9 plots the percolation limit in volume percent of whisker versus the length to diameter ratio, on a log-log basis. For long whiskers with an aspect ratio of 100, the structure is percolated with just a few volume percent. This is confirmed by polymers filled with carbon nanotubes (10 μm length and 11 nm diameter, or $L/D = 909$) giving conductivity at 1 vol%.

The grain or particle coordination number N_C explains the onset of percolation and the changes outlined above. The coordination number corresponds to the number of touching grains for each grain. For long-range conduction, the probability p of contacting a conductor must satisfy the following inequality,

$$N_C \, p \; > 1.5 \tag{6.1}$$

For a loose single sized powder the coordination number at 60 % density is about 7, implying percolated with about 21 % of the contacts being the same phase, close to the value given previously. For compacted, 100 % dense powder of a single size, the coordination number is 12–14. Assume the lower value of 12 contacts per particle, then the resulting critical percolation condition is 12 %. Thus, a random mixture with more than 12 vol% conductor particles, is expected to be conductive when taken to full density.

Percolation concepts identify the powder characteristics needed to meet performance attributes. The illustrations here are for conductor-nonconductor combinations, but as mentioned already, the concepts are equally important to thermal, mechanical, wear, fracture, and other characteristics. The transfer of heat, stress, or cracks depends on continuous pathways associated with phase connectivity. On one hand, a fully connected structure might increase strength yet at the same time the same connectivity might provide an easy fracture path to decrease fatigue life.

Particle Characteristics

Examples of Powders

Powders are available in a broad array of sizes, shapes, purities, and other attributes, including porous particles and hollow particles. Handbooks detail how to work backward from the powder characteristics to identify the particle fabrication process. In this section we define the needed powder characteristics and identify the best powder fabrication routes to deliver that powder. To understand and specify a powder requires a language to convey the characteristics. An important idea is in the classic saying—"a picture is worth a thousand words." A potent means to characterize a powder is via the scanning electron microscope (SEM). The SEM operates over a wide magnification range to provide three-dimensional images that capture and convey information on the particle size, shape, and agglomeration. Additionally, many SEM instruments are equipped with detectors for analysis of the X-rays generated by the electron beam, allowing chemical analysis in parallel with imaging.

A few example SEM images are given in Fig. 6.10, contrasting a tellurium elongated (whisker) particle with an angular titanium particle and an agglomerated silicon nitride powder. Differences in particle attributes are evident. Other artifacts, such as internal pores, are readily identified by microstructure cross-sections.

Particle Size Specification

In dealing with powders, a first concern is with the particle size and its distribution. Just as humans have a range of heights, particles likewise have size variations, even within a single container. This size distribution is usually expressed in terms of the cumulative percent or fraction of particles smaller than a given size. The cumulative measure can be based on the mass, volume, or number of particles. Assuming

Fig. 6.10 Along with several other images in this chapter, this figure shows the diversity of possible as constituents for particulate composites: (**a**) milled tellurium whiskers, (**b**) irregular titanium, and (**c**) agglomerated silicon nitride

spheres, the number of particles N relates to the mass of particles M and theoretical material density ρ as follows:

$$N = \frac{6\,M}{\rho\,\pi\,D^3} \tag{6.2}$$

where D is the particle diameter. For example, 2 g of 10 µm aluminum powder with a theoretical density of 2.7 g/cm^3 amounts to 1.4 billion particles. In most automated particle size analyzers this mass to number transformation is reversed; the number of particles are measured at each size and the mass distribution is calculated. Almost all particle size analyzers collect data by measuring the number of particles versus their size.

It takes many small particles to equal the mass or volume of one large particle. Hence, in conveying particle size data, it is necessary to specify which parameter is being used. Particle size data obtained using microscopy is based on the number of particles while particle size data taken from settling or screening is based on the mass of powder. The resulting distributions are not the same. The population distribution from microscopy is always skewed to favor smaller particles. To rephrase, the particle size distribution skews toward larger particles for the mass distribution while the particle size distribution skews toward smaller particles for a population distribution, even for the same powder.

Fig. 6.11 A scanning
electron micrograph of the
woven wire screen
corresponding to 200 mesh,
where the openings are
nominally 75 μm across
[courtesy L. Campbell]

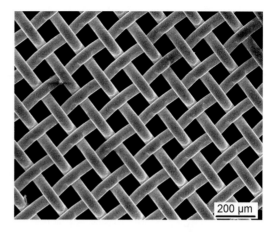

200 μm

Historically, particle size data was obtained using the mass or weight of particles vibrated through a stack of woven mesh screens. A sieve mesh is shown in Fig. 6.11. Mesh designates the number of wires per inch in the weave, so a high mesh implies smaller openings. The screen with the largest opening, or lowest mesh number, is placed on the top of the stack, followed by progressively higher mesh screens with smaller openings. A pan is placed on the bottom to capture the very small particles. After 30 min of vibration the powder is collected off each screen, assuming the size is too large to pass through that opening, so it is retained by the mesh with openings smaller than the particle size. The weight of powder retained on each screen is measured. When summed together this gives the cumulative mass distribution, obtained by adding the amount retained on each screen to that on each of the smaller screens (divided by the total mass).

Sieve analysis is generally applied to particles over 45 μm, corresponding to 325 mesh (a weave of 325 wires by 325 wires in each inch). Other example openings are 60 mesh = 250 μm, 100 mesh = 150 μm, and 200 mesh = 75 μm. The mesh designation increases as the weave becomes tighter, so the particle size able to pass through the mesh is smaller. Particles retained on top of a screen are designated as "+" and powder passing through a screen are designated as "−". Thus, −100 mesh indicates the powder is below 150 μm. Such information is not very useful since the particles might actually be 140 μm or 10 μm or even below 1 μm, yet are still by definition −100 mesh. On the other hand, a mixed designation, for example −200/+325, indicates powder smaller than 75 μm but larger than 45 μm. By employing a stack of screens with progressively larger mesh numbers and smaller opening sizes, a sense of the particle size distribution emerges. Smaller mesh sizes are employed, such as 635 mesh, but the particles stick to the screen and the sieve analysis is not meaningful.

Since screen analysis is not accurate for smaller powders, numerous techniques emerged for measuring powders in the −325 mesh size range. Some of the approaches give only the characteristic or average size for the powder based on measurement of the surface area or gas permeability. Surface area is measured using absorption of nitrogen on powder chilled to liquid nitrogen temperature. Measuring the amount of gas absorbed gives the specific surface area S in m^2/g, from which is calculated an equivalent average particle size D using the material theoretical density ρ as,

$$D = \frac{6}{\rho S} \qquad (6.3)$$

Convenient units are μm for particle size, g/cm^3 for density, and m^2/g for surface area, then the conversion factors cancel. This relation assumes the particles are spheres. For example, an alumina powder with $3.8\ m^2/g$ surface area and $3.96\ g/cm^3$ theoretical density corresponds to 0.4 μm particle size.

Other techniques for measuring particle size rely on settling rates in fluids such as air or water. According to Stokes law the smaller particles sink slower. Particle size is measured by the time for various amounts of powder to settle out of suspension, such as by X-ray attenuation or by measuring the mass settled. More typical are automated devices based on light scattering. In this approach, the particles are passed through a laser beam. Based on light diffraction or scattering around the particle, it is possible to instantaneously capture a signature of the particle size from the angle and intensity of scattering. These data are collected on a photodiode detector array. To ensure that only one particle is being measured at a time, the particles are dispersed in a fluid using high shear or ultrasonic agitation. The test is rapid, requiring only seconds to measure millions of particles. Computers accumulate and reduce the scattering data to provide the calculated particle size distribution. One of the key advantages of laser analysis is the size measurement capability extends from the millimeter to nanometer sizes.

At least 65 other approaches exist to measure the particle size, including image analysis in microscopes, light blocking (large particles block more light), settling in centrifuges (the large particles settle fast and the small particles are slow), and electrical conductivity changes in a capillary tube (large particles give a larger change). They need not be detailed here, since laser scattering now dominates practice.

Once the cumulative distribution is measured the size data are analyzed for three important characteristic particles sizes denoted as D10, D50, and D90. These are the particle sizes corresponding to the 10, 50, and 90 % points on the cumulative size distribution. The D50 is the median particle size, where half of the particles are larger and half are smaller. These three points are marked on Fig. 6.12, which has the cumulative particle size distribution for a milled iron-boron intermetallic powder. In this case about 70 % of the particles are below 10 μm.

Inherent to particle size analysis is the assumption that each particle is a dense sphere. As evident already, this is often a gross simplification. However, the advantage is that spheres allow for a single size measure—the diameter. Accordingly, size data are reported in terms of the spherical equivalent particle size. Each device measures one parameter associated with the powder, such as the particle volume, mass, length, or surface area. To calculate the equivalent spherical

Fig. 6.12 Cumulative particle size distribution based on particle volume for an irregular iron compound powder. Three points are typically reported based on the particle sizes corresponding to 10, 50, and 90 % of the particles, designated as D10, D50, and D90, corresponding to 1.9, 6.4, and 12.8 μm in this case

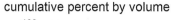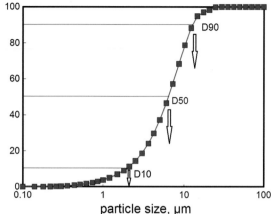

diameter, that measure is converted into particle size assuming spherical particles. For example, if the size measurement is performed in an optical microscope, giving the projected area A, then the equivalent spherical diameter based on projected area D_A is calculated by setting the measured projected area to the equivalent area of a circle, giving,

$$D_A = \left[\frac{4A}{\pi}\right]^{1/2} \tag{6.4}$$

In the same manner, if the particle volume V is measured, then the equivalent spherical volume diameter D_V is given as,

$$D_V = \left[\frac{6V}{\pi}\right]^{1/3} \tag{6.5}$$

Finally, if the data are based on the external surface area S, then the equivalent spherical surface diameter D_S is given as,

$$D_S = \left[\frac{S}{\pi}\right]^{1/2} \tag{6.6}$$

In these formulations, A is the projected particle area, D_A is the equivalent spherical diameter based on projected area, V is the particle volume, D_V is the equivalent spherical diameter based on volume, S is the external surface area, and D_S is the equivalent spherical diameter based on surface area. For a nonspherical powder, these values are not the same, so it is most important to specify the parameter measured to avoid confusion.

Fig. 6.13 A scanning
electron micrograph of the
irregular iron-boron
compound powder
measured for cumulative
particle size distribution
using laser scattering in
Fig. 6.12

Thus, the important part of particle size selection with respect to particulate composites is in specification of the size and its distribution, as well as the means for assessing that distribution. For example a small nonspherical powder is imaged in Fig. 6.13. The laser measured particle size distribution for this powder based on particle volume is shown in Fig. 6.12. The three points at D10, D50, and D90 occur at 1.9, 6.4, and 12.8 μm. The distribution is termed log-normal, as is typical to naturally formed powders. In turn, the mean or average particle size at 7.0 μm is slightly larger than the median particle size (D50) at 6.4 μm, and the mode, or most frequent size, is 8.6 μm. For comparison, the mean size based on surface area is 3.7 μm. In this case the particles differ in composition and density, so it is not possible to extract the mass distribution; hence, particle volume is used for the plot.

Particle Packing

Particles come in a wide variety of sizes and shapes, usually reflecting differences in fabrication technique. In turn the particle packing depends on the size and shape. Spherical particles usually fill a container at 60–64 % efficiency, leaving about 40 to 36 % of the container as interparticle voids called pores. As long as the particles are small, the pores are smaller than the particles. As an estimate the pore size is about 10–15 % of the particle size. Higher packing densities are possible if small particles are mixed to help fill the voids between larger particles. Generally a ratio of 70 % large and 30 % small proves optimal. Likewise, mixed powders with customized broad size distributions are capable of doing the same thing, with the smaller particles filling the voids between the larger particles. Depending on the size distribution apparent densities over 80 %, and even 95 % of theoretical, are possible using mixtures of different particle sizes.

Fig. 6.14 Transmission
electron micrograph of a
nanoscale silver powder,
showing long, chain-like
clusters that lead to a low
coordination number and
poor packing

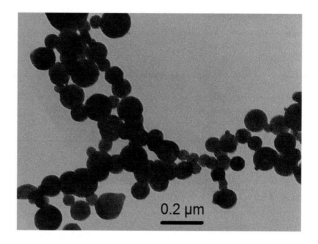

0.2 µm

For nanoscale particles considerable bridging occurs, so the packing density declines to as little as just 4 % of theoretical. This is because the high surface area causes the particles to agglomerate into long chains as evident in Fig. 6.14. The coordination number averages nearly two contacts per particle so the entanglement of the chain-like structure inhibits packing. At these low apparent densities, the pores between the chains are now larger than the particle size. The difficulty in packing small particles arises because the surface forces between nanoscale particles are large compared to the gravity force acting to slide the particles into closer proximity. Large particles are the opposite since gravity acts to ensure closer packing and the surface forces are generally low and do not inhibit particle positioning to fill in larger voids.

Particle packing is assessed using density, defined as mass divided by volume. The value is expressed in convenient units of g/cm^3 or kg/m^3, or if divided by theoretical density it is expressed as a fraction or percentage. Two packing densities are used, apparent density and tap density. The apparent density corresponds to filling of a container without vibration. As the particle shape departs from smooth and spherical, the apparent density declines. Furthermore, as the particle size decreases, especially below 10 µm, surface forces interfere with packing and the apparent density decreases.

Tap density is measured by vibrating a powder until it settles to an asymptotic high packing density. The test is usually performed in a graduated cylinder subjected to 3000 taps. The mass of powder in the cylinder divided by the settled volume gives the tap density. Like the apparent density, the tap density might be given as a ratio to the theoretical density. If the theoretical density is not known precisely, there is a pycnometer test to measure the actual solid volume. Then the measured mass and volume are used to calculate the theoretical density. Pycnometry relies on a sequence of gas pressure-volume exposures, effectively measuring the volume of gas displaced by solid in a known test volume.

Fig. 6.15 Plot of the fractional apparent density for straight whiskers versus the length to diameter ratio. Long whiskers exhibit inhibited packing, and an even lower density occurs if the whiskers are not straight

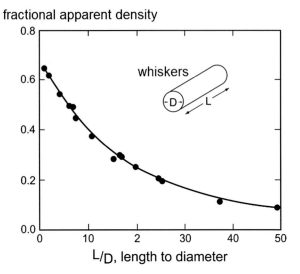

Nonspherical particles tend to pack less efficiently, as evident with elongated particles [18]. Whiskers are a good example. They are elongated but short particles usually from 1 to 25 μm in diameter with lengths between 3 and 150 times the diameter, a parameter known as the aspect ratio. The preference for elongated or whisker particles comes from the improved load transfer between phases [19]. As plotted in Fig. 6.15, the larger the aspect ratio, the lower the apparent density. These data are for straight cylindrical rods. Lower apparent density occur with curved or kinked whiskers.

Another parameter related to particle packing is the flow time measured with the Hall flow test. The simple test reports the seconds required for 50 g of the powder to pass through a 60° funnel with a 2.5 mm diameter exit hole. The related Carney test uses the same 60° funnel with a 5 mm diameter opening. What is reported is the time for discharge of 50 g through the funnel. If the powder fails to flow, it is termed not free flowing abbreviated as NFF. This inability to flow is a precursor to handling difficulties as commonly encountered with high aspect ratio or small particles.

Mixed Particle Packing

Particulate composites often start by mixing powders. The initial packing density for the mixture depends on four factors:

1. size ratio of the particles; one powder is larger and one is smaller
2. composition; ratio of how much of the larger or smaller powders
3. inherent packing of each powder; how they pack by themselves
4. homogeneity of the mixture; are the small particles uniformly dispersed.

Random packing of monosized smooth spheres gives 60 % apparent density and 64 % for the vibrated tap density (these are percentages of the theoretical density for the material). For example, stainless steel at 8 g/cm^3 should pack as spherical particles to about 4.8 g/cm^3 apparent density. Corresponding to this density is an average of six to seven contacting particles around each particle, known as the coordination number. Ideal packings of monosized spheres produce packed arrays with a coordination number of 12 and density of 74 %. Such ideal ordered packing is rarely seen with small particles, but commonly is observed with cannonballs and other large spheres.

Higher packing densities arise by mixing particles of different size. Ponder this situation for improved packing. First, fill a large container with spherical particles. The interstitial holes between the particles are then filled with smaller particles. Optimal packing gains occur when the smaller particles are at least one-seventh the diameter of the large particles. The highest packing density occurs near 70 % large and 30 % small, known as a bimodal mixture. On each side of this peak density, again for particles of large size difference, the specific volume is linear with composition. The specific volume is the inverse of the density. For example, if 20 vol% of small powder with a packing density of 50 vol% is added to a large spherical powder with a 60 vol% packing density, then the overall density is 60 % for the large powder and the remaining 40 vol% voids is filled with small powder at 50 % packing density. This gives an additional 20 % small powder mass (40 % times 50 %) in the same volume, so the overall mixture has a mixed density of 80 %.

Composition determines the resulting packing density [14]. For this case, the specifics of the individual phases are ignored by relying on relative (percent or fractional) packing density. The highest packing density f^* occurs at a critical volume fraction of large particles designated X_L^*. Figure 6.16 plots how packing density changes with composition for mixed powders with a large difference in particle size. Two-dimensional sketches are included to show the relative structural change with composition. The highest packing density depends on the packing of the individual large and small powders f_L and f_S, then,

$$X_L^* = \frac{f_L}{f^*} \qquad (6.7)$$

with the packing density at the optimal composition f^* given as,

$$f^* = f_L + f_S (1 - f_L) \qquad (6.8)$$

If both powders are the same size, then there is no packing density gain from the mixture. For spherical powders with a large size difference, with each packing at 64 % density, the volume fraction of large particles for maximum packing is 0.73, corresponding to a mixture consisting of 27 % smaller particles. The expected packing density is 87 %. Similar concepts are employed for multiple mode mixtures, trimodal and higher [14, 20]. Likewise, nonspherical particles exhibit similar gains in packing at intermediate compositions. However, if the particle sizes are not

packing density

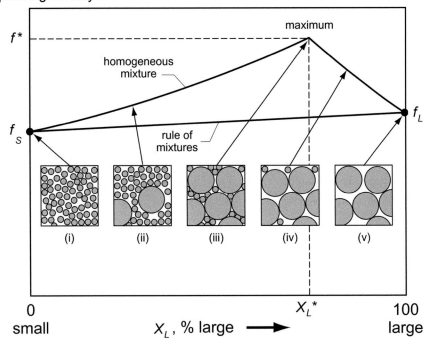

Fig. 6.16 A plot of fractional packing density versus composition for bimodal mixtures of large and small spherical particles. The plot is both apparent density and tap density. The sketches show how the packing density improves up to the critical point where the large particles are closely packed and the small particles fill the interstitial voids. The maximum packing density requires about a sevenfold difference in particle size

significantly different, then the packing behavior is little improved since there is no ability for small particles to fill the voids between large the particles.

The mixing of whiskers and spheres leads to some packing gains, if the whisker has a small diameter. Figure 6.17 plots packing density versus composition for whiskers with a length to diameter ratio of 10. Three variants are shown using sphere diameter to whisker diameter ratios of 4, 10, and infinite. The latter is simply an upper bound case. The peak in packing density occurs at about 25 vol% whisker for the best case, that being spheres much larger than the whiskers.

Powder Modifications

Motivations to modify a powder prior to consolidation are dominated by concerns for improved processing. This comes from changes to the particle size, particle shape, or purity. Size changes include milling or classification (removal). In milling

Fig. 6.17 Packing density
for mixtures of whiskers
and spheres versus the
relative volume percentage
of whiskers. For these plots
the whisker length to
diameter ratio is fixed at
10 and the sphere diameter
to whisker diameter is 4, 10,
or infinite. Peak density
occurs at relatively low
whisker contents

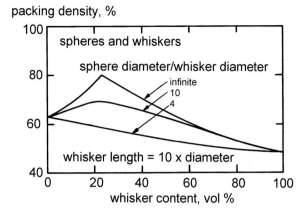

the particle size is reduced by impact attritioning. On the other hand, classification removes the undesirable particle size using screens. It is performed using screens for particles larger than about 45 μm or 325 mesh and air classification for smaller particles.

Particle shape modifications are possible by chemical or mechanical routes. Chemical effects are less desirable due to waste disposal issues. Mechanical modifications might be through milling or tumbling. An example is to tumble a powder in a rotating horizontal cylinder. If the rotation rate is balanced, the particles roll over one another, giving rounded shapes similar to how rocks are rounded in a river bed by the rolling action. The process is accentuated by the addition of about 10 vol% large, hard balls. Typically the added ball size is about 30-fold that of the powder size, giving sufficient mass to impact the particle surface without causing fracture. Impurity removal is usually accomplished by heating in a protective atmosphere. After evaporation or reduction of the impurity, the particles often require a milling step to deagglomerate the lightly sinter bonded particles.

Because of the exposed surface area, particles naturally attract each other to form agglomerates. Also during synthesis small particles preferentially sinter bond together into clusters. An example of a chain-like structure was shown previously in Fig. 6.14. Weak attractive forces are a cause of agglomeration, and it becomes noticeable for particles smaller than 325 mesh or 45 μm. Such agglomeration is reversed using dispersion techniques. For example ultrasonic agitation with a surface active agent (surfactant) is one means to deagglomerate powder. With the proper solution pH control, a surface charge forms on nonconductive particles to cause particle repulsion. The adsorption of small polar molecules on the particle surface acts to sustain the dispersion. As little as 0.1 wt% additive is often sufficient to induce dispersion.

Agglomeration might also be intentional to produce enlarged apparent particle sizes. In common practice, the particles such as WC and Co are mixed and bonded to one another to form large clusters. Agglomerates are easily handled versus small particles. So for automated conveyance and rapid forming steps, it is desirable to agglomerate the powders, hopefully in a well-mixed arrangement. Spray drying is a productive, large-scale process for intentional agglomeration. It starts by mixing powder, binder, and solvent into a slurry. That slurry is sprayed into a heated chamber where the solvent evaporates as the droplets freeze during flight. The resulting spray dried agglomerates are particle clusters that easily flow and pack, yet the individual particles remain intact. This is evident in the image shown earlier in Fig. 6.1. The polymer, often just paraffin wax (with heptane as the solvent), is removed by heating of the powder after consolidation. If the agglomerate proves too fragile, they can be partially sintered to form thermally agglomerated particles, properly termed aggregates.

Aggregates appear similar to agglomerates, but differ since the aggregate is sufficiently hard to resist dispersion. Chemical bonds, beyond simple surface attractive forces, make aggregates resistant to deagglomeration. Intense milling is required to break aggregates back into particles.

Another approach to forming composites is to coat each particle with the other phase; one phase is the particle captured in a shell of the other phase. Coated powders provide an idealized dispersion. Additionally, the core-shell starting structure avoids contact between the central particles, even at high concentrations where normally both phases would be connected or percolated. In a common example, copper is electroplated onto the core particle, such as tungsten, molybdenum, or silicon carbide. After consolidation, the distribution of ingredients is an idealized dispersion down to the particle level. Figure 6.18 is a fracture surface of a particulate composite formed in this manner. In this composite, central 1 μm Al_2O_3 grains are embedded in a WC-10Co matrix. In spite of a high alumina content, there is no connectivity of this brittle phase since the WC-10Co coating separates the hard core particles.

Fig. 6.18 Fracture surface of Al_2O_3-WC-Co composite. The composite was constructed using 46 vol% alumina core grains with WC-10Co coatings on each grain, thereby avoiding low toughness associated with direct grain contacts between the hard alumina grains [courtesy W. Li]

Summary Comments on Powder Characteristics

Powder characterization is a scientific activity in itself. In the advanced approaches, robust computer routines are applied to particle images to generate mathematical descriptions of particle size and shape. Various measures are used to categorize powders, such as simple parameters—aspect ratio, surface area, or particle packing density, but also approaches via Fourier series descriptions of surface topography. Besides the outside size and shape characteristics, it is helpful to microscopically examine the particles using polished cross-sections to reveal internal voids or other undesirable defects.

Usually the characterization parameters are not independent. For example, a high aspect ratio particle also tends to have a low apparent density. A small particle size also has a high surface area. Thus, only a few parameters need to be measured and the other attributes tend to be predictable. Still a minimum characterization battery is necessary to ensure a powder properly fits with the intended function. This is best achieved by specification of the median particle size, D50, and possibly D10 and D90 to control the width of the size distribution. Sometimes very large or very small powders are scalped from the powder lot for a different application, and this is detected by using all three sizes in a specification. Other useful parameters include the particle packing density. Both the apparent and tap density are used, in part because the ratio of tap density divided to apparent density is a simple signature on particle shape. Tests on the powder might also include determination of agglomeration, evident from the particle size and packing density, and even scanning electron microscopy for confirmation of the desired characteristics. Usually, process control requires information on three powder characteristics—particle size distribution, particle shape, and powder agglomeration.

Powder Fabrication Routes

Early powder processing relied on naturally formed particles and these were usually mined and milled. Particles generated by comminution or milling tend to be large, irregular, and impure in contrast with modern powders that are smaller, rounded, and high in purity. Four general routes are recognized in powder fabrication [21]:

1. **mechanical comminution**, milling, self-impaction, jet milling, or attrition routes—best applied to materials with low ductility and low toughness, such as SiC, B, TiC, TiN, AlN, Al_2O_3, TaC, TiH_2, and similar materials,
2. **electrochemical precipitation**, were deposition occurs in an electrolytic cell to form a sponge or dendritic particle—the powders are pure metallic elements, such as Ag, Au, Cu, Pt, Pd, Ni, and other transition metals,
3. **thermochemical reaction**, including routes such as precipitation, reaction, reduction, oxidation, carburization, nitridation, and related composition changes—the powders formed by thermochemical reactions include pure

elements such as Cu, Mo, and W, and reaction products in the form of compounds such as TiO_2, WC, Si_3N_4, and TiB_2,

4. **phase change**, particle production routes based on phase change usually involve liquid to solid transformations without reactions, the powders are formed by melt spraying (atomization), melt centrifuging, and steps such as plasma spraying or vapor condensation.

Mechanical Comminution

Brittle materials, including diamond, are milled via mechanical routes. In the simple form, the raw material is milled, or otherwise compacted and impacted with high velocity particles, paddles, balls, or hammers, to generate progressively smaller particles with extended milling time. Due to random fracture events, the typical particle shape is angular with a wide particle size distribution. Ductile materials flatten during milling to form flakes, but brittle materials fracture like glass, leaving sharp corners and irregular particle shapes. Figure 6.19 is a scanning electron micrograph of a brittle Fe-Cr-C compound, showing a characteristic angular particle shape created by the random fracture events. This powder is used in forming welding rods.

Milling can be applied to a ductile material by first making the material brittle. Oxides are one example, where the material is oxidized, milled, and then reduced to remove the oxygen. Ductile copper is very responsive to size reduction when it is oxidized, so the oxide is milled and subsequently reduced back to copper. Likewise, hydrogen forms brittle compounds with many materials known as hydrides, and nitrogen is useful for embrittling other materials. The embrittling element is removed after milling. This is the basis for powder production using similar processes of hydride-dehydride, oxidation-reduction, or carburization-decarburization.

Fig. 6.19 Image of an iron compound powder formed by milling. The particles exhibit characteristic sharp edges and angular shapes as is associated with brittle materials after milling [courtesy T. Young]

The milling effect on particle size depends on the milling energy. Progressive size reduction links milling input energy W to the change in particle size using an empirical relation:

$$W = g \left[\frac{1}{D_F^A} - \frac{1}{D_I^A} \right] \qquad (6.9)$$

In this form the equation is set to determine the energy required to form D_F as the final particle size, where D_I is the initial particle size, g is a constant that depends on the mill design, and the exponent A is typically between 1 and 2. Most mills are inefficient, so much energy is wasted in the form of heat and noise.

It is also possible to jar mill a ductile material using large, hard balls to form flakes. Here a cylindrical jar filled with a mixture of powder and milling media is rotated to cause the balls to roll inside the mill. The balls are lifted by the combination of centrifugal force and rotation, but then roll or fall to impact the powder. The result is formation of flakes, but continued milling hardens the material to then induce fracture, giving progressively smaller particles.

Attritor milling is a more intense, higher energy milling variant where a vertical tank is filled with balls and powder. The balls are large and constitute the majority of the charge. Mechanical attrition is induced by rapidly spinning a central rotator known as an attritor or impeller. The particles are forced to weld together, then the welded particles work harden until reaching a condition where they fracture, leading to a progressive weld-fracture lamination sequence over time, termed mechanical alloying. Insoluble powders cold-weld to form laminated agglomerates this way. One example is pictured in Fig. 6.20, showing powder milled from an insoluble mixture of silver, cadmium oxide, and polytetrafluoroethylene. This composite is used to form sliding electrical contacts for powering X-ray tubes in medical computer tomography scanners. The silver provides conductivity, cadmium oxide provides hardness and arc resistance, and the polytetrafluoroethylene provides low sliding friction.

Fig. 6.20 A composite powder formed by mechanical alloying. These particles are designed for use in sliding electrical contacts. Each particle consists of three phases—silver for electrical conductivity, cadmium oxide for wear and arc erosion resistance, and polytetrafluoroethylene for a low sliding friction

Characteristically, milled powders are angular, and are often contaminated by the milling process. Some applications form the mill and balls from a similar chemistry to minimize contamination. For reactive species, milling needs to be performed under inert conditions, such as argon gas or nonpolar solvent (hexane, heptane). More milling results in smaller powders, so selection of effective milling conditions is an important aspect of success.

Jet milling is used for production of particles in the micrometer size range. Here the particles to be milled are entrained in a high velocity gas stream and impacted on similar particles in a cycle of compression and impaction. Literally the particle streams are shot at each other to induce high velocity impact without any milling media. Since it is essentially particle-particle collisions with minimized contamination, jet milling is also called self-impact attritioning. It is used to form particles with sizes in the 1–10 μm range. When coupled with separation cyclones, the small particles are removed while the large particles are returned for further comminution. Considerable quantities of process gas are consumed, adding to the expense so preference is given to lower cost nitrogen as the milling atmosphere. Figure 6.13, shown previously, is a picture of a jet milled powder. One variant relies on impaction of the particles against stationary hard targets, but this approach suffers from contamination due to wear of the impact target.

Electrochemical Precipitation

An older approach to forming elemental powders (chromium, iron, cobalt, nickel, copper, palladium, silver, gold, platinum) relies on electrochemical reactions from a solution. Direct electric current is employed to drive a plating or precipitation reaction. Normally, electrodeposition is intended to form a uniform surface coating, but for powder production the electrodeposit is intentionally disrupted to form a porous product. In turn the sponge deposit is cleaned, dried, and milled into powder. Historically electrochemical processes were a standard means to form high purity metal powders from metals with easily dissociated oxides. Subsequently the powder could be reacted to form an alloy or compound.

Depending on the chemistry of the electrolytic batch as well as electrical conditions and mechanical agitation during formation, a range of powder characteristics are possible. Two characteristic shapes are sponge and dendritic powders. The sponge powders also are formed in other chemical reactions routes such as oxide reduction, so a sponge powder shape is not distinctive, but dendritic powder is an easily identified shape from electrochemical production. Figure 6.21 is a dendritic copper powder. A key advantage is the particle purity, but this is offset by limited composition and particle size flexibility.

The disposal of waste solutions is an environmental difficulty, so largely electrolytic powder production is in decline as concerns over waste disposal result in discontinued operations.

Fig. 6.21 Dendritic copper powder formed by electrolytic chemical deposition

Thermochemical Reaction

Many powder fabrication techniques rely on chemical reactions. The reactions involve options of gas-solid, gas-liquid, liquid-solid, and even solid-solid combinations. The gas solid reactions are commonly employed to react compounds to form metals or to react metals to form compounds. Early powders of silver, copper, and gold were precipitated from solutions by making changes to the solution chemistry to nucleate and grow particles. Often these reactions were performed at room temperature.

An approach to ceramic particle synthesis is the sol-gel route, referring to a broad group of solvation-gelation reactions used to form particles. The raw material is first taken into solution. By adjusting the solution chemistry a colloidal particle, smaller than 1 μm, is precipitated to give a liquid-solid suspension. After removal of excess liquid, the remaining gel structure is filtered and dried to collect particles. Final particle deagglomeration occurs by ultrasonic disruption. The sol-gel route results in significant waste liquid. One of the great successes is to form thin films such as the abrasive layer on sandpaper.

Thermochemical reactions involve a solid and gas reacting at an elevated temperature to produce a powder. Several of the reactions remove a species; reduction reactions remove oxygen, decarburization reactions remove carbon, and dehydriding reactions remove hydrogen. The opposite are the additive reactions employing oxygen, carbon, hydrogen, or nitrogen, usually from gaseous sources such as O_2, CH_4, CO, CO_2, H_2O, H_2, N_2, NH_3 or related species that might even include halides such as chlorine. These have the obvious names of oxidation, nitridation, and carburization. One approach arises by cycling between the two states, leading to spontaneous pulverization—hydriding followed by dehydriding is used to form powders from zirconium, titanium, niobium, uranium, tantalum, and other metals.

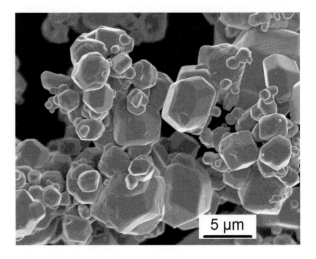

Fig. 6.22 Tungsten powder formed by oxide reduction in hydrogen to produce a grain size below 5 μm. The grains sinter to give agglomerated clusters averaging 20 μm in size

Oxidation of most metals forms a brittle compound. The oxide is easier to mill into a small particle. Subsequent reduction reverts the oxide back to metal. This is an example of a thermochemical reaction for the production of small metal particles. For example, WO_3 is a brittle oxide easily milled into small particles. When the oxide is heated to about 1000 °C (1273 K) in gaseous hydrogen, it reduces to produce a sponge tungsten powder and steam (H_2O):

$$WO_3 + 3H_2 \rightarrow 6W + 3H_2O$$

The reduced tungsten powder is agglomerated since the particles sinter bond together during reduction, leading to the structure evident in Fig. 6.22. That agglomeration is eliminated by subsequent milling. To produce tungsten carbide WC, this same tungsten powder is carburized by heating in methane, or more typically graphite is mixed with the tungsten, and on heating the carburization reaction produces a WC particle.

Chemical reactions provide a host of possible control variables, enabling a diverse range of compositions, particle sizes, particle shapes, purities, and packing characteristics.

Related techniques are vapor condensation reactions relying on precursor compounds that decompose to precipitate a powder. Historically the reactions are in furnaces but more recently include initiation by plasma torches or lasers. The application is most common for the transition metals, such as nickel, iron, and cobalt. A famous example is the carbonyl reaction involving a first step of reacting solid metal with carbon monoxide to form a volatile compound. That compound is distilled and manipulated (pressure, temperature, and catalytic agents) to decompose back into the metal, typically precipitating a small particle. Continued dwell of the particle in the reactor chamber leads to layer upon layer of deposition, giving the onion microstructure shown in Fig. 6.23. In vapor processes the particles grow by

Fig. 6.23 Cross section through a carbonyl iron powder. The onion ring structure is produced by successive waves of vapor phase deposition during growth. The coalescence and bonding of particles is also evident. This powder averages 5 μm in size

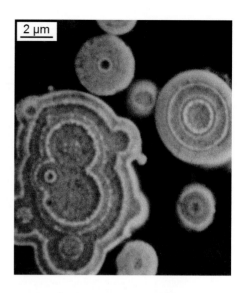

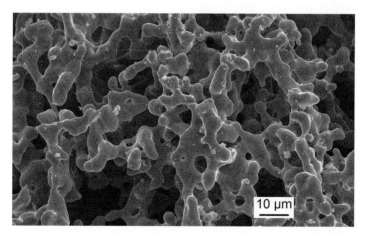

Fig. 6.24 Sponge titanium powder produced by reacting titanium tetrachloride (TiCl$_4$) with molten sodium (Na). After the reaction the NaCl is dissolved in water to leave behind this porous titanium (Ti) powder

continued deposition on the nucleus, and by particle collisions and coalescence to form aggregates. Both are evident in this cross-section micrograph.

Exotic variants of vapor processes involve halides. For example titanium sponge powder is formed from TiCl$_4$ vapor by reacting with molten sodium. The reaction product is NaCl (salt) and titanium. The porous titanium cake contains dissolved sodium chloride that is removed by water washing to give the powder shown in Fig. 6.24. This is subsequently crushed and milled to form a rounded, but low

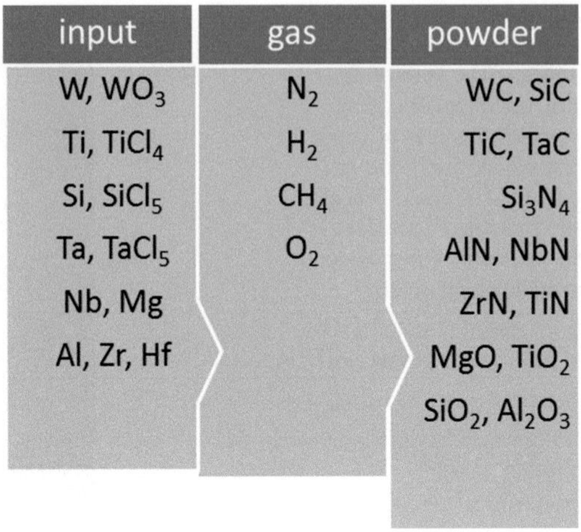

Fig. 6.25 A condensed roadmap to the production of compound particles based on reactions with simple gases, such as hydrogen, nitrogen, methane, and oxygen. Literally hundreds of options exist, but this categorization of the input and output chemistries shows many important cases

input	gas	powder
W, WO$_3$	N$_2$	WC, SiC
Ti, TiCl$_4$	H$_2$	TiC, TaC
Si, SiCl$_5$	CH$_4$	Si$_3$N$_4$
Ta, TaCl$_5$	O$_2$	AlN, NbN
Nb, Mg		ZrN, TiN
Al, Zr, Hf		MgO, TiO$_2$
		SiO$_2$, Al$_2$O$_3$

packing density powder. Alternatively, TiO$_2$ is formed from the same TiCl$_4$ vapor by heating in oxygen. This is a massive industry since the titania powder is the basis for white paper, white paint, and even food coloring.

Reactions involving a gas species results in a wide range of variants as outlined in Fig. 6.25. Hundreds of related examples are possible, such as Fe$_3$O$_4$ reacting in hydrogen to form sponge iron—known as reduced iron and a common ingredient in flour and cereal. The starting material comes in the form of compounds or elements, and the products likewise are compounds or elements. Reaction processes have produced a wide variety of metals, carbides, oxides, and nitrides. Unfortunately, residual contaminants are a difficulty, so special effort is required to purify the powder.

In spite of much study, the vapor phase reactions are not necessarily the fastest, cheapest, or best ways to make powders. Other chemical reactions involve solid-state routes. In this approach two ingredients are heated in a reaction chamber. When an initiation temperature is reached an exothermic reaction takes over, further heating the phases to produce the product. Probably one of the earliest versions were in the production of SiC and WC in the 1800s using electric discharges to provide the initial heating. Several compounds are fabricated in this manner, including examples such as WC, MoSi$_2$, TiC, TiB$_2$, NiAl, Nb$_5$Si$_3$, TaC, TiN, VC, SiC, and NiTi. Typically the reaction is applied to species widely separated on the periodic chart—for example Nb and Al, W and C, or Ti and B. Since heating is spontaneous once the reaction is initiated, the materials tend to sinter bond and require pulverization or milling to deagglomerate the particles after the reaction.

Phase Change and Atomization

The production of powders by a phase change dominates metallic systems. One example is the change from vapor to solid. Small particles form when the ingredients are heated to the vapor state, then that vapor is chilled to nucleate small solid particles. Vapor condensation is a favorite means to form nanoscale powder and is widely used for silver as added to antibiotic ointments. For nanoscale powder production, heating is usually in vacuum, where the vapor travels away from the heat source to cool and condense as a fog of small particles, usually smaller than 1 μm. Due to sinter bonding between contacting particles, vapor condensation synthesis tends to form agglomerates and chain-like particle structures [22]. An example is shown in Fig. 6.26 for vapor phase silica with a median particle size near 0.025 μm (25 nm). This image is generated using transmission electron microscopy, and curiously the smaller particles are semitransparent to the electron beam.

The phase change from liquid to solid is associated with atomization. Starting with molten metal a stream of melt is disintegrated into droplets by high velocity jets, high pressure water, rapid spinning, or plasma spraying. The key is to break the molten stream into small droplets prior to solidification. Because of lower cost construction materials, most atomization is applied to materials that melt below about 1700 °C (1973 K). Higher temperature materials are available, but the extra cost in refractory materials for system construction add to the cost. Most atomization is geared to particle sizes larger than 10 μm. If the melt is formed from an alloy, then the atomized powder is homogeneous in the alloying ingredients. Effectively, each particle is a microscopic casting containing the proper ratio of ingredients.

Water atomization is used to form larger particles, usually with a broad particle size distribution. Some oxygen contamination occurs during water atomization since the molten metal decomposes water. As a side reaction, the liberated hydrogen can be an explosive problem so proper care is needed to properly vent the atomization event.

Fig. 6.26 Silica particles, termed fumed silica, produced by evaporation and condensation of SiO_2. The nanoscale particles naturally bond together during synthesis as evident in this transmission electron micrograph

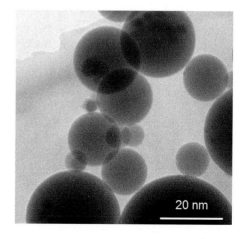

20 nm

Fig. 6.27 Water atomized stainless steel powder illustrating a typical rounded but irregular particle shape

100 μm

Particles are formed in water atomization by impacting a free-falling liquid metal stream with high pressure water jets. The hot metal stream is immediately disintegrated by the cold water. Molten metal droplets are ejected and rapidly freeze, in about 10^{-6} s, so the particles are irregular in shape. Figure 6.27 is a picture of a typical water atomized powder. Adjustments to the powder composition are performed in the melt prior to atomization. If the melt is a pure metal, such as copper, then the powder is copper with possibly some oxygen contamination, often near 0.1 wt%. If the melt is a complex alloy, then the particles likewise are the same alloy. Pressure is the main process control variable in water atomization; higher pressures produce smaller particles. The technology is ideally suited to 50–150 μm powders. Practical limits hinder making particles below about 1 μm diameter and these are more rounded in shape. Water atomization forms powders at rates from 1 to 150 kg/s. In large scale production water atomization is one of the lowest cost powder fabrication routes.

Gas atomization performs similar melt stream disintegration, but relies on high pressure gas instead of water. Common gases are air, argon, nitrogen, or helium. Most gas atomizers are vertical, consisting of a melting chamber on top. The melt is fed into the nozzle to induce disintegration. Accelerated droplets emerging from the pressure nozzle undergo flight in the chamber where they spheroidize and solidify. Figure 6.28 is a cross-section schematic of an inert gas atomizer such as used to form aerospace and medical alloys. Many compositions are made by gas atomization, including iron, steel, stainless steel, copper, brass, nickel alloys, bronze, aluminum, titanium, and cobalt alloys, and generally any material that melts below 2000 °C (2273 K). For high purity powders, the entire atomization event and subsequent handling take place in inert gas to avoid oxygen contamination. As evident in the scanning electron micrograph in Fig. 6.29, the gas atomized particles are spherical and smooth. Consequently, gas atomized particles exhibit excellent packing and flow characteristics. This contrasts with the lower packing density and retarded flow common to water atomized powder.

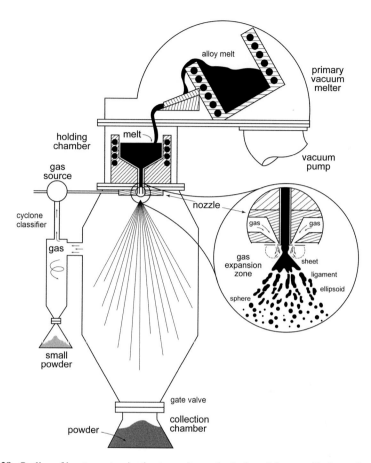

Fig. 6.28 Outline of inert gas atomization to produce spherical particles, usually from alloy melts. The hot liquid is impacted by high pressure gas to generate droplets that freeze to be collected as spherical particles at the bottom of the atomizer

Fig. 6.29 Spherical particles formed by inert gas atomization, with an occasional splat or satellite particle adhering to the particles. This is a nickel-base superalloy powder [courtesy R. Iacocca]

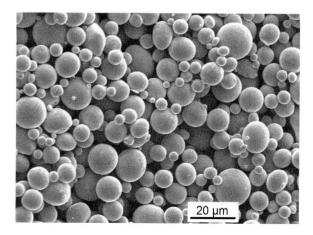

Variants on water and gas atomization include use of high pressure oil as the atomization fluid. Another option is to use centrifugal force to disintegrate the powder. A simple means is to feed the molten metal stream onto a spinning cup or disk, where the melt is thrown off as small droplets. It is also feasible to dip the side of a spinning disk into the melt to pick up melt, freeze it onto the disk. With sufficiently high velocities, the melt sticks only momentarily and then centrifugal force causes the frozen melt to detach. A textured disk results in particles with custom shapes, such as fibers, ligaments, tadpoles, or even staples. Most commonly the centrifugal particles are spherical and large, ranging near 200 μm.

One other option used for higher melting materials, such as titanium, is to feed wire into a plasma torch. The wire melts and the high velocity gas passing through the plasma torch shears and discharges small droplets that subsequently solidify into spheres. Plasma atomized particles are typically larger in the 50–300 μm range. Due to the large size, high fabrication temperature to evaporate impurities, and avoidance of contact with crucibles, the plasma atomized powders are high in purity, pack very efficiently, and show excellent flow.

Mixing, Testing, and Handling

To form a particulate composite involves mixing phases, with at least one being a powder. Due to the wide range of applications, from candy and ice cream to cement and paint, a large body of literature exists on the fundamentals [23]. Mixing refers to intermingling of phases with differing chemistries. On the other hand, blending refers to a combination of two powders of the same chemistry, but possibly different production lots, particle sizes, or particle shapes. Both rely on the same rotational devices to shear and disperse the ingredients, although the mixing and blending devices look the same. The formal phrase is mixing when two different powders are combined, for example iron and copper powders, but it is blending when two lots of powder of similar attributes are combined, such as when two gas atomized iron particles are combined.

Mixing and blending remove variations in the powders, although segregation naturally occurs after mixing. To avoid natural separation, a polymer process control agent is commonly employed and these are the same polymers as used for molding, extrusion, and other binder-assisted shaping steps [24, 25]. Both wet and dry mixing and blending are in use. For dry mixing, the raw powders are combined without a liquid and forced to shear past one another; intermingling occurs along the shear interface. The smaller the particle the more mixing time required to achieve homogeneity. Wet mixing implies a solvent or liquid polymer is added to coat the particles. Mixing is usually a batch process, where the ingredients are loaded into a mixing chamber, possibly heated or evacuated, and after sufficient rotations the mixture is discharged. Occasionally continuous mixing is employed, where ingredients are added at one end of a mixer. After a combination of shear,

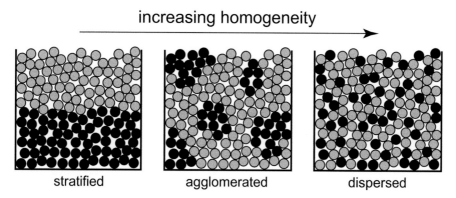

Fig. 6.30 Sketches of the possible homogeneity variations. In the first case the two powders are stratified and show poor homogeneity. In the middle case the particles are clustered and lack homogenous dispersion. The third case represents homogeneous mixing to disperse the two ingredients

separation, extrusion, and compaction events the mixture exits the mixer. Often the continuous mixer relies on rotating screws and external heat to lower viscosity.

The homogeneity of a powder mixture is tied to the randomness of how the two ingredients are dispersed; a homogeneous mixture has total randomness. The best quantification of homogeneity is from composition measurements taken from several randomly selected spots. A homogeneous mixture shows the same ratio of ingredients in all regions of the container. An inhomogeneous mixture has composition variations between spots. In the two-dimensional sketch of Fig. 6.30, the inhomogeneous case is represented by the stratified condition on the left. A partially homogeneous case is shown in the center where clusters exist. A fully homogeneous case is on the right. The homogenization is evident in making a chocolate milkshake. At the start the chocolate ice cream settles while the milk rises, but with intense shear from a spinning blade the two ingredients are evenly dispersed, giving a uniform color.

The sample size used to measure composition influences the measured homogeneity. If the sample consists of thousands of particles, then the segregation or agglomeration might be missed; the composition test would always give the same result. If the test is at the range of just a few particles, then composition variations are evident. The scale of scrutiny, effectively the sample size relative to the particle size, is an important parameter in determining homogeneity.

Quantitative determination of mixture homogeneity relies on statistical analysis of several samples randomly taken from widely spaced areas within the powder lot. The composition fluctuation is used to calculate the variance S^2. This is compared to the variance anticipated for perfectly mixed random samples $S_R{}^2$, and the variance for the initial unmixed condition $S_I{}^2$ as follows:

$$M = \frac{S_I^2 - S^2}{S_I^2 - S_R^2} \qquad (6.10)$$

In this scheme, the mixture homogeneity M varies from 0 to 1, with unity representing an ideal mixture, assuming a small scale of scrutiny. Test precision varies with the square root of the number of samples taken to compute the variance, so many samples are helpful. Prior to mixing, the ingredients are segregated with an initial variance given as:

$$S_I^2 = X_P (1 - X_P) \qquad (6.11)$$

where X_P is the concentration of the major component. The variance for a homogeneous and well-mixed system is zero, so ideally $S_R^2 = 0$. Consequently, the mixture homogeneity is given by a simplified relation,

$$M = 1 - \frac{S^2}{S_I^2} \qquad (6.12)$$

Powder ingredients are usually mixed in tumbling or rotating containers. These are cylinders, V-shaped containers, Y-shaped containers, and might have internal baffles, blades, or stirring rods; Fig. 6.31 is an illustration of just one of many types of mixers. It is filled only 20–40 % of the internal volume. On rotation the powder tumbles and repeatedly splits and recombines. Inside the chamber are spinning blades to disrupt the powder flow, creating chaotic agitation that also induces deagglomeration. Both wet and dry mixing are common, and in the latter a fugitive liquid is added that is later evaporated. Water, gasoline, kerosene, or ethanol are example liquids. For cases with polymers added to the mixture, hot mixing is an option for melting and spreading the polymer.

During mixing, the shear planes of particles flowing over one another induce particle interdiffusion to give homogenization. Initially the mixture homogeneity M increases exponentially with mixing time t,

$$M = M_I + \alpha \, exp[C + K \, t] \qquad (6.13)$$

where M_I is the initial mixture homogeneity, α, C and K are constants for any specific mixing conditions; these parameters vary with mixer design, operating conditions, powder characteristics, and powder surface condition. Eventually the homogeneity reaches an asymptotic value where the rates of mixing and segregation are balanced. The homogeneity is often less than 1 and is attained within 10–20 min. Beyond these times there is little benefit from longer mixing. Unfortunately, it is common for mixed powders to segregate during discharge from the mixer and during subsequent handling. This is one reason for adding transient liquids or polymers to coat and agglomerate the particles. Several process control agents were listed earlier in Table 6.1.

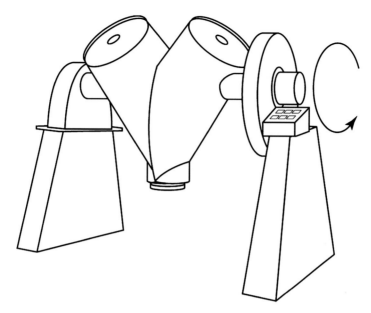

Fig. 6.31 A schematic of a V-shell powder mixer. As the chamber rotates the powder is split into two sections and then recombines a half rotation later. When a set of spinning blades are included inside the mixer, the particles are rapidly dispersed and homogenized. Discharge is through the bottom of the V-shell

Best mixing is performed when only 40 % of the mixer is filled with powder and rotation is at modest rates, ensuring the powders shear and tumble. Too much powder and there is no room for tumbling, and too little powder is inefficient. If mixer rotation is too fast, the particles are held by centrifugal force on the container wall and do not intermix. Slow rotations do not introduce shear and prove inefficient as the particles simply roll without turbulence. Mixing is most difficult when the powders differ in size, shape, or density; unfortunately these differences are inherent to particulate composites.

Discrete Element Simulations

Computation tools are used to predict the structure and behavior of particle composites. The most basic approach relies on atomic level molecular dynamics simulations focused on the interaction of individual atoms as dictated by nuclear attraction and repulsion energies. This is especially useful for understanding interface behavior. It is also used to simulate small particle sintering. At the particle level, discrete element analysis (DEA) allows for computer experiments where each particle is an independent calculation cell [26]. These particle-scale calculations provide predictions of the packing structure, such as connectivity for mixed

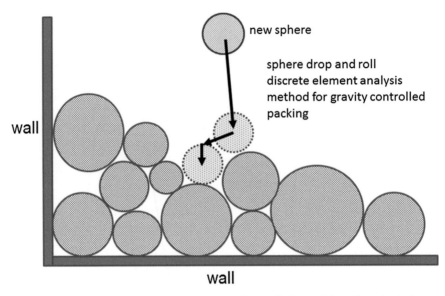

Fig. 6.32 The concept of computational drop and roll as applied to particle packing simulations in discrete element analysis. In this case, the particles are bunkered against container walls

particles of differing sizes [27]. At larger scales, particulate composite microstructures are analyzed via finite element analysis to understand load transmission and fracture path.

The DEA concept was developed by Newton in 1697, but was not implemented until 1971 when computation tools advanced sufficiently to handle many particles at once. The early treatments examined soils to assess building foundation stability and landslides. For particles, the calculations start with random assembly using the drop and roll technique. A schematic of the approach is given in Fig. 6.32. Each particle is added to the assembly under the action of gravity. The particle is randomly assigned a size, x-y-z position, and initial velocity vector. Depending on the height from which the particles fall, it is possible to have considerable bounce, tumble, and roll. Mixtures of different phases require assignment of density to each particle. Millions of particles are computationally assembled and allowed to respond to agitation, gravity, stress, container walls, and other forces [28]. Resting locations are determined based on geometric stability. Subsequent interactions between particles are governed by effects ranging from elastic/plastic behavior, momentum, friction, adsorbed films, or electrostatic forces. Compaction is allowed using the constitutive response for each individual particles. Sinter bonding is induced by shrinking the distance between particles, giving flattened faces.

An example of the DEA structure is pictured in Fig. 6.33 using a transparent box. This represents spherical particles with a D90/D10 size ratio of 1.6 in a loose assembly. The packing density is 58.4 % with an average of 6 contacts per particle for the case where the container is 44 times larger than the average sphere. Packing density increases as the container size is larger than the particle size.

Fig. 6.33 Computer simulated particle packing for spheres with a range of particle sizes. This image represents the particles packed against a transparent wall for a structure where the container is 44 times the average particle size, giving 58.4 % packing density

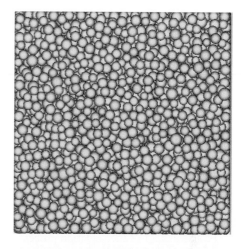

Fig. 6.34 A discrete element analysis output for a two phase mixture of differing sphere sizes. The structure has a tendency for the larger particles to collect at the top. Such stratification is a natural aspect of particle motion during assembly, leading to dilations that let the small particles fall, effectively pushing the larger particles upward

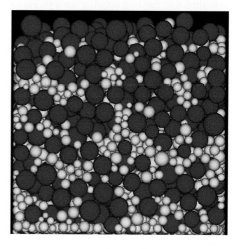

The DEA simulations can handle transients such as powder flow, milling, separation and segregation, agglomeration, and compaction. The predicted properties include packing density, grain coordination and connectivity, pore size distribution, compaction deformation, and sintering shrinkage.

Mathematically, DEA relies on equations for particle motion from mechanics. Packing stability is determined by gravity acting to induce particle contact with at least three supports from the container or other particles. The simulations allow for different particles with differing sizes, shapes, and densities. This is illustrated with two different powders computationally poured into an invisible container as imaged in Fig. 6.34. Here both particles form continuous networks, although the larger particles tend to rise to the top. Thus, DEM calculations provide insight into issues relevant to the selection of the particle size inherent to particulate composites. The approach contrasts with finite element methods, where a continuum is assumed;

discrete element methods handle heterogeneity, critical events, randomly mixed phases, and predict defects, microstructure, grain connectivity, and other features relevant to property predictions. The topic is covered in a stand-alone course supported by commercial software, so this treatment is only intended to introduce the field.

Selection Summary

Particle selection involves a wide range of considerations. Cost and availability are always concerns [29]. Generally smaller powders are more costly. The exceptions are silica, titania, and carbon black because of widespread use in cement and food, tires, plastics, and paints.

From the start of the selection process a vision is needed of the final composite structure. On the one hand, a hard phase bonded with a tough phase might improve toughness by separating the hard grains in a continuous matrix—diamonds bonded with cobalt are intentionally fabricated this way. On the other hand, a heat sink with low thermal expansion and high thermal conductivity requires both phases be interlinked. Thus, besides cost and availability, powder selection considers the particle size distribution, particle shape, and desired phase connectivity.

A wide range of powder fabrication routes confuse selection since often there are too many offerings. Accordingly, discrimination arises by examining different powder fabrication options in terms of agglomeration, purity, and ability to form the desired structure. In addition, consideration is needed to determine how the powders will be mixed or modified. Since several particulate composites are well established, a first step in powder selection is to study what has worked in the past. From that base thoughtful improvements arise.

Study Questions

6.1. The mixing of particles with large size differences results in low homogeneity. Make a schematic plot that compares homogeneity versus time for the mixing of particles of the same size (differing in composition) and the comparable behavior when mixing of particles with a large difference in size and composition.

6.2. Composite properties depend heavily on interface strength. In some cases the strength increases dramatically, reaching to fivefold more than the strength of the major phase, but in other cases just 20 vol% additive reduces strength to half that of the major phase. Identify a particulate composite with high apparent interface strength?

6.3. What are some options to modify the shape of a powder, such as the removal of sharp corners?

6.4. Percolation is critical to several properties. For equal sized spheres consolidated to full density, what is the composition range over which both phases are percolated?

6.5. Assume the percolation threshold occurs at a critical second phase composition of $V_{2c} = 0.2$ (V_C is the volume fraction of second phase). Theory says the change in conductivity near the threshold depends on $(V_2-V_{2c})^{1.2}$. Plot the conductivity versus composition for full density composites of copper and alumina from 18 to 22 vol% copper where both grains are the same size.

6.6. A carbon nanotube (CNT) is a hollow carbon fiber with atom thick walls. The percolation threshold for CNT when mixed with epoxy is 0.004 (0.4 vol%) because of a length to diameter ratio of 1000. Graphene sheets are reported to have a diameter to thickness ratio of 10,000. At what concentration of graphene would a plastic be expected to be conductive?

6.7. One early means to form large spherical particles was based on the rotating electrode process, using a sacrificial electrode and welding arc to throw off large droplets that spheroidized. Figure 6.35 is a micrograph of an Al-SiC particle made this way. Suggest any possible reason this process has not been overly successful.

6.8. Samples are taken after 10 min of mixing for two powders. The composition of each sample is measured for ten small samples, giving minor phase concentrations of 11.0, 10.1, 11.7, 10.6, 10.5, 8.2, 10.1, 8.6, 9.9, and 10.2 wt%. What is the best estimate of the nominal bulk composition?

6.9. Use the data from Study Question 6.8 to calculate the homogeneity.

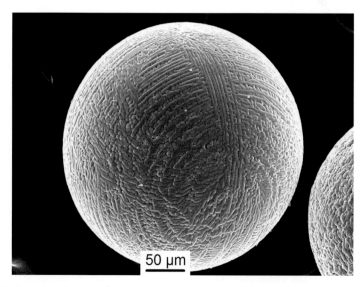

Fig. 6.35 Scanning electron micrograph of a powder made by the rotating electrode process from an aluminum alloy composite containing 4 % silicon carbide [courtesy R. Yamanoglu]

6.10. The powder in Study Question 6.8 is mixed for an additional 20 min with the hope of improving the homogeneity. The resulting ten samples give 9.7, 10.6, 9.6, 11.5, 9.2, 10.1, 9.3, 11.5, 10.5, and 9.8 % of the minor phase. Was the added mixing time useful in improving the homogeneity?

6.11. A particulate composite for high wear applications is formed using a mixture of 4 µm alumina (Al_2O_3) and 150 µm high alloy steel. The alumina has a tap density of 1.66 g/cm^3 and theoretical density of 3.96 g/cm^3 while the steel powder has a tap density of 4.96 g/cm^3 and theoretical density of 8.00 g/cm^3. If only tap density is a concern, what composition gives the highest fractional packing density?

6.12. Using the alumina-steel powders detailed in Study Question 6.11 to estimate the tap density in g/cm^3 for a mixture using 40 wt% alumina.

References

1. J. Khambekar, B.H. Pittenger, Understanding and preventing metal dust hazards. Int. J. Powder Metall. **49**(4), 39–47 (2013)
2. A. Jillavenkatesa, S.J. Dapkunas, L.S.H. Lum, *Practice Guide Particle Size Analysis: Particle Size Characterization*, Special Publication 960-1 (National Institute of Standards and Technology, Gaithersburg, 2001)
3. R.M. German, S.J. Park, *Handbook of Mathematical Relations in Particulate Materials Processing* (Wiley, Hoboken, 2008)
4. M.F. Ashby, *Materials Selection in Mechanical Design*, 4th edn. (Butterworth-Heinemann, Oxford, 2010)
5. A. Lazzeri, Y.S. Thio, R.E. Cohen, Volume strain measurements on $CaCO_3$/polypropylene particulate composites: the effect of particle size. J. Appl. Polym. Sci. **91**, 925–935 (2004)
6. H.M. Jang, W.E. Rhine, H.K. Bowen, Densification of alumina-silicon carbide powder composites: II, microstructural evolution and densification. J. Am. Ceram. Soc. **72**, 954–958 (1989)
7. M.H. Bocanegra-Bernal, Review: hot isostatic pressing (HIP) technology and its applications to metals and ceramics. J. Mater. Sci. **39**, 6399–6420 (2004)
8. J.L. Johnson, Economics of processing nanoscale powders. Int. J. Powder Metall. **44**(1), 44–54 (2008)
9. S.Y. Fu, X.Q. Feng, B. Lauke, Y.W. Mai, Effects of particle size, particle/matrix interface adhesion and particle loading on mechanical properties of particulate-polymer composites. Compos. Part B **39**, 933–961 (2008)
10. A.S. Ioselevich, A.A. Kornyshev, Approximate symmetry laws for percolation in complex systems: percolation in polydisperse composites. Phys. Rev. E **65**, (2002). paper 021301
11. J. Gurland, An estimate of contact and continuity of dispersions in opaque samples. Trans. Metall. Soc. AIME **236**, 642–646 (1966)
12. J.M. Montes, F.G. Cuevas, J.A. Rodriguez, E.J. Herrera, Electrical conductivity of sintered powder compacts. Powder Metall. **48**, 343–344 (2005)
13. C.W. Nan, Y. Shen, J. Ma, Physical properties of composites near percolation. Annu. Rev. Mater. Res. **40**, 131–151 (2010)
14. R.M. German, *Particle Packing Characteristics* (Metal Powder Industries Federation, Princeton, 1989)
15. W. Haller, Rearrangement kinetics of the liquid-liquid immiscible microphases in Alkali Borosilicate melts. J. Chem. Phys. **42**, 686–693 (1965)
16. D. He, N.N. Ekere, Effect of particle size ratio on the conducting percolation threshold of granular conductive-insulating composites. J. Phys. D Appl. Phys. **37**, 1848–1852 (2004)

17. D. Stauffer, A. Aharony, *Introduction to Percolation Theory*, 2nd edn. (Taylor and Francis, London, 1992)
18. J.V. Milewski, in *Packing Concepts in the Use of Filler and Reinforcement Combinations*, ed. by J.V. Milewski, H.S. Katz. Handbook of Reinforcements for Plastics (Van Nostrand Reinhold, New York, 1987), pp. 14–33
19. J.E. Spowart, D.B. Miracle, The influence of reinforcement morphology on the tensile response of 6061/SiC/25p discontinuously-reinforced aluminum. Mater. Sci. Eng. **A357**, 111–123 (2003)
20. L.Y. Yi, K.J. Dong, R.P. Zou, A.B. Yu, Radical tessellation of the packing of ternary mixtures of spheres. Powder Technol. **224**, 129–137 (2012)
21. R.M. German, *Powder Metallurgy and Particulate Materials Processing* (Metal Powder Industries Federation, Princeton, 2005)
22. K. Lu, *Nanoparticulate Materials Synthesis, Characterization and Processing* (Wiley, Hoboken, 2013)
23. A.W. Nienow, M.F. Edwards, N. Harnby, *Mixing in the Process Industries*, 2nd edn. (Butterworth-Heinemann, Oxford, 1997)
24. K.H. Kate, R.K. Enneti, S.J. Park, R.M. German, S.V. Atre, Predicting powder-polymer mixture properties for PIM design. Crit. Rev. Solid State Mater. Sci. **39**, 197–214 (2014)
25. B.R. Sundlof, C. Perry, W.M. Carty, E.H. Klingenberg, L.A. Schultz, Additive interactions in ceramic processing. Ceram. Bull. **79**, 67–72 (2000)
26. S. Luding, Introduction to discrete element methods. Eur. J. Environ. Civil Eng. **12**, 785–826 (2008)
27. D. He, N.N. Ekere, L. Cai, Computer simulation of random packing of unequal particles. Phys. Rev. E **60**, 7098–7104 (1999)
28. C.L. Martin, D. Bouvard, S. Shima, Study of particle rearrangement during powder compaction by the discrete element method. J. Mech. Phys. Solids **51**, 667–693 (2003)
29. J.L. Johnson, Opportunities for PM processing of metal matrix composites. Int. J. Powder Metall. **47**(2), 19–28 (2011)

Chapter 7
Fabrication

The fabrication sequence employed to make a particulate composites varies with the constituents, composition, available equipment, and component details. The common fabrication routes rely on either particle-liquid or mixed particle combinations. Consolidation is in a single step, such as casting liquid with entrained particles, or in two steps, such as pressing mixed particles followed by sintering. The pressing and heating can be combined using hot pressing or hot isostatic pressing. Maps show the trade-offs in processing parameters of particle size, time, temperature, and pressure.

Introduction

Particulate composite fabrication is possible using either solid-solid or solid–liquid approaches. The decision on how to manufacture the composite considers the composition, component design, production quantity, material properties, as well as the powder characteristics. The usual goal is full density, but in some applications porosity is retained for lubrication, filtration, insulation, energy absorption, or reduced strength. For example in bearings, frictional heat causes lubricating oil stored in the bearing's pores to expand to form a lubricating film for reduced wear. Likewise, frangible practice ammunition relies on pores to induce fracture when the bullet strikes a target, thereby avoiding ricochets. However, most applications demand the performance level that comes from full density. Accordingly, fabrication approaches focused on densification are emphasized here.

Sufficient work must be performed on the composite to induce densification. The same level of densification is often possible using different combinations of processing parameters. In this chapter an array of options are detailed, recognizing each has advantages and disadvantages.

Particle densification is assisted by heat, since almost all materials soften as temperature increases. Glass is an example, where molding relies on heating to soften the glass to enable shaping without fracture. A high temperature reduces the required pressure and shortens the molding time. Usually a balance is struck

© Springer International Publishing Switzerland 2016
R.M. German, *Particulate Composites*, DOI 10.1007/978-3-319-29917-4_7

between the benefits of high heats to soften the material and the offsetting reduction in tool life that accompanies high temperatures. The fabrication options fall into three clusters [1–13]:

- **Liquid processes**, such as molding, casting, or infiltration where solid particles and liquid are combined and then the liquid solidifies to form the second phase; a variant relies on vapor infiltration into a sintered porous solid.
- **Simultaneous pressure-temperature processes**, involving combinations of plastic flow and diffusional creep to consolidate mixed powders by hot pressing, hot isostatic pressing, hot extrusion, or spark sintering.
- **Two-step processes**, mixed powders are first shaped near room temperature using pressure, followed by a second stage where heat is applied to sinter bond and densify the shaped powder structure.

Individual processes are detailed in this chapter and newer exothermic reaction routes are introduced as a novel means way of forming composites.

Mechanistic Background

Particle Compaction

Compaction corresponds to pressure induced densification, also termed pressing. When pressure is applied to powder, the density increases by particle sliding and deformation. However, the rate of densification declines as pressure increases since resistance to compaction increases as pores are eliminated and particle contacts increase in size. Typical composite compaction behavior is illustrated in Fig. 7.1 for pure aluminum and a mixture of aluminum (soft) and steel (hard) particles mixed at a 50–50 vol% ratio [14]. The rate of density change with pressurization is steeper at the early portion of the compaction curve, but as pores are removed the rate of densification decreases. The relation expressing fractional density dependence on applied pressure is as follows [11, 15, 16]:

$$f = 1 - (1 - f_O) \, exp[B - \theta P] \qquad (7.1)$$

where f is the pressed fractional density starting with a powder at a fractional density f_O and P is the applied pressure. Effectively, the starting powder density is the apparent density. The dimensionless parameter B accounts for particle rearrangement at low pressures and is between 0.01 and 0.04. For example, spherical particles generally undergo about 4 % rearrangement at the start of pressing, from 60 to 64 % density, so B would be 0.04. The parameter θ has units of inverse pressure and reflects the particle's resistance to deformation. Hard particles resist densification with a corresponding low θ. For this reason, compaction is easier on mixtures that contain a high content of softer particles. Indeed, if

Fig. 7.1 Compaction behavior for pure aluminum powder (*upper curve*) and for the same aluminum powder mixed with 50 vol% hardened steel powder. The composite starts at a slightly higher density but exhibits inhibited compaction compared to the soft aluminum powder. Both curves show decreasing rates of densification as pressure increases

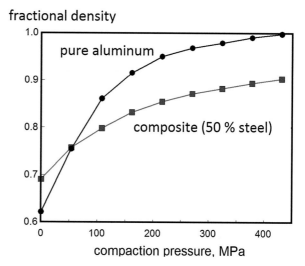

possible it is common to intentionally soften particles to improve compactability. For metals this is performed using annealing treatments.

During pressing, particles initially rearrange by sliding and slipping over neighbors. Subsequently, as pressure increases the particles interlock. If the material is ductile, plastic deformation occurs at particle contacts to form flat faces. With higher pressures, the particles harden and become more resistant to further deformation [17]. Thus, continued deformation requires higher and higher pressures to further deform the progressively harder particles. Brittle materials cannot deform so they fracture and the fragments then repack to give limited densification at high pressures.

The resistance to densification increases as pores are eliminated, so full density by pressing requires ultrahigh pressures, sometimes up to 6 GPa or 60,000 atmospheres pressure [18]. Such high pressure presents considerable difficultly since that stress exceeds the strength of any affordable tool materials. Other problems hinder ultrahigh pressure compaction, including compact cracking on ejection (due to stored elastic energy), trapped impurities, and limitations in equipment capacity. Generally only flat, thin structures that look like hockey pucks or poker chips are produced at these high pressures because of these limitations. Further, safety issues arise since if the tool fractures, the released elastic energy can generate fragment projectiles that can do great harm. Thus, practical issues limit the most powder compaction to 800 MPa or less.

Soft particles deform easily in compaction giving a higher relative density; thus, composites with high contents of soft particles are responsive better to compaction. Figure 7.2 plots mixed steel-aluminum powder data, giving the pressed density versus composition for a constant 280 MPa compaction pressure [14]. As the hard particle content increases the system is less responsive. Effectively, higher pressures are required to press powder mixtures with a high content of hard particles.

Fig. 7.2 Plot illustrating
how compressibility is
degraded by hard particles.
These data are for
aluminum powder with
mixed steel (hard particle)
in various concentrations.
Compaction is at 280 MPa
for all mixtures

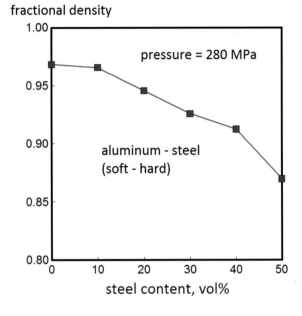

Sintering

Besides pressure, heat is a potent powder consolidation option. Further, the simultaneous action of pressure and heat provides another option. Sintering refers to a
heating cycle used to induce atomic motion to bond and densify particles [19]. With
small nanometer to micrometer sized particles sintering is often sufficient to cause
densification without external pressure [20]. Sintering densification occurs because
random atomic motion, as induced by high temperatures, gradually acts to reduce
surface energy by the pulling the particles together. Figure 7.3 is a scanning
electron microscope picture of the sinter bonds formed between spherical particles
during sintering. These bonds grow due to diffusion over the particle surface and
along the interface between particles. Diffusive atomic motion is sensitive to
temperature. For example, the rate of surface diffusion behaves as follows:

$$D_S = D_{SO}\exp\left[-\frac{Q_S}{R\,T}\right] \tag{7.2}$$

where D_S is the surface diffusivity with units of m^2/s, D_{SO} is the frequency factor
also with units of m^2/s, Q_S is the activation energy for surface diffusion with units of
J/mol or kJ/mol, T is the absolute temperature in K, and R is the gas constant equal
to 8.314 J/(mol K). Diffusion is possible by transport though the lattice, along the
surface, and along grain boundaries (subscripts V, S, and B), with the latter being
most important to densification [21]. Each process has a functional form similar to
Eq. (7.2), but with different frequency factors and activation energies. In most cases

Fig. 7.3 Neck growth between contacting particles is inherent to sintering. This scanning electron micrograph is taken during early stage sintering, showing necks growing between the spherical particles. With more sintering the bonds grow, pulling the particles together, thereby filling void space to give densification

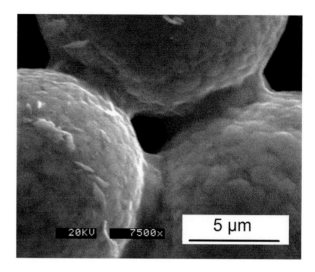

the activation energy is proportional to the material's absolute melting temperature. For example, gold melts at 1063 °C (1336 K) and has an activation energy for grain boundary diffusion of 110 kJ/mol, while lead melts at 327 °C (600 K) and has an activation energy for grain boundary diffusion of 68 kJ/mol. Usually, surface diffusion happens at lower temperatures compared to grain boundary diffusion and volume diffusion is active only at very high temperatures.

The area over which diffusion occurs and the diffusion distances depend on the particle size, resulting in faster sintering with smaller particles. As particles sinter bond, grain boundaries form in the contacts, enabling densification by grain boundary diffusion. In cases where a liquid forms between the particles, even faster sintering is possible. Thus, liquid phase sintering is a common means to sinter particulate composites. After sintering the solidified liquid forms a separate phase.

For sintering an easy basis for tracking densification is the sintering shrinkage. A generic model is effective in relating the adjustable process parameters in a predictive scheme. Shrinkage is defined as $\Delta L/L_O$, corresponding to the change ΔL in a dimension from the size prior to sintering L_O. The convention is to drop the negative sign for shrinkage. An example of sintering shrinkage versus hold time during sintering is plotted in Fig. 7.4 for two composites—Fe-20Cu sintering at 1120 °C (1393 K) and W-20Ni sintering at 1500 °C (1773 K) [19]. Both systems are solid–liquid mixtures at the respective peak temperatures. On a log-log plot the sintering shrinkage is described by a one-third time dependence. Expressed mathematically as [19, 22]:

$$\left(\frac{\Delta L}{L_O}\right)^3 = \frac{g\,D\,\delta\,\Omega\,t}{R\,T\,G^4} \tag{7.3}$$

where D is the diffusivity (similar to Eq. 7.2), g is a geometric constant near 20, δ is the diffusion path width estimated as five atomic diameters, Ω is the atomic volume,

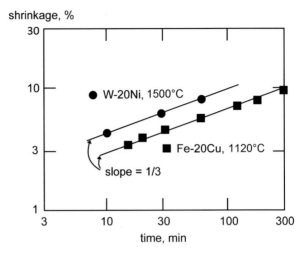

Fig. 7.4 Two examples of sintering shrinkage versus time during composite densification. The scale is log-log. The *upper curve* is for W-20Ni at 1500 °C (1773 K) and the *lower curve* is for Fe-20Cu at 1120 °C (1393 K). Both systems follow diffusion controlled densification

t is the sintering time, R is the gas constant, T is the absolute temperature, and G is the grain size. Grain size enlarges during sintering, as discussed below. Although originally developed for single phase materials, this equation is useful for composites [23]. Often not all of these parameters are known, so a few experiments are performed to identify the sensitivity to time and temperature. Once the shrinkage model is created, then it is possible to predict the changes resulting from new processing conditions. Although the approach shown here is based on monitoring shrinkage, most properties improve in proportion to sintering shrinkage. Accordingly, it is possible to track hardness, strength, elastic modulus, thermal conductivity, magnetic saturation, and related properties, although shrinkage is the most common monitor for sintering behavior.

Sintering repositions mass one atom at a time, resulting in an increased density. The one-dimensional form is the shrinkage, but shrinkage occurs in all three dimensions. An illustration of the net change is given in Fig. 7.5 for a latch. The upper image is after sintering and the lower version is prior to sintering. Note that in a few instances where a strong chemical reaction occurs during sintering, the opposite is seen and the component swells.

Densification from the initial fractional density f_O to the sintered density f follows:

$$f = \frac{f_O}{\left[1 - \frac{\Delta L}{L_O}\right]^3} \tag{7.4}$$

This assumes no mass loss during sintering. The actual density is obtained by multiplying the fractional density by the theoretical density. For example, W-10Cu has a theoretical density of 17.3 g/cm^3, so at 95 % density the measured density is 16.4 g/cm^3.

Fig. 7.5 A practical example of the shrinkage associated with sintering. The lower components is as-molded with binder in the pores. Once that binder is removed and the component is sintered, a much smaller but similar shaped part results as shown on the top

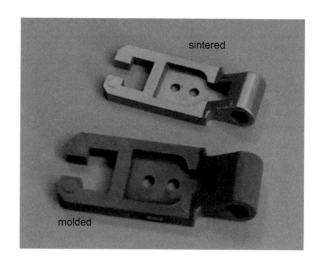

Grain growth, or coarsening, reflects the increase in the average crystal size that occurs as part of sintering. It is accompanied by a decrease in the number of grains. Grain growth depends on atomic diffusion, just like sintering, thus densification and microstructure coarsening go hand in hand. Grain growth occurs such that the mean grain volume increases linearly with time, expressed as follows [24],

$$G^3 = G_O^3 + k\,t\,exp\left[-\frac{Q_G}{R\,T}\right] \qquad (7.5)$$

where G is the grain size, so G^3 is proportional to the grain volume, G_O is the starting grain size (usually equal to the particle size), k is the growth rate constant, t is the hold time, and Q_G is the activation energy for grain growth, R is the gas constant, and T is the absolute temperature. A plot of mean grain volume (grain size cubed) versus sintering time is given in Fig. 7.6 for tungsten grains in a W-Ni-Fe composite. The behavior agrees with Eq. (7.5).

Many composites undergo extensive grain enlargement during sintering, where thousands of initial particles coalesce to form a single sintered grain. This is especially true in liquid phase sintering when the liquid wets the solid particles. The capillary force pulls the grains together to induce coalescence while the liquid provides a rapid atomic diffusion path. Sintering with a liquid phase produces a composite structure, since the microstructure consists of the solid grains surrounded by an interpenetrating network of solidified liquid, as evident in Fig. 7.7. This cross-sectional image shows the structure of stainless steel with a chromium boride phase. Each of these grains is composed of approximately 125 initial particles; the final 50 μm grain size is five times larger than the initial 10 μm particle size, giving a volume ratio 5^3 or 125 particles consumed in making each grain.

The composite phases respond to pressure and temperature in different ways. As an example, Fig. 7.8 plots the final density for alumina with added silicon carbide,

Fig. 7.6 Mean grain size cubed (proportional to grain volume) versus hold time for W-15.4Ni-6.6Fe sintering at 1507 °C (1780 K). The linear behavior corresponds to diffusion controlled coarsening; effectively the same mass transport events that give densification also give grain growth

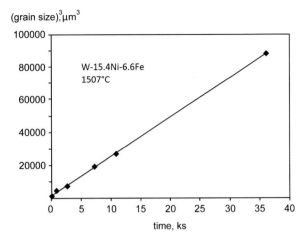

Fig. 7.7 A composite microstructure formed by liquid phase sintering. The majority phase is stainless steel with extensive sinter bonds. The second phase is solidified liquid, consisting of chromium and boron. The solidified liquid forms between the stainless steel grains. The *black spots* are chromium-silicon oxides resulting from impurities in the powder

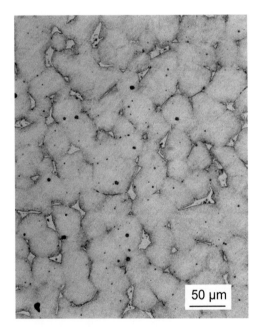

pressed and sintered at 1800 °C (2073 K) for 60 min (1 h), starting in all cases from 0.56 fractional density [25]. Plotted here is the final density versus composition, showing how silicon carbide degrades the alumina densification. The mixture response depends on a host of factors, including composition, particle sizes, and phase connectivity. If compaction or sintering data are known for each of the individual constituents, then the composite behavior is usually intermediate between the two extremes. As illustrated by the two straight lines in the plot with a transition near 10 vol% SiC. At higher SiC contents the response is little changed

Fig. 7.8 Sintered density versus composition for composites of Al$_2$O$_3$ and SiC. Mixtures of 0.7 μm alumina and 1.0 μm silicon carbide were pressed to 56 % density and sintered at 1800 °C for 1 h. Sintering densification is significantly impeded by the silicon carbide

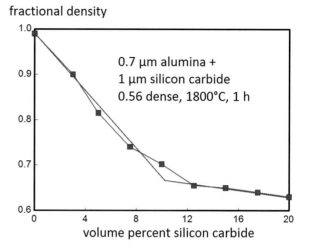

as the SiC content is sufficient to resist densification at all compositions. Effectively, compaction and sintering both exhibit behavior intermediate between that of the pure constituents, complicated by many specifics of the particular system.

Pressure Amplification

Hot consolidation options include various combinations of temperature and pressure. Densification by compaction relies on a high applied pressure. The applied pressure is amplified at the particle contacts. If P_A is the applied external pressure used for consolidation, then the effective contact pressure P_E acting at the particle contacts is estimated knowing the fractional density f and starting fractional density f_O as follows [3]:

$$P_E = P_A \frac{(1 - f_O)}{f^2 \, (f - f_O)} \tag{7.6}$$

Figure 7.9 plots the ratio P_E/P_A to show how the effective (local) contact pressure declines during densification while the applied pressure is constant. Both bond size enlargement and the coordination number increase cause the change. The calculation shown here applies to spherical particles starting from an assumed initial density of 60 %, or 0.6 fractional density. Only at full density is the effective pressure equal to the applied pressure. Otherwise the applied pressure is amplified at the particle contacts to induce localized deformation. The applied pressure might be below the yield strength of the material, but the amplified contact pressure might exceed the yield strength sufficiently to induce plastic flow.

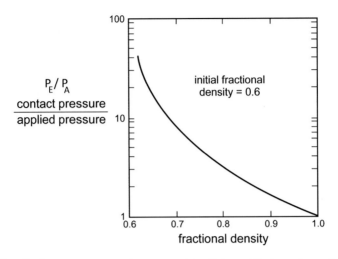

Fig. 7.9 The effective contact pressure between particles P_E is amplified over the applied pressure P_A during particle consolidation. Only at full density are the two equal. This plot gives the contact pressure divided by the applied pressure for spherical particles with an initial density of 60 %

Besides pressure, strain rate, reflecting the loading or deformation rate, influences powder densification. Powders subjected to high strain rates, as attained in forging or extrusion, are more resistant to densification. This is because rapid deformation causes entanglement of crystal defects that effectively harden the particles. Even at elevated temperatures it takes time to relax the entangled defects. Low strain rates, such as those associated with diffusional creep processes like hot isostatic pressing, avoid this hardening.

A difficulty with pressure consolidated composites comes from residual stresses. The two phases in the composite are usually different in elastic modulus and thermal expansion coefficient. On cooling stresses arise because of these differing properties, sometimes leading to deformation or cracking. For example, this is evident in magnesia-graphite composites where intermediate holds are required during cooling to relax residual stresses.

Plastic Deformation

Almost all materials soften as temperature increases, the common exceptions being graphite and nickel aluminide (Ni_3Al). With sufficient heating even brittle materials are deformable. Permanent deformation from stresses above the yield strength is caused in crystalline materials by dislocation motion, termed plastic flow. It is a means to quickly densify particles, since dislocations move at the speed of sound. Thus, with heat and high pressure a powder densifies rapidly. Heating the structure reduces strength, making it easier to densify. Once plastic flow ends, continued densification relies on much slower atomic diffusion events.

Whenever particles prove resistant to pressure-based densification, then a higher temperature is a means to soften the powder, making it easier to compress the particles. Densification is especially rapid when a powder is heated close to its melting range. Plastic flow densification continues as long as the contact stress exceeds the material's yield strength at the consolidation temperature. The fractional density f attainable by plastic flow depends on the applied pressure P_A as follows [19]:

$$f^3 = \left[\frac{P_A (1-f_O)}{1.3 \, \sigma_Y} + f_O^3\right] \qquad (7.7)$$

In this equation, f_O is the starting fractional density, so $(1-f_O)$ is the starting fractional porosity, and σ_Y is the yield strength of the material at the compaction temperature. The loss of strength on heating is termed thermal softening as illustrated for Al-20SiC composites in Fig. 7.10. These data are for 1 μm grain size composites formed by hot isostatic pressing [26]. In this case the yield strength approaches zero at the aluminum melting temperature. If thermal softening data are not known, the protocol is to assume a linear decline in yield strength from room temperature to zero at the melting temperature. This simple model is illustrated, showing over-estimated strength at intermediate temperatures.

At high temperatures and fractional densities over 0.9, a modified relation is applied to estimate the fractional density. This relation ignores the starting density to estimate the fractional density as a function of the applied pressure P_A and yield strength σ_Y:

$$f = 1 - exp\left[-\frac{3 \, P_A}{2 \, \sigma_Y}\right] \qquad (7.8)$$

Equation (7.8) says full density (f near 0.99) requires an applied pressure about three times the yield strength at the consolidation temperature. Plastic flow is the dominant fabrication process in forming particulate composites by hot extrusion, hot forging, and dynamic or explosive compaction. They all operate at high pressures, leading to practical difficulties from rapid tool wear due to the combined stress and temperature required for fast consolidation.

Besides temperature and pressure, there is also a dependence on particle size. Larger particles exhibit less work hardening during densification, so they respond with a higher density during plastic deformation. On the other hand, diffusional processes are enhanced by small particle sizes. Plastic flow is first but diffusional contributions become important as plastic flow ends. Thus, large particles are favored in approaches relying on high pressures, but smaller particles are favored in approaches relying on high temperatures.

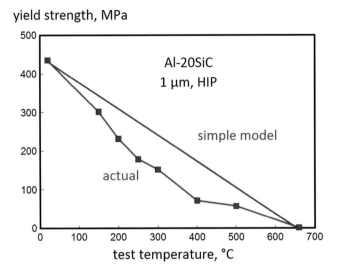

Fig. 7.10 Thermal softening makes is easier to deform a material at high temperatures. This plot of yield strength versus test temperature is for Al-20SiC consolidated by hot isostatic pressing. A simple linear model for thermal softening is compared to the data; that model links the room temperature strength to zero strength at the onset melting temperature

Diffusional Creep

Diffusion is by random atomic motion at high temperatures. It is accentuated by stress gradients in a process known as creep. The combination of external stress and high temperature allows the atomic motion to relax the stress. Although atoms move randomly, subtle differences in energy between regions under tension and regions under compression lead to cumulative motion to remove the stress. Just as water flows downhill due to a gravity gradient, atomic motion is biased to remove stress.

For powders, a change in component shape or density at high temperatures is mostly by grain boundary diffusion, with some involvement from volume diffusion. Creep models are good means to explain powder consolidation [3, 4, 19, 27]. With a small grain size the higher grain boundary area naturally increases grain boundary diffusion. The stress arises from the external pressure and from pore curvature. While heat increases atomic motion the simultaneous stress biases the atomic motion to fill pores. The progressive pore filling increases density and is the fundamental process occurring in hot pressing and hot isostatic pressing [4].

The one-dimensional measure of densification is the shrinkage. As mentioned, a decrease in compact size is expressed as a positive shrinkage. Shrinkage rate is the change in dimensions over time normalized to the starting size. When diffusional creep is controlled by volume diffusion, the time dependent (t) change in length (L) normalized to the initial length (L_O) is given as follows:

$$\frac{dL}{L_O\,dt} = \frac{g\,D_V\,\Omega\,P_E}{R\,T\,G^2} \tag{7.9}$$

where g is a geometric term equal to 13, D_V is the volume diffusivity, Ω is the atomic volume, R is the gas constant, T is the absolute temperature, and G is the mean grain size. Diffusivity is sensitive to temperature, in an Arrhenius behavior similar to that shown above for surface diffusion, for example in volume diffusion the expression becomes,

$$D_V = D_O\,exp\left[-\frac{Q_V}{R\,T}\right] \tag{7.10}$$

Again R and T are the gas constant and absolute temperature, and Q_V is the activation energy for volume diffusion and D_O is the frequency factor; both factors are tabulated for many materials in handbooks.

Assuming isotropic densification in each dimension gives the density from integration of the shrinkage rate over time. Since mass remains constant, the change in dimensions, termed the shrinkage $\Delta L/L_O$, results in a density increase according to Eq. (7.4).

Also important is diffusion along grain boundaries. For that process the relation between parameters is given as,

$$\frac{dL}{L_O\,dt} = \frac{g\,\delta\,D_B\,\Omega\,P_E}{R\,T\,G^4} \tag{7.11}$$

where the geometric term g is 48, and δ is the width of a grain boundary, usually assumed to be five atoms wide. Grain boundary diffusion is more sensitive to grain size when compared to volume diffusion, so grain growth during densification is a factor in determining the dominant diffusion mechanism. Grain boundary diffusion D_B also has an Arrhenius temperature dependence, but the activation energy Q_B for grain boundary diffusion is lower versus volume diffusion. Thus, at typical consolidation temperatures grain boundary diffusion tends to be faster when compared to volume diffusion.

Computer routines are used to solve these equations versus time, temperature, pressure, density, particle size, grain size, and other parameters [3, 22]. The output is a prediction of sintered density versus the fabrication parameters [28]. Since plastic flow is nearly instantaneous, most of the attention is focused on the time-dependent diffusional creep contribution. The densification rate slows as full density is approached and the effective stress declines. Further microstructure coarsening in terms of grain growth or pore coarsening results in large diffusion distances and slower densification.

Computer Simulation

Densification work concepts realize the energy required to densify a composite structure involves interplay between several independent parameters of time, temperature, pressure, green density, and particle size [29, 30]. Smaller particles are harder to compact but easier to sinter, and higher temperatures and pressures assist densification; atomic motion increases and the material softens to become more responsive to external stress. High temperature densification often requires a few hours to reach full density. However, many combinations of time, temperature, and pressure provide equivalent levels of densification. Accordingly, the parameter combinations that deliver full density are swept into a work of densification concept used in computer simulations. Effectively the densification work evaluates how much mechanical and thermal energy is needed to consolidate the powder based on a tradeoff of pressure, temperature, and hold time.

As an illustration, consider the role of particle size. Small particles produce an attractive capillary force that induces densification. For many materials the surface energy is near 1–2 J/m^2 and the microstructure scale is on the order of 0.1–20 μm. The combination of surface energy and curvature pulls the particles together by capillary attraction via a sintering stress. The surface energy divided by the microstructure scale factor gives a compressive stress from 1 to 20 MPa at the particle contacts. An external pressure supplements the inherent sintering stress to accelerate densification. If temperature is increases, the strength of the powder and its resistance to densification decreases. What is needed is an assessment of what an increase in temperature provides as benefit versus an increase in applied pressure. Models provide assessments for density and its sensitivity to these overt fabrication parameters. Curiously, such a work of densification approach is divorced from the equipment and technological details; densification is a reflection of the input work. That work can be supplied by a variety of conditions.

Software captures the relevant body of mathematics into simulations that predict density for a set of conditions [3]. Subsequently, commercial software added shape complexity, nonisothermal cycles, tool friction, gravity, and factors important to processing, costs, component distortion, and identification of potential defects [28]. The simulations apply the models introduced here for plastic flow, diffusion, and creep; sometimes in the form of viscous flow by assuming the composite is similar to a hot polymer or glass. Indeed, in 1905 Einstein conjectured diffusion and viscosity were the inversely proportional to one another, even though atomic structure and atomic diffusion were not accepted.

The simulations predict processing effects in the form of a densification map. Figure 7.11 is a densification map for Ti-10TiC at 870 °C (1143 K) with a particle size of 50 μm and initial density of 56 % of theoretical. Experimental data are indicated by the square symbols for two times (1 and 0.25 h) and three pressure options. Once trained using experimental data, the simulation is an accurate guide to full-density consolidation parameters.

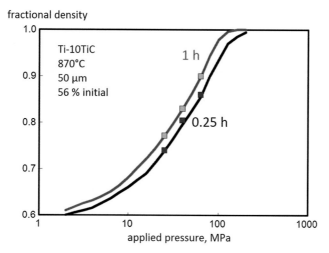

Fig. 7.11 A map of density versus pressure at a constant temperature of 870 °C (1143 K) for Ti-10TiC composite consisting of 50 μm particles at a starting density of 56 %. Two hold times are shown, 1 and 0.25 h. Over this pressure range, densification is dominated by diffusional creep. The results from confirmation experiments are included as the symbols

Densification maps provide a means to assess the interplay between processing variables. Once a calculation is validated, it is turned to the question of what combinations of parameters results in full density. For the Ti-10TiC composite, the answer is full density occurs in 1 h at 100 MPa and 870 °C (1143 K). If a lower pressure of 50 MPa is required, then the consolidation temperature needs to increase to 1120 °C (1393 K) or much longer, near 24 h is required at 870 °C (1143 K). Typically long cycles are discouraged due to cost, so a solution near 1 h is preferred. From a practical standpoint, large structures heat slower, so time is extended to ensure uniform heating in such cases.

Further calculations are associated with predicting final component size and shape. Distortion occurs in complex geometries due to density gradients, gravity, tooling, wall friction, temperature gradients, and component design features such as holes or thick sections. For example, heat transfer is poor in a loose powder. Temperature gradients arise between thick and thin sections during rapid heating, leading to differences in densification. The consequence is component warpage, especially for large components. Finite element calculations predict the heat flow and when coupled to densification models, allow for preform design to offset distortion. As an example, Fig. 7.12 provides the cross-section profile of a thick walled container processed by hot isostatic pressing, comparing the starting and final shapes. The container corners cause compact distortion. Computer simulation techniques are important in optimization of tooling and process design.

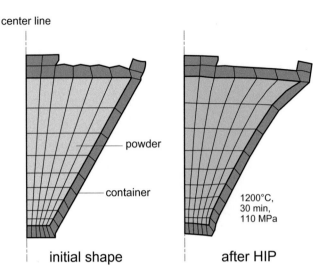

center line

powder

container

1200°C,
30 min,
110 MPa

initial shape   after HIP

Fig. 7.12 Finite element analysis of distortion during hot consolidation, in this case for hot isostatic pressing (HIP), where the differential heat flow and stiffness of the container causes distortion near the corners

Options in Forming Composites

As introduced earlier, processing options for a particulate composite rely on three variants:

- **Liquid processes**; including semisolid casting, molding, or infiltration as outlined in Table 7.1.
- **Simultaneous pressure-temperature processes**; a single densification step using combined heating and pressure as outlined in Table 7.2.
- **Two-step processes**; such as first shaping and then sintering; pressure and heat are applied in sequential steps as outlined in Table 7.3.

There are several minor variants, such as vapor infiltration or reactive synthesis, but the vast majority of particulate composite processing falls under these three headings.

One factor in the selection of a specific fabrication pathway involves the number of components to be made. Those processes with high tooling costs naturally are best justified when the production quantity is large. Another factor is the number of features and the tolerances, and general component geometry such as symmetry, long, flat, round, or thin. Selection of the best fabrication option involves analysis of the natural fits and options. For example, binder-assisted shaping by tape casting is ideal for thin, flat structures. Extrusion is best applied to long, constant cross-section structures. Injection molding is applied to high quantity production of three-dimensional shapes. Slurry casting is suitable for complex, thick and large components, usually produced in low quantities (100s). Slip casting is desirable for

Table 7.1 Key elements of direct fabrication using particles dispersed in liquid phases

• Forming performed with higher melting temperature solid particles dispersed in lower melting temperature liquid phase
• Composition limited by viscous flow of slurry, generally less than 60 vol% solid
• Direct shaping process requires heat resistant custom mold
• Product is composite with limited bonding of solid particles, liquid solidifies to form continuous matrix phase
• Volume change of liquid on solidification produces solidification pores
• Examples processes—casting, extrusion, molding, slurry filtration
• Example systems—nylon-glass, aluminum-silicon carbide, polypropylene-graphite, silver-titania, steel-titanium carbide, steel-titanium boride
• Key control is viscous flow of slurry

Table 7.2 Key elements of direct fabrication using mixed powders

• Consolidation and densification of mixed solid particles via simultaneous application of temperature and pressure to shape and bond
• Lower temperatures require higher pressures, initial densification is generally by plastic flow requiring short times due to thermal softening
• Higher temperatures require lower pressures and densification is generally by diffusional creep requiring longer times due to limited plastic flow
• Applicable to all compositions
• Example processes—forging, hot pressing, hot isostatic pressing, spark sintering, spray forming, hot extrusion, explosive compaction
• Example systems—steel-copper, cobalt-diamond, titanium-titanium carbide, cobalt-tungsten carbide, aluminum-silicon carbide, tool steel-titanium carbide, silicon carbide-diamond
• Key control is via temperature-pressure–time combination in terms of plastic flow and diffusional creep

Table 7.3 Key elements of two-step shaping and sintering

• Mixed powders, first step is associated with pressure to shape the powders and second step is associated with heating to bond and sinter densify the powders
• Lower forming pressures rely on added binder (polymer) to enable shaping; examples include injection molding, extrusion, slurry casting, tape casting
• Higher forming pressures achieve high pressed density with less added binder; examples include uniaxial die pressing, cold isostatic pressing, roll forming
• Second step is thermal treatment, usually by sintering at high temperature, potentially forming a liquid phase; one variant is to infiltrate melt into shaped powder skeleton
• Example systems—tungsten carbide-cobalt, tungsten-copper, silicon nitride-silicon carbide, alumina-glass, steel-copper, titanium-hydroxyapatite
• Key controls are particle size, particle size ratio, pressure in the first step, and temperature in the second step

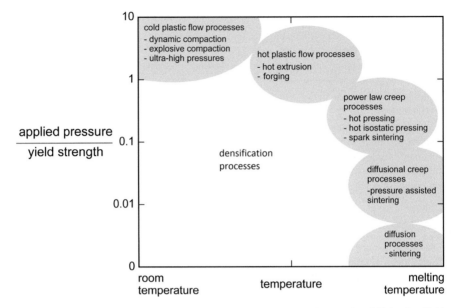

Fig. 7.13 Outline map of relative roles of temperature and pressure in reaching full density. Cold processes employ high stresses to attain densification by plastic flow and high temperature processes require low stresses for densification by diffusional processes. The densification rate is slower as stress decreases. The yield strength corresponds to the strength at the consolidation temperature

thin-wall structures where precision is less important. In these options the specific forming variables are adjusted to the composition and component geometry.

A layout of consolidation options based on relative temperature and relative pressure is given in Fig. 7.13. This conceptualization outlines the processes into clusters based on the governing densification step. To span over a wide range of materials, the plot normalizes temperature to the absolute melting temperature, called the homologous temperature. Likewise, pressure is normalized to the yield strength of the material at each temperature. These normalizations apply to the major phase in the composite. Once a decision is made on the type of consolidation process, then certain fabrication details emerge. Additional factors include practical issues such as equipment availability, tooling cost, and production rate.

Much more of the logic for assessing options and selecting a best fabrication approach is the subject of manufacturing textbooks and can only be outlined here [31]. Very often the final decision is based on cost, both for set up and tooling as well cost per component. A classic dialog starts with the question "Can you make this component?" followed by "How much will it cost?" Obviously the first answer must be yes, but usually the answer to the second question determines process selection.

Liquid–Solid Shaping Processes

Casting, Molding, Extrusion

When one phase forms a liquid, then the fabrication of a particulate composite is a semisolid forming step similar to traditional liquid shaping. On cooling the liquid solidifies to form the second phase in the composite. Options along these lines include casting, molding, or extrusion [32–35]. Each is limited to a narrow composition range. For example, as solid particles are added to a liquid to form a slurry, the mixture viscosity increases. Too much powder stops flow and leads to voids. On the other hand, with too little powder the solid and liquid separate and fail to hold shape. Any difference in solid versus liquid density promotes phase separation. Accordingly, there is a practical range where liquid–solid forming processes prove successful—typically between 40 and 60 vol% solid [35].

Several approaches are used to shape the paste. One option is very similar to casting. The solid particles are mixed and entrained in heated liquid. That semisolid mixture is cast into a mold. No pressure is required, although vibration helps overcome flow resistance. The density difference between the solid and liquid results in segregation if the mixture is not frozen quickly. Low solid contents produce heterogeneous structures due to density differences between the phases, leading to a layer of pure liquid.

The viscosity η of a solid–liquid slurry depends on the volume fraction of solid particles ϕ as a ratio to the maximum or critical volume fraction of solid ϕ_C as follows [36]:

$$\eta = \frac{\eta_L}{\left[1 - \frac{\phi}{\phi_C}\right]} \tag{7.12}$$

where η_L is the viscosity of the pure liquid. Equation (7.12) is based on volume fraction. For spherical powders the critical volume fraction corresponds to the tap density, near 0.6. For irregular particles with asperities the tap density is lower and the critical volume fraction lower, more typically 0.4 or even smaller. In nanoscale structures, the particle-particle friction is very high so the critical volume fraction is as low as 0.10–0.05. Liquid viscosities are low, often similar to water, but a high particle content makes the mixture similar to toothpaste, resisting flow. Indeed, toothpaste, paint, and ice cream are mixtures of particles in liquids. Attaining a dispersion of nanoscale particles is a large challenge, leading to much attention to mixing technologies [37].

Injection molding is a forming technique that shapes polymer-powder mixtures at relatively low temperatures (below 200 °C or 473 K) and pressures (about 15 MPa) [36]. The polymer is molten under these conditions to act as a carrier for the powder. For injection molding, the ideal viscosity at the molding temperature is about 100–200 Pa · s. For comparison, the viscosity of water is 10,000 times lower at about 10^{-3} Pa · s and the viscosity of peanut butter is about 250 Pa · s. This viscosity is determined by both molding temperature and solids content in the binder. For injection molding, the solid consists of about 50–60 vol% of the feedstock [9, 10, 38].

The idea behind injection molding as a component shaping approach is outlined in Fig. 7.14. The powder and polymer binder are mixed to form granules. Those granules are heated in a molding machine that holds rigid tooling. A small opening exists in the tooling for pressuring hot feedstock into the cavity, and a small opening at the other end allows air to vent out of the tooling. A runner path inside the tooling guides delivery of hot feedstock into the cavity through a gate. Once the mold is filled, pressure is maintained while the binder freezes, after which the chilled (solid) component is ejected from the tooling. The approach is restricted to compositions where the mixture viscosity matches the equipment capability. Usually the polymer is a low cost ingredient since it is sacrificial, and paraffin wax is a common ingredient since it allows shaping at a relatively low temperature. A few products leaves the binder as part of the molded shape. One example are bonded magnets, where a rubbery polymer remains to provide electrical insulation and flexibility.

Predominantly injection molded structures rely on polymer binders, but lead, aluminum, zinc, and related liquids are possible. A variant known as reaction injection molding employs two nozzles to inject and mix polymer ingredients that cross-link in the mold. Solid particles are included to form a composite. Injection molding options are limited by tool erosion. For example, molding molten titanium laden with particles is possible, but tool life is short due to attack by the liquid metal.

Related ideas are employed in shaping long, uniform cross-section shapes by extrusion. It relies on pushing heated feedstock through a die to create a constant cross-section structure rod, tube, honeycomb, or spiral. The molten polymer imparts viscous flow characteristics to the mixture, allowing particles to slide past each other in the forming tool. A cross-section illustration of the process is given schematically in Fig. 7.15a and an example flexible flat powder cloth is pictured in Fig. 7.15b. After extrusion the solidified binder holds the particles in place. In one variant, the binder is removed and the particles are sintered to density. In another variant, the binder is part of the final composite, for example to form pencil lead which is an extrusion of graphite. A composite of this sort forms the flexible magnetic seal on refrigerator doors, consisting of rubber and magnetic particles.

A related idea relies on a polymer-powder mixture extruded onto a flat table to form a thin deposit. The table moves to sweep out a two-dimensional layer that is coordinated with a computer image of the desired structure. The motion of the extrusion nozzle deposits a molten bead, layer by layer. Each layer builds on the previous layer to generate a three-dimensional object. Sacrificial support material is added to temporarily fill hollow spaces, cantilevers, or overhangs. Of course hollow objects require some means to extract the sacrificial material.

Fig. 7.14 Injection molding relies on hot mixtures of molten binder and solid particles. The paste-like mixture is pressurized to fill the tool cavity before the binder solidifies. After cooling, the tool cavity opens to eject the shaped component and the cycle repeats

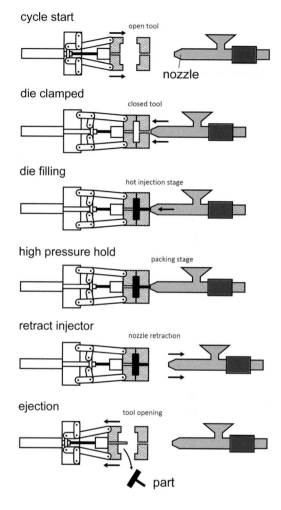

cycle start

die clamped

die filling

high pressure hold

retract injector

ejection

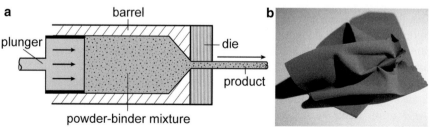

Fig. 7.15 A powder-binder mixture is extruded through a reduced cross-section die to produce a long, thin structure such as a tube, rod, or ribbon; (**a**) the mixture is heated to melt the polymer binder, which solidifies after extrusion to hold the entrained particles into the desired shape, and (**b**) an example of a flexible powder cloth produced by extrusion

Infiltration

In composites with solid contents exceeding about 60 vol%, the powder-liquid viscosity is too high for direct forming. For about 60–85 vol% solid, an option is to infiltrate liquid into the pores between the shaped solid particles [39]. For liquid infiltration, the particles are held in a mold or compacted into a preform. Subsequently liquid (polymer, glass, or metal) is introduced into the pores to create the composite. If the liquid wets the pores, no pressure is required as the solid sponges the liquid into the pores. For nonwetting systems an external pressure is required to force the liquid between the particles. Both phases are connected at the particle concentrations associated with infiltration. Figure 7.16 is a cross-section contrast of pore structures (black) associated with (a) open pores and (b) closed pores. Open pores link to the exterior surface and exhibit an elongated, nonspherical character. On the other hand, closed pores exist as spherical cavities that do not connect to the surface and appear circular in cross-section. Infiltration requires open pores. Due to wetting differences, copper infiltrates tungsten without pressure, but aluminum is pressurized to infiltrate silicon carbide. Infiltrated composites are characterized by both phases being interconnected and intertwined.

The capillary pressure pulling a wetting liquid into the pores depends on the pore size, contact or wetting angle, and the surface energy. The capillary pressure with respect to atmosphere pressure ΔP varies with the inverse of the pore diameter d_p as follows:

$$\Delta P = \frac{2\,\gamma_{LV}\,\cos\,(\theta)}{d_P} \tag{7.13}$$

where γ_{LV} is the surface energy of the liquid–vapor interface and θ is the wetting or contact angle. Small pores are associated with a small particles, resulting in a higher capillary pressure. If the liquid is not wetting, then external pressure is required to force liquid into the pores. Once the liquid is infiltrated, it solidifies on cooling to deliver the composite. Most liquids contract on solidification, so some porosity forms during cooling. To preserve component dimensions, infiltration cycles are fast, thereby avoiding swelling or shrinkage reactions. An excess of infiltrant leads to liquid bleeding or compact distortions, so it is common to slightly under fill the pores to avoid leaving a surface residue. Properties are usually highest when the infiltrant liquid wets the solid.

As illustrated in Fig. 7.17a, liquid infiltration is similar to how water seeps into a sponge. In some applications, sintering and infiltration occur in a single heating cycle using infiltrant stacked on top of the pressed powder perform. Cooling solidifies the liquid to form the second solid phase as shown in Fig. 7.17b. This microstructure corresponds to steel powder processed to about 80 % density with infiltration of the remaining pores using a copper-based alloy near 1100 °C (1373 K). As an illustration of common usage, Fig. 7.17c shows a picture of deadbolt door lock components fabricated this way.

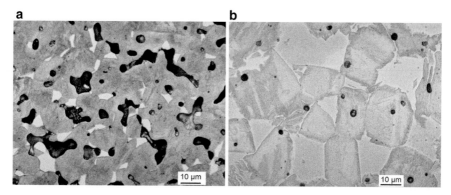

Fig. 7.16 Open pores and closed pores differ with regard to their connection to the exterior surface. Open pores are able to pass fluids, but closed pores are sealed. Infiltration requires open pores evident by a characteristic worm-hole shape such as seen in (**a**). Closed pores are spherical and lack connections to each other and the external surface and are characterized by the circular shapes in cross-section as seen in (**b**) [courtesy Y. Wu]

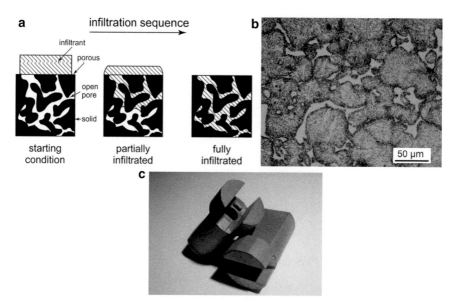

Fig. 7.17 Infiltration of open pores relies on melting the infiltrating phase to allow capillary forces to pull the liquid into the open pores, as diagrammed in (**a**). On cooling the infiltrant solidifies to give an interconnected composite structure, such as shown by the steel infiltrated by copper in (**b**) [courtesy of E. Klar]. A common product are lock components as seen in (**c**)

Related composite systems rely on copper, aluminum, nickel, silicon, or silver to infiltrate steel, tool steel, tungsten carbide, silicon carbide, titanium carbide, tungsten, diamond, molybdenum, cadmium oxide, aluminum nitride, or other higher temperature materials. Infiltration was an early approach to forming high performance particulate composites such as titanium carbide and tool steel, giving a composite termed Ferrotic. The process is called "sinter-casting" to reflect the hybrid between casting and sintering.

A lower temperature variant of infiltration fills the pores with molten polymer. The liquid might be a monomer that polymerizes after infiltration. Although most infiltration relies on liquid, vapor infiltration is possible using a vapor species that decomposes inside the pores. For example, methyltrichlorosilane (CH_3SiCl_3) vapor is used to form silicon carbide in pores at 1200–1400 °C (1473 –1673 K). Pore filling by vapor infiltration is performed slowly to avoid closing surface pores. Otherwise residual pores remain due to incomplete infiltration. Chemical vapor infiltration is reserved for high value composites used on space vehicles or military aircraft.

Simultaneous Temperature-Pressure Processes

Many materials are sintered at high temperature without pressure. Composites with a high level of second phase often resist sintering densification, so simultaneous temperature and pressure are used for consolidation. In essence, materials designed to be strong at high temperatures inherently resist densification. For example, plotted in Fig. 7.18 are comparative data for 0.4 µm ZnO showing sintered density versus temperature, and the same powder with added 12 µm SiC at a composition of ZnO-6SiC (10 vol% SiC) [40]. These data were collected for both compositions during constant heating rate sintering at 4 °C/min. The composite exhibits retarded sintering densification in spite of slightly improved packing. Consolidation cycles with simultaneous pressure and temperature are a means to compensate for the impeded sintering densification. A variety of approaches are in use, generally characterized by pressure acting as a supplemental force while heating is used to soften the composite.

Slow Densification Processes

Densification is difficult when both phases in the composite remain solid. At low pressures densification is by diffusional creep, depending on the relative temperature and pressure. In such cases, full density requires holds at the peak temperature from 15 to 120 min, mostly dependent on the harder and higher temperature phase densification rate.

Hot pressing is an effective means to consolidate mixed powders. It was adopted for widespread use in the 1940s. Subsequently hot isostatic pressing, hot extrusion, and related heating and pressing ideas emerged. The dimensional change is large since the starting condition is loose powder. Hot pressing applies the pressure along one axis, so the dimensional change is all along that direction due to radial

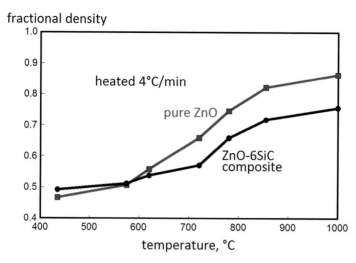

Fig. 7.18 Sintering data from constant rate heating experiments for pure 0.4 μm ZnO and a composite consisting of the same ZnO with 6 wt% SiC added. Sintering is retarded by the inert SiC addition. To compensate for the decreased sintering densification, pressure-assisted techniques are used with simultaneous temperature and pressure

constraint from the tooling. In such a case the axial dimensional change is large, near 40 % in the compression direction.

Hot pressing is a conceptually similar to die pressing, but the tooling is heated and the cycle is long. A cross-section of the hot pressing concept is given in Fig. 7.19. Heat is supplied by a furnace or induction coil outside the tooling. Graphite is the most common tool material. It requires protective atmosphere or vacuum to avoid oxidation. Heating softens the material and pressure induces diffusional creep to densify the compact. It is widely used to consolidate materials that are unstable at high temperatures, such as diamond-metal composites.

On initial pressurization, densification starts with particle rearrangement followed by possible deformation at the particle contacts if the powder is ductile. As the contacts enlarge, the effective stress falls and further densification depends on diffusional creep. The maximum pressure is limited by the strength of the graphite tooling, usually about 10 MPa, but possibly up to 50 MPa. Temperature is an important means to accelerate densification, but small grain sizes and high pressures also aid densification.

Cycle times in hot pressing are measured in hours. Two factors make the cycles long. One is simply the thermal mass where the tooling and compact must be heated and cooled in each cycle. The other actor is the slow creep rate due to the limited stress that can be imposed by the tooling. Maximum temperatures reach up to 2200 °C (2473 K). Because vacuum or inert gas is required to avoid oxygen reactions, leak tight chambers surround the hot zone leading to added equipment expense.

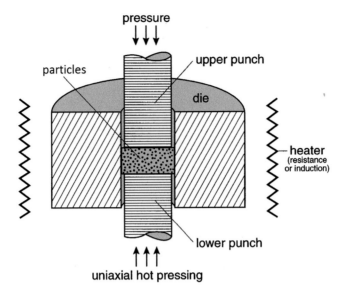

Fig. 7.19 Cross-sectional diagram of uniaxial hot pressing. The particles are contained inside a die cavity (usually graphite) with an upper and lower punch acting to pressurize the powder. Heat is supplied by an external heater or induction supply. The assembly is often contained in vacuum to protect from oxidation

Hot isostatic pressing (HIP) is a variant where temperature and pressure are applied simultaneously using heated inert gas [12]. Densification usually is under diffusional creep conditions. In one approach, powders are poured into an oversized container designed to shrink to the desired final size. On heating the container softens and deforms at high temperature and common options are titanium, stainless steel, iron, or glass (if heated prior to pressurization). After powder loading into the container, it is baked to evaporate contaminants, then evacuated and sealed. Within the HIP chamber the combined action of heat and pressure cause powder consolidation. Figure 7.20 shows a schematic of hot isostatic pressing using a thin-walled container. In this illustration the powder is encapsulated and evacuated prior to HIP consolidation. After the cycle, the container is removed. A containerless option is available for presintered compacts if they start with about 95 % density, corresponding to the closed pore condition.

Densification requires the application of both heat and pressure, desirably with independent control of the temperature-pressure cycles. Most desirable is a cycle where first the container is softened by heating, then pressurized. It is common to add gas to the chamber at a low pressure and rely on heating to help drive pressurization. Peak temperatures of 2200 °C (2473 K) and pressures of 200 MPa are possible. The HIP pressure chambers come in various sizes up to 1.5 m diameter and 2.5 m high. Graphite heating elements are common, but refractory metals including molybdenum and tungsten are options.

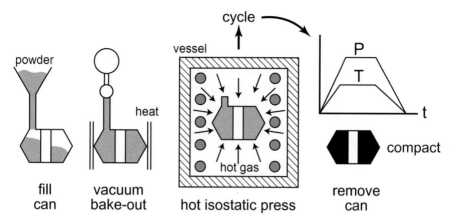

Fig. 7.20 Hot isostatic pressing (HIP) relies on temperature and gas pressure to densify powder sealed inside a container. Cycles involve either simultaneous heating and pressurization, or sequential heating followed by pressurization. After densification the outer container is removed from the compact

A significant advantage for HIP is that the pressure is applied from all directions so it is a hydrostatic forming process. In contrast to uniaxial hot pressing, the hydrostatic conditions result in less particle-particle shear, but more uniform dimensional change. Since there is less shear at the particle contacts, any surface contamination on the particles is likely to interfere with particle bonding. This sometimes leads to precipitation to outline the prior particle boundaries with weak interfaces. Hot pressing is uniaxial with nonuniform densification, leading to more interface shear. In both hot pressing and hot isostatic pressing, the outside compact surface is contaminated by the tooling or container. These surfaces are removed after consolidation by chemical dissolution, machining, or abrasion. Since final dimensions are attained in a post-consolidation step these are termed near-net-shape approaches.

Containerless hot isostatic pressing uses the same device, without the necessity of powder encapsulation. For this to succeed, the powder preform is first sintered to about 95 % density in vacuum. At this point the pores are closed and will not allow gas penetration between the particles. From that point in the consolidation cycle, gas pressure is applied to the hot compact. The integration of sintering and HIP into a single cycle is called sinter-HIP. A schematic of the temperature and pressure curves is illustrated in Fig. 7.21. Heat is first applied under vacuum, and pressure is applied after substantial sintering to collapse any residual pores. Typical cycles are a few hours. Sinter-HIP is used for WC-Co composites to ensure high performance. It is also possible to first sinter, cool, and then move to HIP for final densification. One advantage of HIP is the possible fabrication of large components, reaching upwards of 1000 kg. On the other hand containerless HIP is applied to small components, often in the 10 g range.

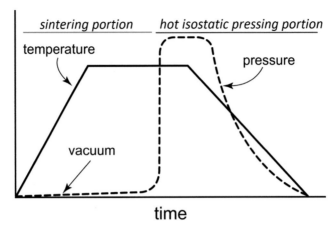

Fig. 7.21 Sinter-HIP is a hybrid densification approach that avoids an outer container. The first part of the cycle is conducted in vacuum to densify the structure to about 95 % density. Then the furnace is pressurized in the second half of the cycle to densify the structure. The benefit is a faster cycle with no loss of heat and no handling between the two steps

Reactions between powders are used to form compounds during hot pressing or hot isostatic pressing. The product microstructure is more homogeneous if pressurization is delayed until after the reaction. For example, TiB_2 is fabricated from a mixture of Ti and B powders. The mixed elemental powders are encapsulated and heated in a HIP to 700 °C (973 K) under a pressure of 100 MPa. At that point, the exothermic reaction initiates to give $Ti + 2B \rightarrow TiB_2$ liberating 293 kJ/mol that causes self-heating. After the reaction, while the compact is hot, pressure is applied to densify the heated compact. A composite of titanium reinforced with titanium diboride, Ti-TiB_2, starts with an excess of titanium [41]. Another composite formed in this manner is $MoSi_2$-SiC. However, in some systems undesirable interface phases form that degrade composite properties. Generally, reactive hot consolidation is difficult to control, so it is mostly used when the geometry remains fixed once the process is stabilized, such as in reactive hot pressing [42, 43].

Spark sintering is a variant of hot pressing where electric current passes through the powder compact to provide direct resistive heating [5, 6]. The idea was used to sinter lamp filaments in the early 1900s and was termed spark hot pressing during the 1950s or simply spark sintering. Both direct and alternating currents are in use. The largest composite application is cobalt-diamond stone cutting tools. A process schematic is shown in Fig. 7.22. Current densities range up to 300 A/cm^2. The peak temperature is limited by the available power, but 2700 °C (2973 K) is possible in special cases. To attain full density, various trade-offs are recognized between current pulsing, hold time, peak pressure, and peak temperature [44]. Commonly, the powder is contained in a graphite punch-die assembly. It may be necessary to insulate part of the assembly to ensure current flows either through the powder or through the tooling; the latter is necessary for high conductivity materials. High current densities enable short consolidation cycles to minimize reactions, phase

Fig. 7.22 Spark sintering involves electrical current discharges through a powder compact during hot pressing. The powder is captured in graphite tooling to ensure electrical conduction for heating. Cycles are very rapid, making it useful for consolidation of unstable materials, such as diamond, and where rapid consolidation is sought to avoid grain growth

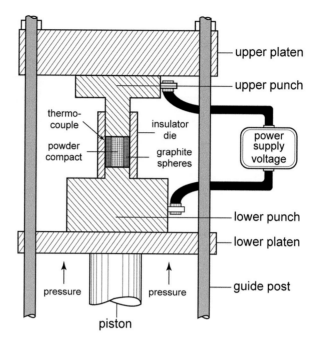

decomposition, and grain growth. Early speculation suggested the combination of direct and alternating currents created a discharge plasma between the particles, but careful observation found this was not the case. Prior to this the process was called spark plasma sintering, but the absence of a plasma results in calling the approach spark sintering or field-activated sintering.

High Stress, Rapid Consolidation Processes

Diffusional creep processes correspond to slow strain rates when compared to plastic flow, accordingly creep requires longer consolidation times. Creep implies the effective pressure falls below the yield strength of the composite during densification. Typically a short burst of plastic flow ends and slower diffusional creep is the dominant process. With sufficient applied pressure plastic flow can reach full density. Clearly, relatively high compaction pressures give rapid densification, as long as the stress exceeds the yield strength.

Hot extrusion is a means to combine temperature and pressure in a fast densification event. The mixed powders are placed in a canister, degassed, heated, and evacuated, then forced through a die using a high pressure ram. The area reduction associated with passage through the die causes the powders to yield, deform, and bond, producing a dense rod or tube. Intense shear disrupts the particles surfaces to improve bonding. Figure 7.23 illustrates how canned powders are pressurized by a

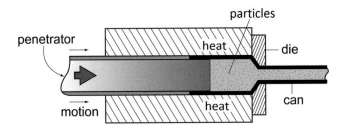

Fig. 7.23 Hot extrusion of particulate composites involves filling a canister with the particles, degassing the canister, and inserting the hot canister into a high pressure hydraulic press. It is necessary to preferentially pressurize the powder (not the can) using a penetrator. The reduction in cross-sectional area in the die is often 25-fold

plunger to force extrusion. Densification corresponds to a high area reduction as measured by the extrusion constant. This parameter is a measure of the difficulty in achieving deformation and flow; the extrusion force F and extrusion constant C relate to one another as follows:

$$F = C A \ln(R) \qquad (7.14)$$

where A is the cross-sectional area of the feed material and R is the reduction ratio, equal to the cross-sectional area of the starting billet divided by the cross-sectional area of the product. It is this area reduction that causes particle shear responsible for interfacial bonding. The force to initiate first flow in extrusion is higher than the force needed to maintain flow, and both forces increase with smaller particle sizes. The force required to sustain flow is recorded as the extrusion force F. The extrusion constantt C declines as temperature increases, but increases as the extrusion strain rate increases. Low melting temperature compositions inherently have lower extrusion constants.

Although material properties influence the ease of extrusion, temperature is the main control variable. Too high a temperature damages the microstructure and shortens the extrusion tool life. Alternatively, a low temperature inadequately softens the material and fails to give densification or requires excessive stress with a short tool life. Generally, the reduction ratio in extrusion must exceed 10, and often a reduction ratios near 25 gives the best composite properties.

A variant on hot extrusion is equal channel angular extrusion (ECAE). Instead of the cross-section change to induce interparticle shear, the equal channel process introduces a 90° bend in the flow path without a cross-section change. A high pressure pressures pushes the composite material through the tooling, as sketched in Fig. 7.24. As the composite turns the corner, the outside and inside edges exhibit large velocity differences that shear the mixture. For systems such as Ti-TiN and Ti-TiB the process generates considerable strengthening, but the gain in strength comes with a loss in ductility [45]. Similar to hot extrusion, the product is long and thin, often in the shape of a square rod.

Fig. 7.24 The equal channel angular extrusion is used to densify and shear mixed powder compacts. The flow around the corner induces intense shear deformation to densify and strengthen the compact, but the shapes are limited to rod-like structures

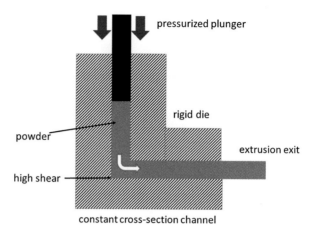

Another route for rapid composite densification is hot forging. Mixed powders are pressed into a preform, sintered, and upset forged to full density [46]. Densification is by plastic flow coupled to thermal softening; the stress necessary to achieve full density decreases as temperature increases. Although similar to hot pressing, the high velocity and pressure employed in hot forging deliver nearly instantaneous densification by plastic flow. In hot pressing the compact remains in contact with the die wall during pressing, while in powder forging there is lateral spread to induce shear prior to die wall contact. The lateral motion, as sketched in Fig. 7.25, improves particle bonding due to shear and differential particle flow. Presintering provides some strength but not enough to withstand large levels of lateral flow, otherwise the compact cracks. Thus, preforms are designed to flow and make early contact with the tooling during consolidation, prior to fracture. Due to the interfacial shear during densification, the composite strength is often higher than obtained with approaches such as HIP. In production the forming rates are at most four forgings per minute. Often tool life limits peak temperatures to about 1200 °C (1473 K) or less.

An alternative to hot forging is granular forging. This concept is illustrated in Fig. 7.26; it is a hybrid between hot isostatic pressing and uniaxial hot pressing that starts with preheated granules (graphite or ceramic) to act as pressure transmission media. An external punch presses on the granules. Prior to loading the powder compact is also heated. Then, the forging stroke presses on the granules with the embedded component in a relatively rapid compact densification stroke. A disadvantage is nonuniform densification resulting from how the granules are pressed in one dimension, but act to give three-dimensional compaction. The granules redistribute the stress, but this is only partly effective, so the compact distorts during densification.

In a comparison of densification routes, those producing shear between the particles result in superior properties. For Ti-20TiB composites, the hardness and elastic modulus from hot pressing were slightly higher than attained using hot isostatic pressing [47].

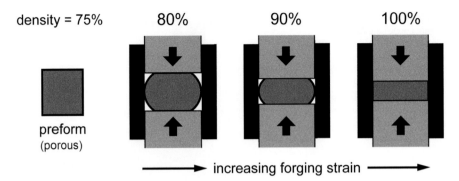

Fig. 7.25 Forging is a means to rapidly densify a heated powder preform. Radial flow of the preform improves interparticle shear. Too much radial flow causing cracking, so by the end of the forging stroke the preform is fully constrained by the tooling

Two Step Press-Sinter Processes

An approach to composite fabrication relies on two steps; pressurization and heating are performed in sequential steps. Shaping is possible with several approaches, but for particulate composites two approaches are dominant—compaction and molding. The former relies on high pressure and the latter relies on lower pressure shaping.

Compaction

Uniaxial die compaction produces a weak powder perform that is a "green" body. In a separate step, the preform is heated to sinter the particles. Usually the first step is performed in rigid tooling. The idea of die compaction is sketched in Fig. 7.27. This same procedure is used to form vitamin and pharmaceutical pills, as well as candies. At the start of the compaction cycle, the powders are fed from a hopper via a feed shoe as it passes over the die cavity. Uniform cavity fill is achieved by feed shoe oscillations over cavity. The deposited powder is at the apparent density. Once the feed shoe swings away, the upper punch enters the cavity and applies pressure. Pressure first causes the particles to rearrange, then deform and bond. As pressure increases the rate of densification slows since particle rearrangement ends and the particles harden from deformation.

Example pressing behavior was plotted in Fig. 7.1, showing rapid early densification and slow densification as pressure increases. Eventually the powder reaches a point of diminishing return where little density change occurs even with large increases in the compaction pressure. Hard particles inherently resist compaction, so the greater the hard particle content the lower the pressed density. Figure 7.28 plots fractional density versus hard particle content for a mixture of lead (soft) and

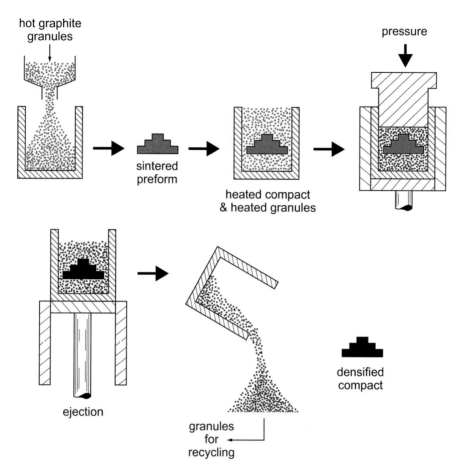

Fig. 7.26 Granular pressing is a means to induce compaction of a shaped preform using uniaxial pressing; the heated granules provide pressure transmission to the component from the applied pressure. Densification is by plastic and diffusional flow. The granules are usually graphite and are reused

steel (hard) particles compacted at 400 MPa, illustrating how hard it is to compact mixed powders.

Depending on the powder and tool material, the peak compaction pressure ranges up to 1000 MPa. A few instances achieve 4500 MPa. Very hard powders and very soft powders are both pressed at the lower pressures. For example, Al-10SiC composite pressed at 150 MPa reaches 90 % density, but harder particle mixtures, such as WC-10Co, pressed at 350 MPa only reach 60 % density. In such cases wax is added to the powder to provide handling strength, since the pressed powder is weak until sintered. Reduced tool wear is the largest factor in determining which polymer addition to employ.

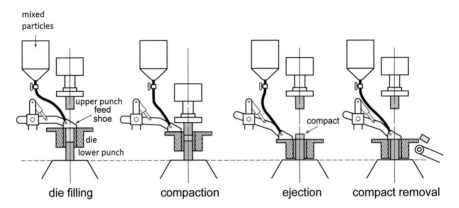

Fig. 7.27 Uniaxial cold compaction is illustrated here, where the mixed particles delivered by the feed shoe fall into the die cavity, then are compacted using a high pressure. The upper punch withdrawals at the end of the compaction stroke, while the lower punch pushes the compact out of the tooling in the ejection stroke. After this, the cycle repeats

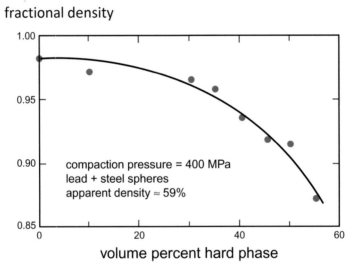

Fig. 7.28 Plot of fractional pressed density versus content of steel spheres as the hard phase. The lead powder is soft and at 400 MPa reaches 98 % density, but with added hard steel particles the composite progressively resists pressing

In high volume production, compaction is fully automated with only one operator for four presses. Pressing rates depend on the press design, ranging from one
compact per minute for large structures to 5000 compacts per minute for small,
simple structures formed in multiple cavity tooling. Die compaction is best applied
to thin structures that lack undercuts. An undercut makes ejection from the tooling
difficult if not impossible. Die compaction for automotive and electronic

applications often exceed millions of compacts per week. Designs compatible with die compaction have holes, imprints, and steps, but generally die compaction is restricted to flat, squat shapes that are easily ejected from the tooling.

Cold isostatic pressing (CIP) is a room temperature variant that relies on a flexible die, properly termed a bag. Like hot isostatic pressing, the pressure is hydrostatic, meaning it is uniform in all directions. Accordingly, the green density is homogeneous within the pressed body. Molds for cold isostatic pressing are formed using flexible polymers or rubbers dipped on a master form to produce the desired shape. Cores and inserts are possible. Figure 7.29 is a cross-sectional schematic of a cold isostatic press for the consolidation of a tube shape billet around a solid core. The particles occupy the annular space between the solid core rod and flexible bag. Pressure works on the outside of the bag to densify the powder. Compaction pressures up to 1400 MPa are possible; however, most CIP is performed at pressures below 420 MPa. A perforated external container might be added to hold the flexible bag alignment during powder loading and pressing. Also, it is helpful to evacuate the bag to remove air prior to compaction.

Wet bag CIP is applied to low production quantities. In this variant, the filled and sealed rubber bag is immersed and pressurized in a fluid chamber. After pressurization the wet mold is removed from the pressing chamber and the compact extracted from the mold. An alternative, known as the dry bag approach, is favored in high-volume production, because the bag is built directly into the pressure cavity. In this case the flexible bag deforms, but is not ejected; end plugs allow powder loading and component unloading with the rubber bag still in the press. As an estimate, wet bag pressing might produce a compact every 15 min, but dry bag pressing might produce six compacts every min.

Cold isostatic pressing is applied to shapes that cannot be die compacted, long or thin structures. Long tubes are commonly produced using CIP. Since the tooling is flexible, cold isostatic pressing is not able to produce high precision components. Also, cycle times are long, especially for wet bag pressing, making cost high compared to die compaction. Thus, CIP processes are reserved for long-thin geometries that are not possible via die compaction.

Various high velocity compaction options exist. The high strain rate and high pressure are outside the range encountered in traditional pressing. The high velocity approaches include explosive compaction, shockwave compaction, and gas gun or high velocity compaction; generically they are variants of dynamic compaction.

Dynamic compaction employs a high velocity shock wave to exceed the material strength during consolidation. Frictional heating between the particles helps to bond the particles. In some cases frictional heating is sufficient to form a melt at the particle contacts. Peak velocities are 10 km/s with pressure pulses up to 30 GPa. Gas guns are not so effective. They impact a high-velocity projectile against a powder preform using a pressurized launcher, reaching up to 500 m/s velocities and 2 GPa peak pressures. Such pulses are sufficient to densify softer powders, such as aluminum. Particle bonding is especially enhanced if the shock wave forms a melt at the particle contacts. Unfortunately, dynamic compaction equipment is expensive, production rates are low, tool life is short, and most powders densify more at

Fig. 7.29 Wet bag cold isostatic pressing relies on a pressure vessel filled with a flexible container around the powder. In this instance the powder is located between a solid core and an outer flexible container. Pressure is applied inside the chamber to compact the powder uniformly in all directions. After compaction the wet bag and core rod are stripped from the compact to form a tube

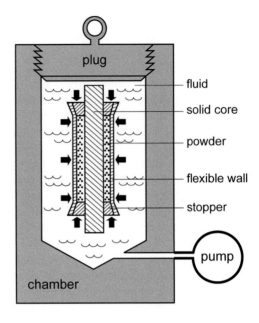

slow strain rates. High strain rates cause strain rate hardening. Effectively, fast pressurization induces more resistance to densification. In other words, similar peak pressures delivered at slow strain rates prove superior. Accordingly, in spite of much attention, dynamic compaction remains a laboratory tool.

Explosive compaction relies on shock waves for densification. Figure 7.30 illustrates how explosive compaction is applied to a flat compact geometry. The pressure wave from the explosive densifies the powder by simultaneous pressurization and heating over a microsecond. The echo from the shock wave often damages the compact, so some means of trapping or diverting the shock wave is necessary. The ratio of explosive mass to powder mass is the primary determinant of density. Small particles are difficult to compact and brittle materials are preheated to minimize fracture. Peak velocities of 8 km/s are reported in pressure pulses of 35 GPa. Product shapes are relatively simple and often contain cracks.

Shaping

Powder shaping processes create a component without deforming the particles. Several options are applied to particulate composites, including injection molding, slip casting, slurry casting, tape casting, and powder-polymer extrusion. The forming pressures are relatively low, below 100 MPa, as compared to die compaction pressures of 350–850 MPa. For complex shapes and high production quantities it is appropriate to rely on automated processes based on powders mixed with sacrificial binders. Those mixtures are known as feedstock. When the heated the

Fig. 7.30 A conceptual outline of explosive compaction. The powder is packed and sealed in the cavity. The shock wave from detonation of the explosive drives the piston into the powder. It is critical to manage pressure echoes to avoid damage to the densified compact

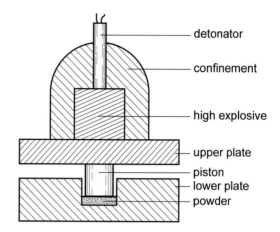

feedstock binder melts to form a slurry with the powder. Under pressure the hot slurry is molded into a cold die, and then frozen to produce the desired shape.

An outline of the injection molding cycle is sketched in Fig. 7.31 [11]. After mixing the feedstock pellets are fed into a heated barrel of the molder where the lower melting binder melts. The viscosity of the mixture is controlled to 100–200 Pa-s, sufficient to flow like toothpaste. For thin-walled components the molding time is just a few seconds, but for thick-walled components molding is much slower. The forming time reflects heat transfer in the tooling, realizing the binder must be frozen before the mold can be opened. As a scaling factor, the cooling increases by about 5–10 s for each mm of component wall thickness. Once ejected from the mold, the binder is removed and the compact sintered.

Common binder systems consist of paraffin wax, or similar low-cost polymer, that melts at about 65 °C (338 K). Other polymers are added to provide strength, such as polypropylene. To reduce the forming pressure, solvents or plasticizers are added; these include gasoline, kerosene, palm oil, stearic acid, or mineral oil, largely to reduce viscosity.

Once the mold is filled and the binder frozen, the component is removed from the tooling. At that point the binder provides handling strength. Subsequently the binder is removed as part of the heating cycle to the sintering temperature, called debinding. Several options exist, but the most common is to slowly heat the component to evaporate the binder. Most of the evaporation occurs between 250 and 350 °C (523 and 623 K). As binder is extracted, the particle structure becomes fragile, so heating continues to the sintering temperature without intermediate handling. Sintering densifies and strengthens the structure, leading to shrinkage. An example of sintering shrinkage is evident in Fig. 7.32. This injection molding approach is widely employed to form small, complex shaped components. For example thread guides, cutting tools, computer hinges, and surgical devices. Ideal applications hover near combinations such as:

Fig. 7.31 In binder-
assisted injection molding,
the composite particles are
mixed with a sacrificial
binder (paraffin wax for
example) at a ratio of
typically 60 vol% particles.
When heated this mixture is
molded in a traditional
plastic molding machine.
The binder is the removed
and the particles sinter
densified

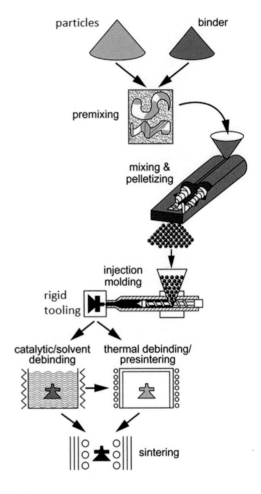

- production quantity of at least 200,000 parts per year
- component mass near 10 g (range is 0.1–150 g)
- longest dimension is near 25 mm (nominally 1–200 mm)
- wall thickness near 3 mm (range is 0.1–120 mm)
- design with about 75 specified features.

Instead of molding into a die, the same powder-binder feedstock can be extruded through a die. This produces a long structure of constant cross-section. One common use of extrusion is in the production of pencil "leads". These are mixtures of graphite (carbon), clay (hydrous alumina-silicate), and water. The ingredients are mixed into a paste that is extruded as a rod, akin to how spaghetti noodles are formed. After drying to remove the water the rod is sintered for 45 min at 1000 °C (1273 K), giving a hardened structure that is subsequently encapsulated in wood. Another application of extrusion is the production of twist drill blanks from WC-Co. The extrusion step relies on the same equipment as used to shape plastics.

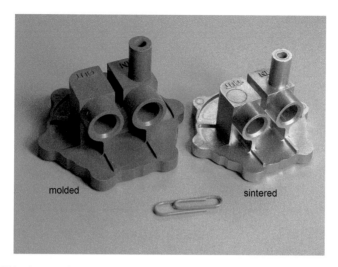

Fig. 7.32 This photograph compares the size of the molded body on the left and the sintered body on the right, the shrinkage corresponds to the void space eliminated after the binder is removed

For flat sheets, the powder-binder shaping operation is performed using tape casting. Here solvent is added to lower the mixture viscosity to the consistency of paint. The solvent evaporates after the sheet is formed. The polymer-powder sheet is flexible, allowing some shaping into curved surfaces prior to sintering.

For hollow objects, slip casting is a typical solution. It also relies on powder-binder-solvent mixtures. After shaping in mold, the solvent is evaporated to harden the binder and provide handling strength. Whereas injection molding changes the binder from solid to liquid using heat, tape casting and slip casting rely on solvent addition/removal to achieve a similar goal. Very low pressures are employed so there is no particle deformation, making the slurry processes insensitive to material hardness.

Sintering

Sintering occurs after compaction or shaping. Heating a powder compact induces atomic motion to repositions atoms into pores. The energy reduction happens by forming bonds between particles, thereby eliminating surface area. Consequences of sintering are a densification and strengthening of the molded structure [19]. Due to surface energy, capillary forces pull the particles together. An external pressure applied during sintering accelerates the densification rate. For small powders (measured in the micrometer size range) sufficient capillary force exists that no external pressure is required to reach full density. Additionally, a liquid phase can accelerate sintering, keeping hold times at the peak temperature to 15–30 min.

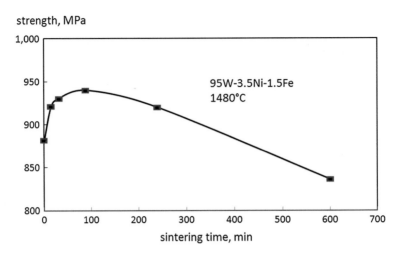

Fig. 7.33 A plot of tensile strength for a 95 W-3.5Ni-1.5Fe composite sintered at 1480 °C (1753 K) for various times. The strength peaks at a short sintering time and declines with longer times. Similar behavior occurs when the sintering temperature is increased over the optimal level, leading to over-sintering

Because liquid phase sintering is a faster process, about 80 % of particulate composites are formed using a liquid phase during the heating cycle.

Sintering processes are thermally activated, meaning input energy is necessary to induce atomic motion. Higher temperatures cause more and faster atomic motion, so temperature control is a critical aspect of sintering. Of course, higher temperatures also induce more microstructure coarsening since densification and coarsening both depend on the same atomic motion. However, high sintering temperatures or long hold times cause grain growth or even swelling, both of which degrade properties. For example, Fig. 7.33 plots the sintered strength for a tungsten-nickel-iron composite, showing peak strength occurs at less than 100 min hold, beyond which strength declines due to grain growth.

One difficulty in sintering mixed powder composites comes from the differing densification conditions for the two phases. Usually one particle is more active at the sintering temperature, meaning the other phase resists densification. The sluggish phase is the higher melting temperature phase. Under conditions where one phase exhibits high sintering shrinkage, the addition of a second phase tends to resist densification, roughly in proportion to the volume concentration. This is seen in Cu-W, where the large melting temperature difference results in retarded sintering as the tungsten content increases. The retardation is because the second phase reduces the number of contacts between sintering particles, resulting in less overall activity. Figure 7.34 depicts how this happens using two particles of different sizes. On the one hand, mixed particles of different sizes have improved packing, but as the sintering shrinkage falls the mixture exhibits degraded shrinkage.

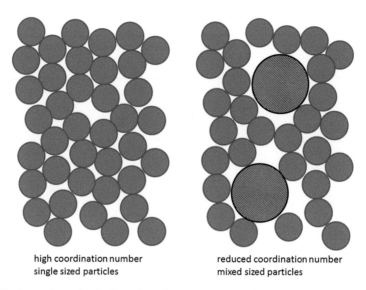

high coordination number   reduced coordination number
single sized particles   mixed sized particles

Fig. 7.34 A two-dimensional schematic on how the presence of large particles reduce the small particle coordination number. Sintering shrinkage depends on bonding at each contact point. If the large particles resist sintering, then shrinkage for the composite declines as the quantity of large particles increases

Final sintered density is a combination of changes in packing and shrinkage, so the peak density depends on the relative gains and losses for each. Figure 7.35 plots data on sintering shrinkage of bimodal mixed large and small particles during constant rate heating to various temperatures. The small particles (0.56 µm alumina) produce 17 % shrinkage on reaching 1600 °C (1873 K) and the large particles (4.5 µm alumina) produces 4 % shrinkage. Sintering shrinkage is degraded as the composition increases in large particle content. The combination of improved packing at intermediate compositions but with degraded shrinkage leads to the density behavior plotted in Fig. 7.36. The sintered density behavior ranges between the two end values. For lower sintering temperatures the peak density occurs at compositions rich in large particles since packing gains are dominant. But for high sintering temperatures, the improved small particle shrinkage delivers the highest density. Such combined effects underscore the difficulty in sintering mixed powder composites.

One approach to improved sintering arises from coated particles. The larger particle is coated with the smaller, more easily sintered phase. Coated particles deliver more uniform sintering, making is easier to densify the composite. For example, ZrO_2 coated ZnO (20 vol%) sintered to 100 % density on heating to 1500 °C (1773 K) but only 88 % using the same composition and particles when simply mixed [48]. Performance is improved because of the homogeneity of the sintered structure.

Sintering is performed in high temperatures furnaces. These might be batch furnaces, where trays of components are stacked in the cold furnace, then the

Fig. 7.35 Sintering shrinkage for mixtures of small (0.56 μm) and large (4.5 μm) alumina particles. The data were collected during constant heating rate 10 °C/min experiments to six temperatures. At each temperature the plots show a decrement in sintering shrinkage as the large particle content increased

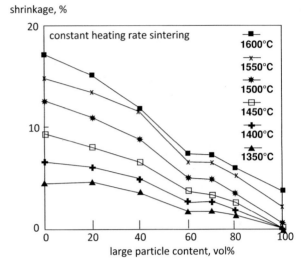

Fig. 7.36 Data for sintered density for the same experiments plotted in Fig. 7.35. Density is plotted versus composition. The bimodal large-small particle mixture leads to a higher green density near 70 vol% large particles, but the large particles degrade sintering. The net effect is a shift in peak density as temperature increases. The highest sintered density comes with small particles alone. Such behavior illustrates the difficulty in sintering particulate composites

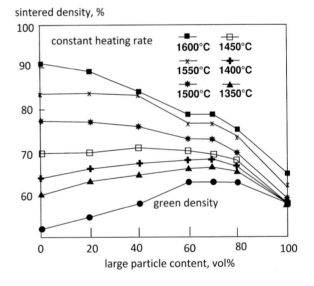

furnace heats in a cycle lasting up to a day. A diagram of a front door loading furnace is diagrammed in Fig. 7.37. A load might consist of 200 kg or more (a few reach 10,000 kg capacity). Depending on the construction materials, peak temperatures for batch furnaces can reach 2500 °C (2773 K).

Alternatively, a continuous furnace is employed for high production rate sintering. A belt, conveyor, or pusher mechanism transports the components through zones of progressively higher temperature, with a long final cooling zone. Atmosphere protection might involve flooding the furnace with hydrogen, nitrogen, or other gas. One design is shown in Fig. 7.38. Upwards of a dozen hot

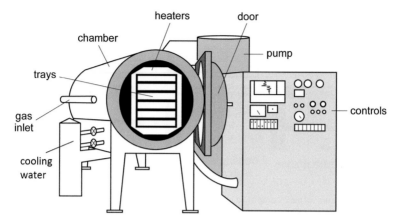

Fig. 7.37 Batch sintering is performed in furnaces using stacked trays holding the components inside the furnace. Once the door is closed the control system cycles through a heating cycle. It is common to operate in vacuum, with loads of 200 kg or more, requiring cycle times of 24 h

zones are employed to properly stage the heating cycle. Reusable trays transport the components through the furnace using automated loading and unloading. Continuous furnaces are designed for peak temperatures of 800, 1120, 1400, 1600, or 1850 °C (1073, 1393, 1673, 1873, or 2123 K), reflecting differences in furnace design, transport mechanism, heating elements, and support trays. Depending on the peak temperature, support trays are fabricated from graphite, ceramics such as alumina, stainless steel, or refractory metals such as molybdenum.

For materials not sensitive to oxidation, sintering is conducted in air. This is the case for oxide ceramics, such as SiO_2, Al_2O_3, MgO, TiO_2, BeO, ZnO, and ZrO_2. Most other materials require protection during sintering, and that is possible in vacuum or artificial atmospheres of inert or reactive gases (nitrogen and hydrogen as examples). Many variants exist on the furnace design and atmosphere. Further, a new trend is to change the sintering atmosphere to induce migration of one phase. This makes it possible to add functional gradients to the structure [49].

Most of the densification in sintering occurs as the peak temperature is achieved. A short hold at the sintering temperature ensures uniform heating of the load, but extended times prove detrimental. Over-sintering refers to the situation where a composition exhibits a loss of properties due to excessive time or temperature. Such behavior is evident in the form of microstructure coarsening and a peak in density and properties.

After reaching the peak temperature, the sintered compact is cooled to room temperature and removed from the furnace. One aspect of sintering is control of the component surface appearance, requiring protection against adverse reactions. For some metals a reducing atmosphere of hydrogen is employed to remove oxides. Hydrogen is potent in this regard and is used at low concentrations in many furnaces. A favorite combination is about 95 % nitrogen with 5 % hydrogen, to avoid explosive danger while providing oxide reduction.

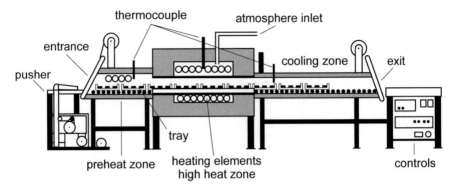

Fig. 7.38 Continuous furnaces employ a sequence of temperature zones, peaking with a high heat zone and ending with a cooling zone. The components are stacked on trays and transported through the zones at a fixed rate. Such a furnace might be 30 m long with capacities from 50 kg/h. Depending on the construction, the peak temperatures might range from 1120 to 2200 °C (1393–2473 K)

Liquid phase sintering is a means to form a composite by either melting one phase or reacting the powders to form a melt, such as eutectic melt. The rate of atomic motion in liquids is hundreds of times faster than in solids, so enhanced sintering significantly reduces the hold time and requires less equipment capacity; sintering costs less using liquid phases. On cooling the liquid solidifies to form a second phase, intertwined with the phase that did not melt. Too much liquid causes component distortion, so liquid phase sintering is usually relegated to composites with high solid contents, over about 85 vol%. This nestles in the ranges where infiltration or casting routes are not effective.

In liquid phase sintering the solid grains are wetted by the liquid at the peak temperature to bond the particles together with a process akin to brazing [35]. Figure 7.39 illustrates the steps, starting with a mixture of powders. During heating the particles bond by solid-state diffusion, but on liquid formation the structure densifies by grain rearrangement induced by the liquid. Solubility of the solid in the liquid induces wetting and densification of the structure. After a short time, maybe 15 min, the grains are reshaped by dissolution into the liquid and reprecipitation as solid to fill pores. As pores are eliminated the grains bond into a solid skeleton with interpenetrating liquid.

During liquid phase sintering the capillary force pulling the structure together is equivalent to a few atmospheres of external pressure. Several events enable densification, including grain rearrangement, solid dissolution, and pore filling. An important event is grain shape accommodation, where the solid grains change shape to better fill space and eliminate porosity. The microstructures in Fig. 7.40 shows grain shape accommodation. The grains are bonded to each other and the solidified former liquid phase penetrates between the grains. Grain shape accommodate ensures a tight fitting together of the grains to eliminate pores. Atomic motion removes corners and small grains preferentially with growth of the large

Fig. 7.39 Liquid phase sintering is widely practiced for sintering particulate composites. The initial state involves mixed particles that heat to form a liquid. That liquid induces rearrangement and repacking of the particles to improve density. Capillary forces combine with diffusion to reshape the solid grains and remove pores. The final densified product is a mixture of solid grains and interpenetrating solidified liquid

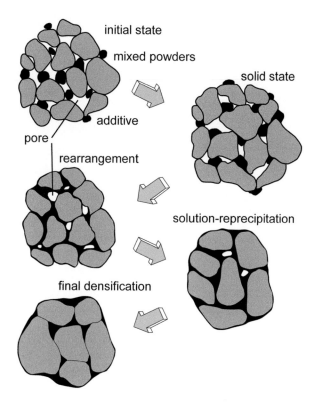

grains. Over time the smaller grains disappear, the larger grains coarsen, while the growing large grains reshape to eliminate porosity [24].

The detailed sintering trajectory depends on many parameters, including factors such as solid solubility in the liquid and peak temperature and hold time. In some instances the liquid even remains glassy after sintering, such as in porcelain. Glass is a viscous liquid at high temperatures.

Sintering improvements are possible using small concentrations of a third phase. An example is W-Cu where 0.5 wt% Co improves sintering. The cobalt promotes tungsten bonding with liquid copper filling the voids. There is no interaction between copper and cobalt as the two metals are insoluble, but cobalt accelerates tungsten sintering.

Novel Fabrication

Novel fabrication methods are gateways to defense agency research funding. Thus, an abundance of ideas emerge to solve leading edge problems, and particulate composites are frequently employed to solve these problems.

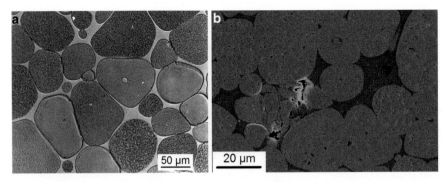

Fig. 7.40 These liquid phase sintered composites shows the characteristic features of grain shape accommodate, solid skeletal bonding, and interpenetrating (but now solidified) liquid. This structure in (**a**) is W-3.5Ni-1.5Fe composition sintered at 1500 °C (1773 K) to give full density with considerable grain growth and (**b**) is Mo-5Ni-2Cu sintered at 1400 °C (1673 K) with a small pore region near the center

One novel approach to forming a composite comes from reacting constituents to generate new phases [50]. Originally termed reactive sintering, the idea is to mix powders and during a reaction produce two or more phases. The reaction is akin to that seen on holidays when children ignite sparklers. Once initiated, the reaction propagates as a combustion wave with the release of heat and light. Mixed powders can react to release substantial heat. For example, a mixture of 68.5 wt% of nickel powder with 31.5 wt% aluminum powder corresponds to the stoichiometric compound NiAl. When mixed in this ratio and ignited at about 600 °C (873 K), the reaction exothermically heats to 1200 °C (1473 K). That heat ignites more of the mixture to propagate without external heating. A composite results when multiple phases are involved. For example, SiC mixed with the Ni and Al results in a NiAl matrix with dispersed SiC reinforcement.

Thousands of variants build on the reactive sintering idea, also known as self-propagating high-temperature synthesis. Handbooks are available detailing the synthesis of borides, carbides, aluminides, silicides, oxides, and nitrides. Pressure applied after the reaction improves densification. However, pressure applied during the reaction causes inhomogeneity by squeezing soft phases out of the component. Reactive sintering has failed to find many successes, largely because control is difficult during the short reaction time. One commercial success is $MoSi_2$-SiC used for high temperature heaters.

A related idea comes from directed metal oxidation, where liquid metal is exposed to oxygen to grow a ceramic phase within the metal. Oxygen diffusion into the liquid metal under controlled conditions, results in oxidation fingers within the melt. For example, aluminum oxide reinforcement inside aluminum as a three-dimensional interconnected composite. An ideal alloy is Al-7 Mg. It oxidizes to give 50 vol% alumina (Al_2O_3) in an Al-Mg matrix giving a density of 3.3 g/cm^3, yield strength of 127 MPa, and ultimate tensile strength of 317 MPa, with an elastic modulus of 159 GPa. The thermal expansion coefficient is 14 ppm/°C and the thermal conductivity is 66 W/(m °C).

Laser processes are attractive for additive manufacturing where limited production seeks to avoid the cost of tooling. A high-power laser, under computer control, directly sinters deposited powder in the x-y plane. Progressive laminate layers create a three-dimensional product [51]. By layering the thin laminates, a three-dimensional solid grows along the z-axis. The laser build delivers about 1 cm^3/h build. Faster builds make for weak bonds between layers and anisotropic properties. A sketch of one additive manufacturing variant is given in Fig. 7.41. Multiple powder feeders can be used to mix powders to form composites. Rapid cooling after heating reduces grain growth versus normal sintering cycles. Prime targets are in custom designed medical or dental devices, tool and die structures, aerospace devices, and shapes that are otherwise difficult to form. Figure 7.42 is an example of the latter where complex arcs are formed in a steel-bronze composite. Molds created using laser sintering enable complex internal (conformal) cooling channels that are difficult to form using machining. However, close tolerances require final finishing by machining, grinding, or polishing. The fabrication of high strength titanium aerospace components is one of the applications reaching widespread production status.

Spray forming is a means to form large, thin-walled shapes, such as tubes [11]. A spray of semisolid droplets is deposited on a substrate as sketched in Fig. 7.43. A second solid powder is co-sprayed to be caught up in the deposit, for example solid ceramic particles are injected into the melt spray. The semisolid droplets trap the ceramic particles to form the composite. When properly conducted, the product is dense with a small grain size.

Plasma spraying provides a related concept involving particle injection into a hot plasma torch. During a brief transient in the torch, the particles heat and accelerate to splash against a substrate. When properly controlled the deposited particles are in a semisolid sate, allowing bonding into a nearly dense deposit. As drawn in the cross-section in Fig. 7.44, the plasma arc is distorted by the high velocity of the injected gas. Mixed particles are injected into that plasma arc, which can reach temperatures of 5000 °C (5273 K). Since the particles are in the arc a short time, they must be sufficiently small to melt without evaporating. Usually particles between 40 and 80 μm work best. A composite such as imaged in Fig. 7.45 results from plasma spraying mixed particles. This image is a section though the spray deposit of mixed stainless steel and alumina.

In spray processes, the desire is for the particle to arrive at the substrate in a semisolid condition so they can spread and bond to one another. Splatter occurs if the particles are too hot. On the other hand the particles bounce away if they are too cold. Spraying in vacuum avoids trapped gas pockets. Spray technologies are best applied to surface coatings in jet engines, power turbines, and wear surfaces.

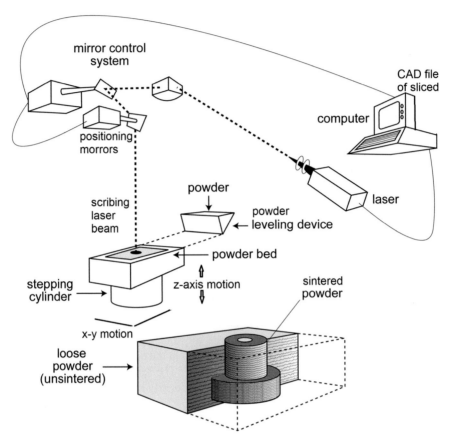

Fig. 7.41 One means to build limited production components is with additive or freeform fabrication. Illustrated here is a computer controlled laser scanner coupled with layered particles. Laser scribing on the top surface the particles induces sintering. By coordination of the repeated layering and the computer image, three-dimensional objects are created by layer-wise sinter builds

Fig. 7.42 This complex component is a demonstration of the interlaced features possible with additive manufacturing, consisting of composite consisting of steel infiltrated with bronze [courtesy A. Hancox]

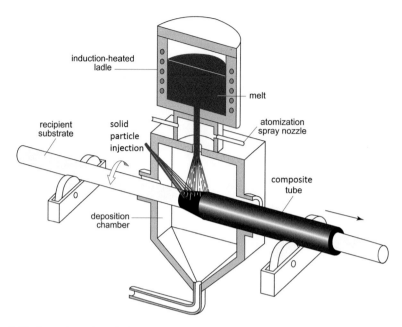

induction-heated
ladle

melt

recipient
substrate

solid
particle
injection

atomization
spray nozzle

composite
tube

deposition
chamber

Fig. 7.43 Spray formed particulate composites involve atomized droplets into which are injected hard particles. The semisolid droplet builds a simple shape by depositing on a sacrificial solid substrate

Manufacturing Defects

Every manufacturing operation has potential defects. For particulate composites, the three most common are:

- pores or voids from incomplete densification, gas reactions, volatile phases, or volume changes (phase changes) during processing
- weak interfaces due to poor wetting, insolubility, or impurity segregation during slow cooling
- cracks associated with improperly aligned tooling, improper handling, or thermal stresses.

Some of the difficulties arise from polymeric binders using in processing, especially during binder-assisted shaping in slip casting, slurry molding, injection molding, or extrusion. On occasion lumps of polymer remain during shaping to then evaporate to form a large void. Normally the binder evaporates during heating, and if not carefully coaxed out of the body the vapor causes bubbles, cracks, or voids.

Other defects are similar to those encountered in related forming technologies and are not unique to composites. A good example are weld lines formed where flow occurs around pins but improperly bonds on the downstream side. Sink marks form near thick sections as indents that arise due to unpressurized feedstock shrinkage during cooling. Powder separation from the binder occurs at high shear

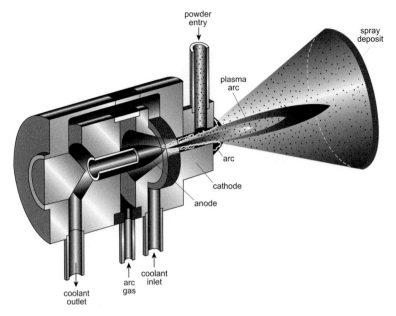

Fig. 7.44 Plasma spraying is a means to build thin wall structures. Particles are injected into a plasma arc where they melt and spray onto a cold substrate. The deposit reflects the ratio of ingredients feed into the spray torch

Fig. 7.45 The composite structure formed by plasma spraying, giving particle layers formed by sequential deposits of semisolid droplets [courtesy of K. Shaw]

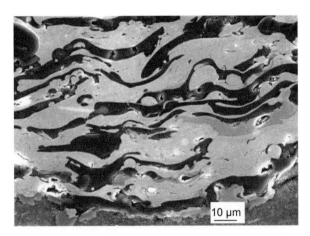

rates during shaping when there are large density differences between ingredients, especially if the feedstock viscosity is too low. These defects occur independent of the powder chemistry.

To avoid common errors, it is best to anticipate the fabrication process during component design. Some of the typical goals are to design round corners to help avoid stress concentrations. Components with thick and thin sections see stresses during processing that cause thin sections to distort. Flat surfaces are beneficial for support during processing. A key justification for process simulations is to identify problems early, prior to final tooling design.

Outline for Fabrication Process Selection

Besides technical capability associated with determining which process can deliver the component, process selection depends on cost and quality criteria. A historical background of prior successes helps guide the process selection. Effectively, the process, material, and component come together to guide fabrication final decisions [31]. Further considerations include the component shape, complexity, and production quantity. A high tooling expense presents a significant hurdle for some production routes, but at high production quantities the tooling is amortized and is less an issue. Consultation with experts on each forming option helps prevent errors.

Listed below are some of the common fabrication routes and their key attributes:

- binder-assisted extrusion shaping followed by sintering, gives long constant cross-section geometries, extruded product is cut to length, intermediate tooling cost, best for rods, tubes, spirals, and similar shapes
- binder-assisted injection molding followed by sintering, gives high complexity, high production quantity in small components, good tolerance capabilities, tooling cost depends on complexity but is generally high, equipment cost is variable, production rates from 2 per minute at 0.1 kg to 1 per second in multiple cavity tooling, widely available process
- casting of solid–liquid paste into mold, liquid solidifies to form second phase, suitable to 40–60 vol% liquid, separation of solid and liquid common problem due to density differences
- cold isostatic pressing followed by sintering, generally low precision, low production quantity technology (except for dry bag variant), low tooling cost, batch process, high aspect ratio shapes, nominally applied to components in the 1 kg range
- die compaction followed by sintering, highly automated, high production rates, not useful for low production quantities, squat components in mass range from 1 g to 0.5 kg, complexity possible on pressing face, ejection from tooling mandates no undercuts or holes perpendicular to the pressing face, tooling cost depends on component design and can be high for multiple featured component, widely available
- dynamic compaction and explosive compaction, high strain rate self-heating compaction, rapid consolidation based shock wave impact on powder, difficult to control, cracking is common, simple shapes, generally applied to flat compacts
- granular forging with high strain rate pressurization of hot granules and embedded preform, shape determined by perform, distortion typical, expensive, limited

availability, no tooling required for forging but component is formed first by other means, requires post-forging surface machining

- hot extrusion of heated powder forced through a die under high pressure, extruded product is constant in cross-section, equipment and process are expensive, applied to long shapes, high tool cost, excellent densification, simple shapes, limited availability
- hot forging using high strain rate hot compaction, shapes are reasonably complex, applied to high volume production using 15 second cycle time, initial tooling and equipment are expensive, tool wear and short tool life are limitations, component requires machining after densification
- hot isostatic pressing using simultaneous pressure and temperature inside gas pressure vessel, containerless variant applied to small components in the 10 g range, sealed container variant applied to components 1 kg or larger using custom designed single use container, cycle times measured in hours, can be applied to single component, expensive equipment
- hot pressing using simultaneous uniaxial pressing and heating, slow process, simple shapes, surface contamination from tooling mandates post-consolidation machining, cycle times of 15 min to several hours, maximum size limited by press capacity, equipment is expensive, can be used for single component
- infiltration of liquid into pores of a solid, liquid solidifies to form second phase, suitable for 40–15 vol% liquid, pressure required if liquid does not wet solid, process can be automated, no significant tooling costs
- plasma or thermal spraying of mixed particles using hot arc to form semisolid droplets, droplets are sprayed onto substrate, limited to surface layers, 1–3 mm thick, equipment cost is intermediate, deposit rates are low, used for wear and thermal barrier coatings
- reactive synthesis involves a variety of options to give simultaneous heating and pressurization, mixed ingredients undergo exothermic reaction, applied to compounds and intermetallics, difficult to control, limited history of success
- slip casting of a powder-binder-solvent slurry is followed by sintering, low cost porous tooling, shapes are complex and thin walled, low equipment cost, slow process, commonly applied to components up to 1 kg or more
- spark sintering using hot pressing with temperature generated by electrical discharge through the tooling or powder, very limited shape and size based on available power, component are circular to avoid density gradients, equipment cost is high, fast heating, production rates range from 1 to 4 per hour
- spray forming is similar to thermal spray where droplets and particles are co-deposited on substrate, low precision, rough surface, may require sacrificial substrate, high equipment cost, slow, not widely available, generally limited to small production quantities
- tape casting of a powder-binder-solvent mixture followed by sintering, used for thin and flat structures, fluidity of feedstock requires evaporative solvent, widely used for electronic substrates and battery electrodes.

Via experience it is possible to quickly identify the merits and cautions for each approach and to rationalize those attributes with the desired component. Classically, fabrication decisions are biased by what worked in the past. Prior success should always be the starting point in exploring options for component production.

Study Questions

7.1. Practical difficulties are encountered in ultrahigh pressure compaction, such as at 2 GPa. Name at least one difficulty with respect to each of the following: tooling, component geometry, and impurities.

7.2. A laboratory press is rated at 60 tons (120,000 pounds) load. If compaction at 3 GPa is to be performed on this press, what is the maximum circular diameter possible?

7.3. A component starts with a fractional density of 0.55 and undergoes uniform 16 % shrinkage in each dimension. What is the final fractional density?

7.4. Laser sintering is a means for rapid heating. A narrow laser beam enables precise structures, but the production rate is just 10 cm^3 per hour. What might be an application for such a technology?

7.5. Pencil lead is a composite of graphite and clay, extruded with water as a binder. If the final diameter for the extruded rod is 0.5 mm, and the extrusion reduction ratio is 20, what is the diameter of the feedstock being fed into the extruder?

7.6. At the temperature selected to consolidate a composite the yield strength (full density) is estimated at 50 MPa. If the composite is at a 90 % density, estimate the required applied pressure needed to reach full density?

7.7. A full density composite is desired from a mixture of 50 vol% Cu and Co. What is the expected density in g/cm^3? Could this be processed by casting?

7.8. A composite with a starting fractional density of 45 % is to be consolidated to full density. What linear but isotropic shrinkage would produce 100 % final density?

7.9. The rate of sintering shrinkage is measured at 915 °C (1188 K) and found to be 0.1 %/min. The activation energy is measured at 250 kJ/mol. What higher temperature would be required to double the sintering shrinkage rate?

7.10. The activation energy for grain boundary diffusion for copper and gold are nearly the same—107 and 110 kJ/mol, respectively. Silver is slightly lower at 90 kJ/mol. Is the lower value for silver sensible? Why or why not?

7.11. The yield strength of bronze-glass composite at 800 °C (1073 K) is measured at 15 MPa. If the powder is filled into a press at 60 % apparent density and pressed to 8 MPa pressure at 800 °C (1073 K) to induce plastic flow, what is the expected fractional density?

7.12. A powder mixture with an apparent fractional density of 0.64 is measured for compaction behavior, giving the following results in terms of pressure and fractional density: 100 MPa—0.72, 200 MPa—0.79, 300 MPa—0.83, 400 MPa—0.87, and 600 MPa—0.92. Fit this compaction behavior to Eq. (7.1). What is the best estimate for the two adjustable parameters?

7.13. For the powder mixture described in Study Question 7.12, calculate the pressure required to reach a compaction density of 0.98. What practical difficulties arise in attaining this pressure?

7.14. A composite of copper reinforced with 20 vol% silicon carbide is formed by hot pressing. The starting powder packs to 4 g/cm^3. What axial shrinkage is expected if the pressing is to reach full density?

7.15. A composite of Ti-20TiB is pressed to 55 % green density and 25 mm diameter and then sintered to 84 % density. Assuming sintering shrinkage is uniform in each direction, what is the sintered diameter?

7.16. One new material for artists consists of powder, binder, and solvent mixed to form a paste. Artists freeform construct items using a syringe to extrude the paste. Since the syringe is plastic and forming is at room temperature, the mixture has low viscosity. The solvent evaporates to cause the structure to harden and it is then sintered. What might be an appropriate volume ratio of powder, binder, and solvent?

References

1. D.B. Miracle, S.L. Donaldson (eds.), *Composites*. ASM Handbook, vol. 10 (ASM International, Materials Park, 2001)
2. S. Abkowitz, S.M. Abkowitz, H. Fisher, P.J. Schwartz, CermeTi discontinuously reinforced Ti-matrix composites: manufacturing, properties, and applications. J. Met. **56**(5), 37–41 (2004)
3. M.F. Ashby, *Background Reading HIP 6.0* (Engineering Department, Cambridge University, Cambridge, 1990)
4. A. Bose, W.B. Eisen, *Hot Consolidation* (Metal Powder Industries Federation, Princeton, 2003)
5. R. Orru, R. Licheri, A.M. Locci, A. Cincotti, G. Cao, Consolidation/synthesis of materials by electric current activated/assisted sintering. Mater. Sci. Eng. **R63**, 127–287 (2009)
6. M. Pellizzari, A. Fedrizzi, M. Zadra, Spark plasma cosintering of hot work and high speed steel powder for fabrication of a novel tool steel with composite microstructure. Powder Technol. **214**, 292–299 (2011)
7. R.M.K. Young, T.W. Clyne, A powder mixing and preheating route to slurry production for semisolid diecasting. Powder Metall. **29**, 195–199 (1986)
8. J. Tian, K. Shobu, Fabrication of silicon carbide - mullite composite by melt infiltration. J. Am. Ceram. Soc. **86**, 39–42 (2003)
9. X. Liu, Y. Li, F. Lou, M. Li, Al/SiC composites with high reinforcement content prepared by PIM/pressure infiltration. Powder Injection Moulding Int. **1**(4), 53–55 (2007)
10. Z.Y. Liu, D. Kent, G.B. Schaffer, Powder injection molding of an Al-AlN metal matrix composite. Mater. Sci. Eng. **A513**, 352–356 (2009)
11. R.M. German, *Powder Metallurgy and Particulate Materials Processing* (Metal Powder Industries Federation, Princeton, 2005)
12. M.H. Bocanegra-Bernal, Review: Hot Isostatic Pressing (HIP) technology and its applications to metals and ceramics. J. Mater. Sci. **39**, 6399–6420 (2004)

13. H. Ye, X.Y. Liu, H. Hong, Fabrication of metal matrix composites by metal injection molding - a review. J. Mater. Process. Technol. **200**, 12–24 (2008)
14. F.F. Lange, L. Atteraas, F. Zok, J.R. Porter, Deformation consolidation of metal powders containing steel inclusions. Acta Metall. Mater. **39**, 209–219 (1991)
15. R.J. Henderson, H.W. Chandler, A.R. Akisanya, C.M. Chandler, S.A. Nixon, Micromechanical model of powder compaction. J. Mech. Phys. Solids **49**, 739–759 (2001)
16. D.N. Smith, Processing modelling in powder metallurgy and particulate materials. Powder Metall. **45**, 294–296 (2002)
17. A. Arockiasamy, S.J. Park, R.M. German, Viscoelastic behaviour of porous sintered steels compact. Powder Metall. **53**, 107–111 (2010)
18. Z.J. Lin, J.Z. Zhang, B.S. Li, L.P. Wang, H.K. Mao, R.J. Hemley, Y. Zhao, Superhard diamond/tungsten carbide nanocomposites. Appl. Phys. Lett. **98**, 121914 (2011)
19. R.M. German, *Sintering: From Empirical Observations to Scientific Principles* (Elsevier, Oxford, 2014)
20. K. Lu, *Nanoparticulate Materials Synthesis, Characterization, and Processing* (Wiley, Hoboken, 2013)
21. H. Tanaka, H. Nakano, Y. Suyama, Grain shrinkage driven by surface and grain boundary energy in $Ba_5Nb_4O_{15}$ powder. Acta Mater. **55**, 2423–2432 (2007)
22. S.J. Park, S.H. Chung, J.M. Martin, J.L. Johnson, R.M. German, Master sintering curve for densification derived from a constitutive equation with consideration of grain growth: application to tungsten heavy alloys. Metall. Mater. Trans. **39A**, 2941–2948 (2008)
23. L. Olmos, C.L. Martin, D. Bouvard, Sintering of mixtures of powders: experiments and modelling. Powder Technol. **190**, 134–140 (2009)
24. R.M. German, Coarsening in sintering: grain shape distribution, grain size distribution, and grain growth kinetics in solid pore systems. Crit. Rev. Solid State Mater. Sci. **35**, 263–305 (2010)
25. H.K. Kang, S.B. Kan, Behavior of porosity and copper oxidation in W/Cu composite produced by plasma spray. J. Therm. Spray Technol. **13**, 223–228 (2003)
26. F.J. Humpherys, W.S. Miller, M.R. Djazeb, Microstructural development during thermomechanical processing of particulate metal-matrix composites. Mater. Sci. Technol. **6**, 1157–1166 (1990)
27. D.S. Wilkinson, M.F. Ashby, Pressure sintering by power law creep. Acta Metall. **23**, 1277–1285 (1975)
28. S.H. Chung, Y.S. Kwon, S.J. Park, R.M. German, Modeling and simulation of press and sinter powder metallurgy, in *Metals Process Simulation*, ed. by D.U. Furrer, S.L. Semiatin. ASM Handbook, vol. 33B (ASM International, Materials Park, 2010), pp. 323–334
29. D.C. Blaine, S.J. Park, R.M. German, Linearization of master sintering curve. J. Am. Ceram. Soc. **92**, 1400–1409 (2009)
30. S.J. Park, P. Suri, E. Olevsky, R.M. German, Master sintering curve formulated from constitutive models. J. Am. Ceram. Soc. **92**, 1410–1413 (2009)
31. G. Boothroyd, P. Dewhurst, W.A. Knight, *Product Design for Manufacturing and Assembly*, 3rd edn. (CRC Press, Boca Raton, 2010)
32. Y. Goto, A. Tsuge, Mechanical properties of unidirectionally oriented SiC-whisker-reinforced Si_3N_4 fabricated by extrusion and hot pressing. J. Am. Ceram. Soc. **76**, 1420–1424 (1993)
33. S.R. Martins, W.Z. Misiolek, Consolidation of particulate materials in extrusion. Rev. Particulate Mater. **4**, 43–70 (1966)
34. S. Turenne, N. Legros, S. Laplante, F. Ajersch, Mechanical behavior of aluminum matrix composites during extrusion in the semisolid state. Metall. Mater. Trans. **30A**, 1137–1146 (1999)
35. R.M. German, P. Suri, S.J. Park, Review: liquid phase sintering. J. Mater. Sci. **44**, 1–39 (2009)
36. K.H. Kate, R.K. Enneti, S.J. Park, R.M. German, S.V. Atre, Predicting powder-polymer mixture properties for PIM design. Crit. Rev. Solid State Mater. Sci. **39**, 197–214 (2014)

37. F. Hussain, M. Hojjati, M. Okamoto, R.E. Gorga, Polymer-matrix nanocomposites, processing, manufacturing, and application: an overview. J. Compos. Mater. **40**, 1511–1575 (2006)
38. F. Ahmad, R. M. German, Evaluation of Metal Composite Mixes for Powder Injection Molding, In *Advances in Powder Metallurgy and Particulate Materials - 2006* (Metal Powder Industries Federation, Princeton, 2006), pp. 9.168–9.180
39. F.J. Semel, K.S. Narasimhan, Steel based infiltration to achieve full density, high performance PM parts. Powder Metall. **52**, 94–100 (2009)
40. M.N. Rahaman, L.C. De Jonghe, Sintering of ceramic particulate composites: effect of matrix density. J. Am. Ceram. Soc. **74**, 433–436 (1991)
41. K. Kondoh, Titanium metal matrix composites by powder metallurgy (PM) routes, in *Titanium Powder Metallurgy*, ed. by M.A. Qian, F.H. Froes (Elsevier, Oxford, 2015), pp. 277–297
42. K. Peng, M. Yi, L. Ran, Y. Ge, Reactive hot pressing of SiC/MoSi$_2$ nanocomposites. J. Am. Ceram. Soc. **90**, 3708–3711 (2007)
43. S. Sugiyama, Y. Kodaira, H. Taimatsu, Synthesis of WC-W$_2$C composite ceramics by reactive resistance heating hot pressing and their mechanical properties. J. Jpn. Soc. Powder Powder Metall. **54**, 281–286 (2007)
44. O. Guillon, J. Langer, Master sintering curve applied to the field-assisted sintering technique. J. Mater. Sci. **45**, 5191–5195 (2010)
45. W. Xu, X. Wu, X. Wei, E.W. Liu, K. Xia, Nanostructured multiphase titanium-based particulate composites consolidated by severe plastic deformation. Inte. J. Powder Metall. **50**(1), 49–56 (2014)
46. Q. Zhang, B.L. Xiao, W.G. Wang, Z.Y. Ma, Reactive mechanism and mechanical properties of in situ composites fabricated from an Al-TiO$_2$ system by friction stir processing. Acta Mater. **60**, 7090–7103 (2012)
47. M.S. Kumar, P. Chandrasekar, P. Chandramohan, M. Mohanraj, Characterisation of titanium - titanium boride composites process by powder metallurgy techniques. Mater. Charact. **73**, 43–51 (2012)
48. C.L. Hu, M.N. Rahaman, Factors controlling the sintering of ceramic particulate composites: II, coated inclusion particles. J. Am. Ceram. Soc. **75**, 2066–2070 (1992)
49. K. S. Hwang, P. Fan, H. Wang, J. Guo, X. Wang, Z. Z. Fang, Fabrication of functionally graded WC-Co using a novel carburizing process, in *Proceedings International Conference on Refractory Metals and Hard Materials*, 18th Plansee Seminar, Reutte, Austria, 2013, paper HM1, pp. 1–7
50. K. Morsi, The diversity of combustion synthesis processing: a review. J. Mater. Sci. **47**, 68–92 (2012)
51. H. Attar, M. Bonisch, M. Calin, L.C. Zhang, K. Zhuravleva, A. Funk, S. Schdino, C. Yang, J. Eckert, Comparative study of microstructures and mechanical properties of in situ Ti-TiB composites produced by selective laser melting, powder metallurgy, and casting technologies. J. Mater. Res. **29**, 1941–1950 (2014)

Chapter 8
Microstructures and Interfaces

Microstructure gives details on the spatial relation of the phases. Small changes in microstructure might have profound influences on properties. Likewise, the interfaces between phases determine the degree of interaction to further impact properties. This chapter introduces the means to quantify microstructure and interface quality, providing examples of possible property variations.

Microstructure Quantification

Microstructure is the arrangement of phases observed using microscopy, usually the characteristic size scale in the millimeter to nanometer range. An important aspect is isolated how microstructure links to property changes in engineering materials. In turn, microstructure analysis is a frequent aspect in analyzing a particulate composite.

Microstructure is often captured in pictures, letting a picture replace a thousand words. Beyond qualitative aspects such as "small" microstructure quantification provides quantitative measurement parameters, often an average value or maximum value for pore size, grain size, grain separation, surface area, or curvature [1]. Measurements occur on two-dimensional images, assuming the cross-section is representative of the three-dimensional structure. The cross-section is polished to remove distortion from the sectioning process, then imaged. Contrast between phases is desirable, and might be achieved using polarized light or electron contrast, stains, or etchants. Sample preparation for microscopy is assisted by handbooks that detail polishing and etching techniques. Image are generated using light or electron microscopes and digital images are easily analyzed and quantified using laptop computers. Automated image analysis software options includes *ImageJ*, a free program supported by the US National Institutes of Mental Health [2]. Measures range from average parameters to statistical distributions; for example average grain size or grain size distribution.

© Springer International Publishing Switzerland 2016
R.M. German, *Particulate Composites*, DOI 10.1007/978-3-319-29917-4_8

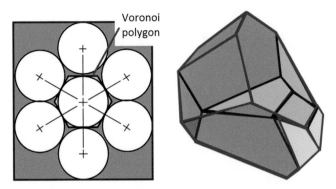

Fig. 8.1 A polygon results from a Voronoi construction to represent grain shape: (**a**) illustrates the idealized two-dimensional structure of circular grains surrounding a central grain. The perpendicular bisectors between grain centers produce a hexagon in this case. An actual three-dimensional polyhedron is more complicated as shown in (**b**) [4]

One construction applied to the microstructure is the Voronoi representation. This is created by drawing lines between the centers of contacting grains. Perpendicular planes are erected on these lines at the points of contact. The result is a polygon representation of the grains with flat faces perpendicular to the contacts; grain-grain bonds are bisected as illustrated in Fig. 8.1a. Here an idealized six grain array is in contact with a central grain. The connectors between centers are shown to illustrate how the hexagon results from the construction. Grain contacts relate to phase connectivity and percolation [3]. A more realistic Voronoi grain is shown in Fig. 8.1b, based on three-dimensional (computer tomography) synchrotron imaging. Unlike the idealized two-dimensional version, the three-dimensional grain is a more complicated, involving different sizes and shapes to the faces, ranging from 3 to 6 edges, corresponding to triangles and hexagons [4]. A Voronoi polyhedron is space-filling and helps is modeling microstructures.

Phase Content

The amount of each phase is a common feature captured by microscopy. Two phases are evident in the micrograph of Fig. 8.2. Overlaid on this micrograph is a 10×10 grid of points. Simple image analysis sorts the points or pixels into bins of dark or light. In this case the black phase is about 10 % of the total. A calculation of the phase content assumes the section plane is representative of the three-dimensional structure. Many pixels and many fields of view enable high accuracy to the assessment. Indeed, computer quantitative microscopy at the pixel level reports the dark phase constitutes 13.5 vol%. By coordinating the measurement with position, it is possible to survey for composition gradients versus location.

Fig. 8.2 A two phase
microstructure with
overlaid square 10 by
10 point array. The volume
fraction of a phase is
calculated by counting the
fraction of points or pixels
that intersect the second
phase

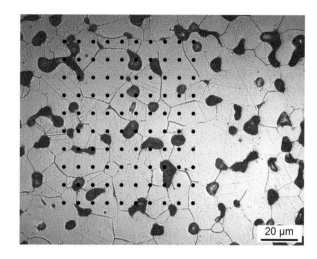

Bulk chemical analysis will not detect composition heterogeneity, where the ratio of phases varies by position. However, chemical testing applied to regions as small as the grain size capture the point-to-point variation. At intermediate size scales, microstructure tests for homogeneity H are possible using a dimensionless parameter (scaled to 100 %), determined from repeated tests spot to spot. Actually the test is possible using chemistry, microhardness, or phase contrast, giving,

$$H = 100 \, \frac{S}{H_M} \tag{8.1}$$

where S is the standard deviations over several repeated tests at different locations, and H_M is the average value. For example, the microhardness test is a means to assess homogeneity. It is performed using indentations at different regions and from the uniformity of the readings is derived a sense of microstructure homogeneity. An alternative uses scanning electron microscopy with energy dispersive X-ray analysis to measure local chemistry in regions as small as 1 μm³. The scale of scrutiny is inherently a part of homogeneity. When testing is performed on many small spots the granularity of the microstructure is evident and the apparent homogeneity decreases. More variation occurs at higher magnifications. Accordingly, tracking the phase variation, say with the standard deviation, versus the magnification provides a measure of the homogeneity versus the level of scrutiny. Coarse areas of analysis appear homogeneous, so the challenge is to assess homogeneity at a size scale reflective of performance. A guideline is the field of scrutiny might be ten times the average grain size.

Grain Size

Grain size indicates the crystal dimensions. Usually it is based on two-dimensional random sections that are used to measure the grain width. Polishing and etching brings out the grain boundaries to more easily delineate individual grains. Since the cross-section is random and not necessarily slicing through the grain center, most grains appear smaller than their true size.

A simple grain size measure counts the number of grain boundaries intercepted by a test line or test circle placed on the magnified image. For full density structures, the two-dimensional grain intercept size G_{2D} is the total test line length L_L divided by the magnification M and number of boundary or interface intercepts N_B,

$$G_{2D} = \frac{L_L}{M\,N_B} \tag{8.2}$$

The three-dimensional grain size G_{3D} is larger than observed using two-dimensional microscopy, namely the usual estimate is $G_{3D} = 1.5\,G_{2D}$ assuming random sections in typical polygonal grains. This conversion from 2D to 3D grain size is invalid for anisotropic microstructures involving whiskers or highly deformed materials.

An alternative approach is to measure the area of each grain in random cross-section. Assuming that grain area is equivalent to a circle leads to calculation of the equivalent grain diameter based on projected area. For solid grains with noncircular shapes, the projected area A and perimeter P are measured on a random cross-section. The calculated diameter of an equivalent circle G_E is,

$$G_E = \frac{4\,A}{P} \tag{8.3}$$

The two-dimensional value from Eq. (8.3) is still not the true grain size. The two-dimensional equivalent grain size must be transformed to a three-dimensional size. To do this the grain is assumed to be sphere and again the 3D version is estimated at 1.5 times the 2D version.

Beyond the average grain size, the grain size distribution is determined by tallying the individual grain sizes and normalizing to the total number of measured grains. The cumulative runs from the size of the smallest to largest, giving 0–100 % of the grains. The median grain size G_{50} and other grain size distribution parameters are calculated from the measurements. This results in the cumulative grain size distribution, such as plotted in Fig. 8.3 [5]. In this plot, the cumulative fraction of grains is expressed as a function of the grain size. The slightly skewed distribution shape agrees with an exponential Weibull distribution. In this form, the cumulative fraction of grains with a size G, pivoting around the median grain size G_{50}, is given as $F(G)$ as follows:

Fig. 8.3 Cumulative grain size distribution based on intercept size [5]. The plot gives the experimental data as symbols and the agreement with a cumulative grain size distribution Weibull function given by Eq. (8.4)

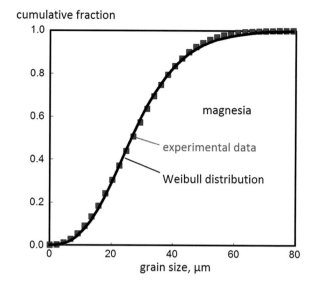

$$F(G) = 1 - exp\left[\ln\left(\frac{1}{2}\right)\left(\frac{G}{G_{50}}\right)^{M}\right] \qquad (8.4)$$

As shown, the cumulative fraction runs from 0 to 1. Typically the exponent M is between 2 and 3; for the data in Fig. 8.3 the solid line represents a best fit to the data using $M = 2.3$ and $G_{50} = 27$ μm. The convergence to a Weibull distribution arises from coarsening, where the small grains coalesce into the large grains over time, reaching this steady-state distribution. The large grains have a slightly lower energy per volume (surface area per volume), so they grow at the expense of the small grains with a slightly higher energy per volume. Although coarsening continues over time, the Weibull distribution best captures the steady state that occurs as some grains are shrinking while other grains growing. The median grain volume increases linearly with time while the number of grains declines; the larger grains grow at the expense of the smaller grains.

Porosity and Pore Size

Like the fraction of phases, porosity is assessed using two-dimensional cross-sections. Porosity is a gross measure of the pore structure as a fraction of the total component volume. For simple geometries, porosity is measured indirectly by determining the mass and dimensions of the component. Density is calculated as mass divided by volume. When the measured density is compared to the theoretical density, any shortfall is attributed to porosity. This is an indirect measure since the pores are never directly identified or measured. On the other hand, microscopy provides a direct measure. If pores are properly imaged, then quantitative image analysis is a direct measure of pores. Microscopy adds knowledge on pore location, pore size, and pore shape.

For microscopy, the pore structure must be preserved during specimen slicing, grinding, polishing, and etching. Distortion by excess polishing smears neighboring material into pores, giving the appearance of low porosity. On the other hand, etching removes mass and enlarges pores, leading to an overestimate of the porosity. The appropriate preparation procedure is to section the material, lightly grind the surface to remove distorted and smeared material, polish, etch, and then fill epoxy into the surface pores. Once cured the epoxy preserves the pores during repolishing, providing an accurate view without distortion.

Point counting or image analysis provides a direct measure of porosity based on the fraction of points or pixels falling on pores. A video image is composed of thousands of pixels, so a digitized image is an effective platform for measurement. This is the same procedure used to identify the phase content based on counting the fraction of points falling on pores. Automated microscopes map large areas quickly, allowing additional assessment of porosity gradients. As just noted, the measured density, such as the Archimedes density, as compared to the theoretical density is a means to estimate porosity. Good agreement between the indirect and microstructure measured porosity is a verification that sample preparation is unbiased.

Pore size and pore shape are measured from two-dimensional cross-sections using microscopy. The aspect ratio is a simple shape index, where the height and width are used to determine if the pore is spherical (aspect ratio close to 1) or elongated. Closed pores are generally spherical. Open pores are akin to caverns and in cross-section appear to have an elongated, wormhole character. Usually if porosity is below about 5 vol% all of the pores are closed with an aspect ratio of 1.0. In optical microscopy, these spherical, closed pores are reflective in the center. On the other hand, with more than 15 vol% porosity the pores are mostly open and not reflective.

Porosity has no units so it is expressed as a percentage or fraction of the total volume. Pore size is distributed, usually following a functional form similar to that of the grain size distribution. This reflects the fact that pores and grains tend to attach to each another. Other tests are possible for measuring pore size distributions, such as gas absorption, mercury porosimetry, gas permeability, and gas-fluid displacement in what is known as the bubble point test. In the latter test, the pores are filled by holding the structure in alcohol or other wetting fluid. While submerged, gas is applied to one side of the porous structure with gas pressure increasing until bubbles emerge from the opposite side. Large pores allow bubble formation at low pressures and small pores require higher gas pressures prior to bubbling. The test is not useful for low porosity structures.

Connectivity

The spatial relation between phases has considerable impact on properties. Consider two composites of the same phases, say 50 vol% each. As illustrated by the two-dimensional sketch in Fig. 8.4, one option is to mix the phases to form a random structure. Accordingly, the grains have a natural variation in the number of contacts with similar composition grains, up to 3 in this two-dimensional section. The alternative, also with 50 vol% of each phase, sketched in Fig. 8.5 is based on surrounding one phase to prevent connections. In this case the central phase has no connectivity. Such isolated phase structures are fabricated using coated particles, leading to a variant imaged in Fig. 8.6. In this case the isolation of the central grain

Fig. 8.4 Two-dimensional grain section of a 50 vol% composite where grain neighbors are determined by random events, leading to a variety of similar composition neighbors for both phases ranging from zero to three in this drawing

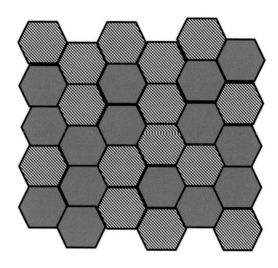

Fig. 8.5 A schematic of a 50 vol% composite created by coated grains to ensure no contacts between the central cores

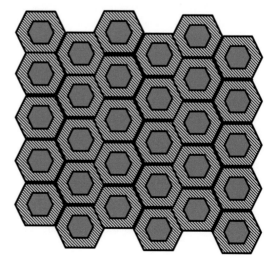

Fig. 8.6 A high contrast
scanning electron
micrograph of a tough
coated hard particle
composite, formed by
coating the hard particles
prior to consolidation
thereby avoiding weak
fracture paths in the
composite

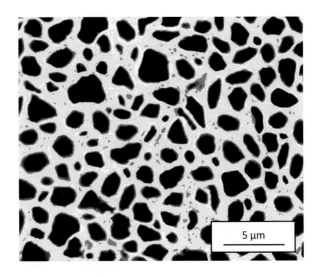

Fig. 8.6 A high contrast scanning electron micrograph of a tough coated hard particle composite, formed by coating the hard particles prior to consolidation thereby avoiding weak fracture paths in the composite

contributes to high hardness and high fracture toughness. The core is a hard, brittle phase that is isolated by a tough surrounding phase. Thus, in spite of similar bulk compositions, the property difference is large. In this case, much wear resistance arises from the central hard grain while toughness is maximized by avoiding easy fracture paths in the brittle phase. Isolated grain structures are invoked in other high performance composites, including soft magnetic components, electrical contacts, wear structures, and frangible bodies [6].

Phase connectivity is a three-dimensional property that links to the grain coordination number. The three-dimensional coordination number is difficult to extract because of opacity. Serial sections spaced close together are one means to capture the connectivity as is microscopic computer tomography. Most commonly, the connections are quantified using two-dimensional measures. Two-dimensional connectivity is the average number of grain contacts visible in cross-section for each grain. It is a distributed parameter, where smaller grains have fewer connections and larger grains have more connections. The calculation starts by selecting a grain, then tracing the perimeter counting the number of similar phases in contact. Contiguity is similar, only it is based on the fraction of grain perimeter that is in contact with similar phases out of the total grain perimeter.

The composite shown in Fig. 8.7 illustrates connectivity analysis. Select a grain and trace around its perimeter, counting the number of bonds with similar grains. In this microstructure the average is 2.0 contacts per grain with a standard deviation of 1.3. Grains on the edge of the viewing field are ignored since some of the contacts are out of view. For this microstructure the distribution is shown in Fig. 8.8, giving an average connectivity per grain C_g of 2.0. If repeated for other views, then a mean connectivity is derived. The higher the volume fraction of a phase the higher the connectivity, reaching 6 for a polycrystalline solid.

Connectivity relates to the underlying three-dimensional grain coordination number N_C,

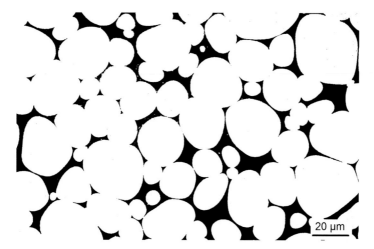

Fig. 8.7 Digital binary micrograph (*black* and *white*) for a composite. The connectivity is determined by traversing along the perimeter of a grain and counting the number of phase changes in completing a circuit. This micrograph averages about 2 contacts per grain in two-dimensions

Fig. 8.8 Histogram plot of the frequency for grain contacts. In the microstructure one grain is selected and the number of contacts with similar grains is counted, with the procedure repeated for several grains to determine the distribution

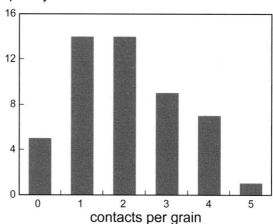

$$N_C = \frac{3\,C_g}{2\,sin\,(\phi/2)} \qquad (8.5)$$

where ϕ is the dihedral angle, a measure of bond size between contacting grains of the same composition. Figure 8.9 shows the interfacial geometry associated with the dihedral angle at a two grain junction. Although surface energies vary with crystal orientation, leading to a dihedral angle distribution, an average is assumed. The interfacial energy balance associates a surface energy vector with each

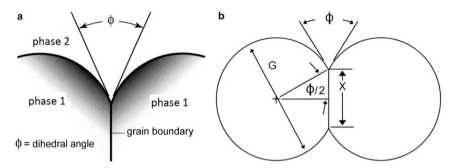

Fig. 8.9 The dihedral angle ϕ is defined by these drawings: (**a**) a sketch of the interface where the grain boundary and surface intersect. The energy balance involves the downward pointing grain boundary energy vector versus the two phase interfacial energy vectors while (**b**) is the definition of the grain contact showing the bond size X and grain size G related to the dihedral angle

interface. When those vectors are balanced in the vertical orientation the resolved vectors give,

$$\gamma_{11} = 2\,\gamma_{12}\,cos\,(\phi/2) \tag{8.6}$$

where γ is the surface or interface energy and the subscript corresponds to the contacting phases, either 1-1 or 1-2. The bond size X between similar grains divided by the grain size G depends on the dihedral angle ϕ,

$$X = G\,sin\,(\phi/2) \tag{8.7}$$

If the neck size X and grain size G are independently measured, then Eq. (8.7) provides a means to calculate the dihedral angle for use in Eq. (8.5).

A related microstructure calculation involves the grain size G, dihedral angle ϕ, and distance between grain centers Y for contacting grains of similar size given by,

$$Y = G\,cos\,(\phi/2) \tag{8.8}$$

Properties such as conductivity, strength, and fracture toughness depend on the level of connection in three dimensions. Intentional dihedral angle manipulation leads to significant property change. This is demonstrated in the Al-SiC composite. In one fabrication approach, a SiC skeleton is formed by high temperature sintering (1900 °C or 2173 K) to create large bonds, then at a lower temperature molten aluminum is infiltrated into the pores [7]. An alternative fabrication approach for the same composition mixes SiC particles into molten aluminum prior to casting. This second approach results in no bonding between the silicon carbide grains. The infiltrated composition with bonded SiC is about 30 % higher in thermal conductivity, but 10 % lower in thermal expansion coefficient. Other lower melting temperature liquids are used in a similar manner, such as magnesium as an infiltrant for oxide ceramics [8].

Contiguity

Another measure of grain contacts is the contiguity, defined as the fraction of grain perimeter in contact with the same phase. The grain perimeter consists of regions in contact with similar grains, pores, and other phases. For example, the grain perimeter in Fig. 8.7 consists of bonds to similar grains and bonds to the second phase. Based on measuring the grain perimeter, the contiguity of phase "1" C_{11} is the perimeter for a grain of phase "1" in contact with similar composition grains, P_{11}, divided by the total grain perimeter P_T,

$$C_{11} = \frac{P_{11}}{P_T} \tag{8.9}$$

In the absence of porosity, the total perimeter P_T is the sum of P_{11} and P_{12}. This latter value is the perimeter for phase 1 in contact with phase 2 contacts.

Measuring the grain perimeter is not necessary, since linear intercepts provide an easier approach. Test lines are randomly overlaid on the microstructure. For either phase, 1 or 2, the number boundaries crossed involving same grain N_{11} contacts is taken as a ratio to the total number of grain N_T contacts. The contiguity for phase 1, designated C_{11}, is determined as follows:

$$C_{11} = \frac{2 N_{11}}{N_T} \tag{8.10}$$

When the test line crosses any boundary (phase 1 with phase 1, phase 1 with phase 2, phase 1 with pore, so on) the total number of contacts increases, while a separate tally is maintained for the number of same grain contacts. The factor of 2 in Eq. (8.10) arises since each same-grain contact is counted once, yet is shared by two grains and so should be counted twice, while each dissimilar grain contact is counted once.

Like connectivity, contiguity depends on volume fraction, dihedral angle, and coordination number. The relation between the volume fraction of a phase, coordination number, and dihedral angle is plotted in Fig. 8.10 [9]. Included with these lines are data from VC-Co composites over a wide range of VC contents; this system has an approximate dihedral angle of 60°. Several studies rely on contiguity to explain fracture toughness, electrical conductivity, strength, and ductility.

Grain Separation

Another measure used to link microstructure and properties is the mean separation λ. It is a microstructure metric applied to pores, grains, and inclusions. The mean separation relates to several factors, including the feature size. For example, the

Fig. 8.10 The variation in three-dimensional grain coordination number (number of neighboring grains of same composition) for VC-Co composites [9, 10]. The *solid circle symbols* are the experimental results and the *lines* show the expected variation with different dihedral angles

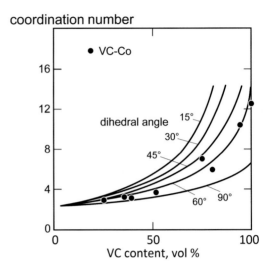

mean grain separation λ measured from one edge of a grain to the nearest edge of the next grain, along a random test line, is given as follows:

$$\lambda_1 = \frac{V_1}{N_G} = \frac{G_1 \left(1 - V_1\right)}{V_1 \left(1 - C_{11}\right)} \tag{8.11}$$

where V_1 is the volume fraction, G_1 is the grain size, C_{11} is the contiguity, and N_G is the number of grains per unit test line length, accounting for the magnification. These calculations provide the mean value. It is also possible to measure the individual separation distances on random test lines to extract the distribution.

The mean separation between pores and phases is often linked to deformation and fatigue life, as well as hardness, ductility, and thermal shock resistance. It is inherently a characteristic of the matrix ligament size between grains. Figure 8.11 plots the mean separation distance versus tungsten content for W-Ni-Fe composites [10]. The greater the tungsten content, the closer the tungsten grains and the thinner the grain separation. Accordingly the strength increases, but ductility decreases [11]. These properties also depend on grain size, volume fraction, and connectivity.

Interface Strength

Interface strength determines how the phases transfer stress, heat, current, or otherwise interact. Interface strength often controls properties; weak interfaces fail under low stress independent of the attributes of the individual phases. In many polymer composites, strength falls as filler particles are added. Such behavior implies the solid particles are effectively voids since there is low interface strength. However, some particulate composites exhibit significant property gains over the

Fig. 8.11 Mean separation of tungsten grains in W-Ni-Fe composites versus the tungsten content [10]. These composites consist of various tungsten levels with Ni:Fe as the balance in a 7:3 ratio, fabricated at 1500 °C (1773 K)

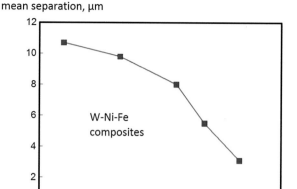

individual phases if a strong interface exists [12]. Indeed, the several composites previously listed in Table 4.3, at nominally 20 vol% second phase, illustrate strength ratios from 0.5 to over 5. At 20 vol% porosity the predicted strength ratio is 0.6. In other words, when 20 vol% second phase drops strength by half, then the additive is not bonding and the composite is the same strength as with 20 % porosity. When a composite exhibits such significant strength degradation, most typically this is associated with one of the following [13, 14]:

- inhomogeneous mixture,
- large grains exceed the critical flaw size (determined by the fracture toughness of the matrix phase),
- poor interface bonding.

Active bonding improves strength, but weak bonding degrades strength. Chemical interaction, as evident by wetting, is a critical indicator of interface strength [15]. For polymer matrix composites, significant strength gain is possible using particles treated to improve adhesion [16]. This involves reactive chemical groups, such as silane (Si-H complexes), to bridge from the added phase to the matrix. Surface treatments for adhesion are very useful in polymer or epoxy composites reinforced with glass or graphite. Likewise, titanium (Ti) or titanate (Ti-O) organometallic compounds provide bonding to metallic particles. Some others are chromium and tin for metal oxide interfaces. Care is needed in handling some of the additives, since they are reactive or toxic.

Tests for interface strength rely on small test geometries. Ideally, pure shear is applied to the interface to measure the fracture strength. Figure 8.12 sketches cross-sections of plate on plate, wire through plate, and tube through plate test options. Other ideas include making thin slices of the composite to allow grains to be pushed through the slice. The fracture force divided by the interface area is the shear strength, a relative characterization of the interface strength. For ductile

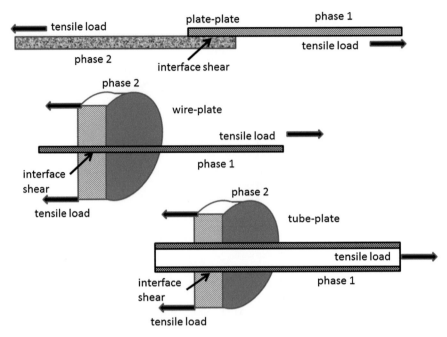

Fig. 8.12 Sketches of three interface shear tests. In the upper drawing, two bonded thin plates are pulled apart while being supported to avoiding bending. In the center drawing, the wire of phase 1 is pulled until it fractures away from the phase 2 plate. In the lower drawing, a tube is used for phase 1, again being pulled (or pushed) until interface fracture. Interface strength is calculated from the failure load divided by the interface area

composites, considerable work hardening occurs prior to interface fracture, usually by cavity nucleation at the interface [17]. Also, the test geometry can bend prior to fracture, implying the fracture load is not pure shear.

Fracture strength and deformation, such as ductility, are used to evaluate interface treatments. Unfortunately, the test results are sensitive to several interactions between the phases, besides just interface strength. Accordingly, tests to compare different interface treatments are should be compared over a narrow range of test conditions, sample geometries, and material combinations. Compression tests are best for brittle materials, to avoid premature failure away from the interface [18]. In cases where the phases react during processing to form low strength interface, it is appropriate to apply protective surface coatings on the minor phase to suppress the reactions [19].

The role of interface doping is illustrated for the Cu-Al_2O_3composite interface in Fig. 8.13 [20, 21]. The interface strength increases with a low concentration of added titanium, but titanium excess attacks the aluminum oxide. The reaction product becomes thicker as the titanium content increases, eventually forming a continuous, weak layer at the interface. A small concentration of titanium provides chemical bridging without forming a new layer.

Fig. 8.13 Bond strength measured for alumina-copper interfaces with different amounts of added titanium to improve wetting [20]. Some titanium improves bonding, but an excess causes a reaction that contaminates the interface

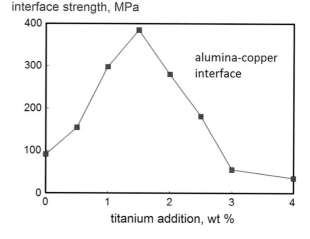

Fig. 8.14 Scanning electron micrograph of a copper-cordierite (MgO-SiO$_2$-Al$_2$O$_3$ compound) after heat treatment, where the nonwetting copper exuded to the composite surface

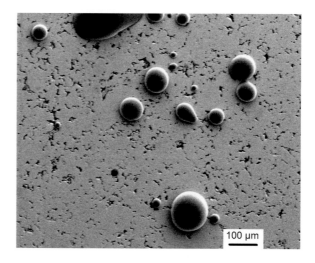

One characteristic of poor wetting is pictured in Fig. 8.14. This scanning electron micrograph shows copper exuded from cordierite (magnesium alumino silicate or MgO-SiO$_2$-Al$_2$O$_3$) after heat treatment. Poor wetting induced copper migration out of the composite where it coalesced on the external surface. Wetting difficulties require modifications, such as adding a bridging agent to the interface. For oxides such as alumina, the interface strength increases with additives that are oxygen active, so titanium proves useful. Generally the peak interface strength measured in metal-oxide interfaces is 600 MPa. For example, in a low interaction case of Cu-Al$_2$O$_3$ the interface strength is 90 MPa, but this increases to 390 MPa with titanium doping. Such doping is performed in rotating reactors where the alumina particles are coated. One approach is to vaporize the titanium and another is to induce chemical reactions using metal chlorides or fluorides.

Improved interfaces come from mechanical or chemical bonding treatments. Interfaces reliant on mechanical treatments use bumps, pores, nodules, or serrations to improve stress transfer. Chemical interfaces rely on bonds across the interface between dissimilar materials. Reactive species are generally most effective [19, 20]. Unfortunately, if the interface depends on diffusion, the process can be slow. For example, in the wetting of copper on alumina at 1100 °C (1373 K), the contact angle progressively falls from 130 to 25° over 120 min. The slow change reflects the limited diffusion at this temperature. Further, diffusion can allow the formation of a new phase at the interface that is detrimental. In Ti-SiC composites, the two phases react to form a brittle interface layer. Likewise, magnesium reinforced with alumina reacts to form $MgAl_2O_4$ spinel on heating at the interface, leading to progressive weakening at high temperatures.

One approach to interface control and wetting comes with nanoparticles. At the nanoscale, nominally with particles below about 20 nm, a low contact angle is common. Spreading requires just split seconds, as short as 5 ms, after contact, so wetting is nearly spontaneous. The use of nanoscale particles to adjust interface bonding is are area of much speculation, but the high cost of nanoparticles and concern over worker health from exposure to these particles are impediments to widespread application.

Interface contamination is a problem in particulate composites. Dramatic shifts in properties occur with formation of reaction products on the interface. Systems where the impurity has some solubility in one phase exhibit the most difficulty. When impurity solubility decreases with cooling, the impurity naturally tends to segregate and interfaces are common segregation sites. Segregation requires atomic diffusion, so rapid cooling to avoid diffusion is a means to reduce problems. Slow cooling lets the decreasing impurity solubility couple to atomic diffusion to produce a segregated interface film or reaction product. The transmission electron micrograph in Fig. 8.15a images impurity precipitates in the form of a string of inclusions. Crack propagation links between the inclusions to greatly reduce strength, ductility, and toughness. In such situations, fracture surface examination shows preference for failure along the interface as evident in Fig. 8.15b. In this particular case, impurities in the consolidated structure diffused during cooling from the hot isostatic pressing temperature. Impurity solubility declined with cooling while the temperature was sufficient to enable diffusion to segregate at the interface. To avoid such problems, changes in processing are required, including purification of the powder, improved powder handling to avoid contamination, interface shear to disrupt surface films, or reheating to dissolve the impurities coupled to rapid quenching to freeze them into solution (avoid segregation).

Interfaces often degrade due to reactions at high temperatures. Table 8.1 lists several composites and the temperature of significant interface degradation [21–26]. The effect comes from formation of a reaction product. In some cases, a low temperature eutectic liquid forms at the interface. Thus, initially the composite appears successful, but on exposure to a high temperature, such as seen in brazing or welding, the composite interface degrades. It is important to test properties under conditions relevant to the application to avoid surprising degradation events.

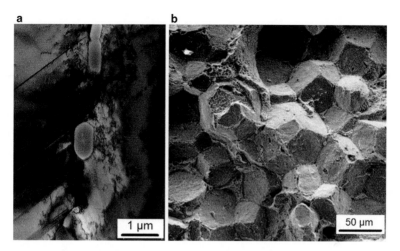

Fig. 8.15 Weak or contaminated interfaces greatly reduce strength. The transmission electron micrograph in (**a**) images inclusions that formed from impurities on an interface boundary (running vertical in this picture). The consequence is evident in the scanning electron micrograph in (**b**) showing how the inclusions provide an easy fracture path to greatly reduce ductility, strength, and toughness

Table 8.1 Onset temperature for interface degradation in several composites [21–26]

Composite	Form of degradation	Temperature, °C
Al-B	Boride formation	500
Al-C	Formation of Al_4C_3 compound	550
Al-Fe	Aluminide formation	500
Ni-Al_2O_3	Compound formation	1100
Ni-SiC	Silicide formation	800
TaC-Co	Eutectic formation	1200
Ti-B	Boride formation	750
Ti-SiC	Compound formation	700
W-Fe	Intermetallic formation	1000
W-Ni	Intermetallic formation	900

Interface Measures

Two parameters provide insight on interface thermodynamics. Both are related to chemical solubility. The first parameter is the wetting or contact angle. It is associated with a solid-liquid-vapor interface, but is applicable to any combination of three phases. As sketched in Fig. 8.16, the typically contact angle θ is measured by the liquid contact along a horizontal solid surface (gravity causes the horizontal spreading). Good wetting is associated with a small contact angle, defined by the horizontal equilibrium of surface energies,

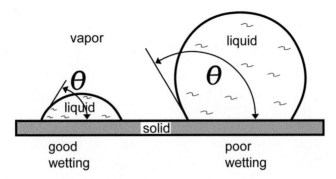

Fig. 8.16 The contact angle (also known as the wetting angle) θ is a reflection of the horizontal surface energy balance. Good wetting is reflected by a small contact angle as shown on the left. Poor wetting is associated with a large contact angle shown on the right

$$\gamma_{SV} = \gamma_{SL} + \gamma_{LV} \cos(\theta) \qquad (8.12)$$

where γ_{SV} is the solid-vapor surface energy, γ_{SL} is the solid-liquid surface energy, and γ_{LV} is the liquid-vapor surface energy. Two cases are depicted, the first for good wetting with a low contact angle and the second for poor wetting with a contact angle greater than 90°. Rearranging the vector balance of Eq. (8.12) gives the contact angle as a function of the relative surface energies,

$$\theta = arccos\left[\frac{\gamma_{SV} - \gamma_{SL}}{\gamma_{LV}}\right] \qquad (8.13)$$

Many metals, ceramics, and compounds have surface energies are in the 1 to 2 J/m^2 range. For example, γ_{LV} for alumina Al$_2$O$_3$ in vacuum is 0.63 J/m^2 at its melting point, aluminum is 1.14 J/m^2, chromia Cr$_2$O$_3$ is 0.81 J/m^2, and chromium is 2.2 J/m^2. In these cases note the oxide is lower in surface energy versus the metal, indicating the oxide coated surface is more stable.

Low contact angles of 30° or less are associated with intersolubility. If a reaction compound forms on the interface, the typical contact angle is 30–40°. Likewise if there is poor intersolubility the contact angle is higher, such as Ag on W (50° contact angle). Higher temperatures improve wetting [21]. For example silver liquid on iron has a declining contact angle on heating, reaching 10° at 1075 °C (1348 K), as plotted in Fig. 8.17. Over this same temperature range the solubility of liquid silver in solid iron increases from essentially zero to about 0.02 wt%. Although that solubility is still small, it is sufficient to induce wetting.

Solubility adjustments are a means to induce wetting, via temperature changes or additives. Figure 8.18 is an example where the addition of copper to silver improves the wetting of iron (reduces the contact angle) at 965 °C (1238 K) [21]. The contact angle falls with an increase in the copper content because of increased dissolution across the interface. A related attribute is the time-dependence of the contact angle.

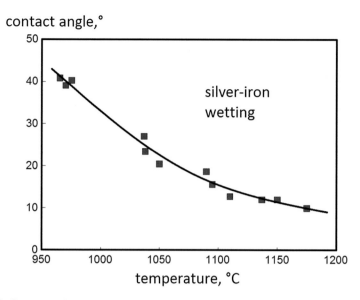

Fig. 8.17 Contact angle versus temperature for liquid silver wetting solid iron [21]. The decrease in contact angle occurs since the solubility of iron in silver and silver in iron both increase at high temperatures

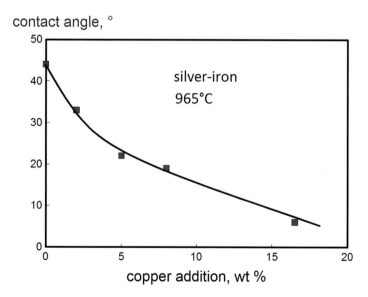

Fig. 8.18 Contact angle data for silver-iron at 965 °C (1238 K) [21]. The addition of copper to the silver lowers the contact angle since it improves intersolubility

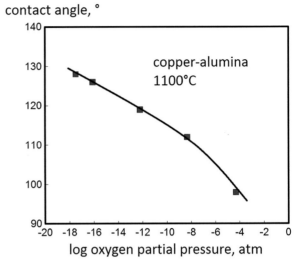

Fig. 8.19 Wetting of copper on alumina at 1100 °C (1373 K) with varying oxygen partial pressures (logarithmic scale) [21, 27]. Oxygen in the process atmosphere improves wetting

Dissolution and diffusion require time, so the contact angle slowly changes until equilibrium is reached. For example at 1100 °C (1373 K) the initial contact angle of molten copper on tungsten carbide is 26°, but over 10 min of isothermal hold it drops to 19° as slight dissolution of WC occurs into the liquid copper. Since dissolution changes the surface chemistry, an advancing contact angle associated with spreading is different from a retreating contact angle such as when liquid is drained from a solid. The result is effectively a hysteresis or memory effect.

Particulate composites benefit from additives that adjust the wetting. The source might even be from oxygen or nitrogen in the process atmosphere. Plotted in Fig. 8.19 is the contact angle of copper on alumina at 1100 °C (1373 K) as a function of the atmosphere oxygen partial pressure [21, 27]. Since oxygen is soluble in copper, the higher oxygen pressure is beneficial for wetting. This is not always the case, and in many situations atmosphere impurities, such as oxygen, increase the contact angle.

Much attention is given to oxide-metal composites, where nonoxide forming metals, such as gold, exhibit contact angles in the 110–140° range when brought in contact with stable oxides. But if the metal forms a weak oxide, then the contact angle drops to 60–90°, and if the metal-oxide system forms compounds, then the contact angle is even lower. Interface reactions depend on diffusion and are often delayed. For example, in wetting SiC with liquid Cu-Cr at 1100 °C (1373 K) in vacuum, the initial contact angle is 120° but falls to 30° after 7 min.

Likewise, metal-carbide interfaces depend on possible reactions and solubility. For the metals of Ni, Co, and Fe, the contact angles with carbides range from 0° with WC and Mo_2C to 35° for contact with ZrC and HfC. Generally the contact angle is lower as solubility of the carbide increases in the metal.

Coupling agents are employed for polymer matrix composites, especially where mineral particles are added to the polymer. Some of the most effective coupling

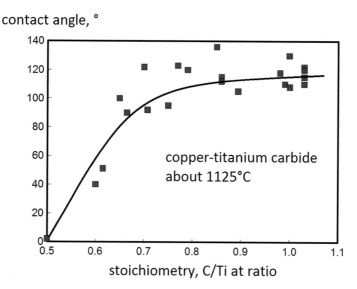

contact angle, °

Fig. 8.20 Copper wetting on titanium carbide at 1125 °C (1398 K), where the titanium carbide stoichiometry in at% is adjusted over a broad range, leading to a high contact angle as the stoichiometric TiC composition is approached [30]

agents are titinates [28]. A large application was to bond magnetic iron oxide particles to polymer surfaces to form magnetic recording media.

Monitoring the contact angle is a means to track interface energy changes [21, 29]. For example, Fig. 8.20 plots of contact angle versus carbon content in titanium carbide, where wetting changes progressively as the stoichiometry moves toward the TiC composition [30].

Since solubility is fundamental to wetting, phases that are intersoluble have low contact angles. Strategies to change wetting in particulate composites can be very effective. Taking the case of copper on alumina at 1200 °C (1473 K), the system without an additive has a contact angle near 125°. With the addition of titanium, the contact angle falls dramatically, passing below 10° with 10 at.% Ti addition to the copper. A weak titanium oxide film forms at the interface.

A related interface parameter is the dihedral angle, illustrated in Fig. 8.9. It is associated with the grain boundary intersection with a second phase. The second phase might be another solid, liquid, or pore. The dihedral angle ϕ describes the balance of surface energies as shown in Fig. 8.21. Two grains form a grain boundary and when that boundary intersects the second phase, a vertical balance of surface energies gives,

$$\gamma_{11} = 2\,\gamma_{12}\,\cos\,(\phi/2) \tag{8.14}$$

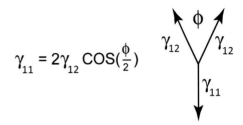

Fig. 8.21 Surface energy balance defining the dihedral angle, in this case the surface energies correspond to the phase arrangement in Fig. 8.9, with the grain boundary energy pointing down

$$\gamma_{11} = 2\gamma_{12} \cos\left(\frac{\phi}{2}\right)$$

where γ_{11} is the interfacial energy (grain-boundary energy) for grains of phase 1, and γ_{12} is the interfacial energy between phases 1 and 2. The interfacial energy balance defines the dihedral angle,

$$\phi = 2\, arccos\left[\frac{\gamma_{11}}{2\,\gamma_{12}}\right] \tag{8.15}$$

The dihedral angle ϕ is a means to assess relative changes in interface energy, since it depends on the ratio of interface energies. In the case of a grain boundary intersecting a vapor the resulting trench is termed a thermal grove. It is small and best observed in a microscope. Temperature changes surface energies, usually resulting in better wetting at higher temperatures. Accordingly, the dihedral angle changes with processing cycles.

Interface Failure

The relative strength of the interface, versus the individual phases, is assessed by fracture path analysis. Fracture follows the weakest path. Quantitative microscopy of fracture surfaces determines which path is preferential for fracture. Generally, composite strength increases with the interface strength. So strength tests are a basis for sensing interface quality and any need for improvement. Fracture test options are discussed in Chap. 3.

A sketch of a fracture path in profile is drawn in Fig. 8.22. The variants include cleavage of either phase as well as separation on grain boundaries or interphase boundaries. Quantitative microscopy applied to the fracture surface allows for identification of a weak path. For example, for the material pictured in Fig. 8.15, the interface phase (seen in 8.15a) is under 1 vol% of the structure, but the fracture path (seen in 8.15b) is over 90 % along the interface. Clearly the interface is the weak link.

For the case of a tough matrix (for now assumed to be phase 1) with no adhesion to the second phase (termed phase 2), composite strength σ_C depends on the matrix, or phase 1, strength σ_1 and volume fraction of second phase V_2 as follows:

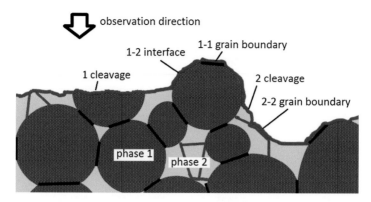

Fig. 8.22 Fracture profile with the key pathways indicated. The sketch is perpendicular to the normal observation direction (from the top looking down on the fracture). For this sketch, both phase 1 and phase 2 are assumed to be polycrystalline with grain boundaries. Traversing across the fracture surface, the failure paths are labeled as phase 1 cleavage, phase 1 and phase 2 interface separation, grain boundary failure for phase 1, phase 2 cleavage, and grain boundary failure for phase 2

$$\sigma_C = \sigma_1 \left[\frac{1 - V_2}{a + bV_2} \right] \tag{8.16}$$

where a and b are empirical parameters, as long as the second phase particles are small, measured in micrometers. The consequence is reduced composite strength with increased concentration of second phase. Several examples of this behavior are tabulated in Table 4.3.

Adhesion and bonding improve with chemical interactions between phases. Good examples are in composites fabricated at high temperatures where solubility and diffusion are active—WC-Co, W-(Ni + Fe), Be-BeO, SiC-ZrO$_2$, Al-Si, Al-TiB$_2$, TiC-Fe, Mo-TiC, Fe-TiB$_2$, and Ti-TiC. These systems are fabricated at elevated temperatures where atomic intermingling occurs at the interface. However, high temperatures reactions can degrade properties. For example, TiAl-SiC composites undergo progressive strength loss during high temperature exposure [31]; interface strength declines from 3500 to 1100 MPa during a 10 h exposure to 1150 °C (1423 K). The behavior is plotted in Fig. 8.23 using the square-root of time as the expected diffusion controlled plotting parameter. Empirical models are offered to explain how interface strength varies with composition using modified versions of Eq. (8.16),

$$\sigma_C = \sigma_1 \left[\frac{1 - V_2}{a + bV_2 + c\,V_2} \right] \tag{8.17}$$

using the same symbols as above (σ_C is the composite strength, σ_1 is strength of phase 1, V_2 is the volume fraction of phase 2, and a, b, and now c are empirical

Fig. 8.23 Interface
strength decay with high
temperature hold time
(in hours) where time is
plotted as a square-root to
illustrate diffusion
controlled interface strength
loss [31]

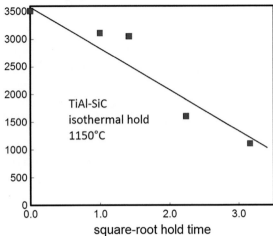

interface strength, MPa

TiAl-SiC
isothermal hold
1150°C

square-root hold time

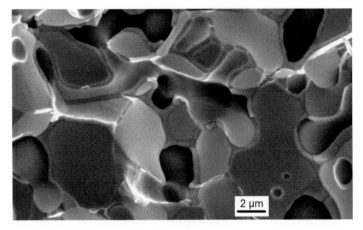

Fig. 8.24 Fracture surface of a porous composite corresponding to a case of little interface
strength

parameters). This form requires input data, so it has no predictive character.
However, it is effective in fitting composite systems, especially polymers with
added mineral, ceramic, or metallic phases that do not bond well to the polymer.

The importance of interface strength is most evident in fracture strength and
related parameters of ductility, fatigue life, and toughness. On the other hand, little
change occurs in elastic modulus and hardness with changes to interface strength.
Microscopy is the critical means to identify weak interface problems. Fracture
along grain faces, as evident in the Fig. 8.24, invokes a need for interface strengthening. Most desirable is the case where fracture propagates through both phases
with little interface failure. The cleavage of both phases, indicative of a strong

Fig. 8.25 Fracture surface
of a W-7Ni-3Fe composite
with desirable cleavage of
both phases, indicating
strong interfaces

interface, is evident in the fracture surface of Fig. 8.25. Failure identifies the weakest link, so fracture surface evaluation guides attention to where the greatest gains are possible.

Study Questions

8.1. For abrasive composites, such as diamond embedded in a metal, why it is not desirable to have bonding between the hard particles?

8.2. The contact angle of gold on titania (Au on TiO_2) is 100° and for gold on alumina (Au on Al_2O_3) is 140°. What possible factor might explain this difference?

8.3. The surface energy for a pure metal in vacuum is 1.1 J/m^2 and for that metal in contact with an oxide ceramic the interface energy is 0.65 J/m^2. The contact angle is 120°, so what is the surface energy for the oxide ceramic in vacuum?

8.4. Fracture surface analysis finds 80 % of the fracture surface consists of interface separation. The composite consisting of 90 vol% of the main phase, with 10 vol% second phase. Is this a case of a weak interface?

8.5. For a dihedral angle of 65° and grain size of 50 μm, what is the expected size of the bond between two contacting grains of the same composition?

8.6. Measured dihedral angle for a composite is 45°. Accurately draw the contact between two grains to reflect this dihedral angle.

8.7. In composite fabrication experiments, it is found gravity causes a "top-bottom" microstructural change, accompanied by differences in grain size, contiguity, and solid fraction. For example a direct comparison of a 78 W

(W-15Ni-7Fe) composition processed at 1500 °C (1773 K) for 120 min on Earth or 120 min in microgravity gives the following comparative properties:

	Earth	Microgravity
Density, g/cm^3	14.93	14.93
Mean grain size, µm	29.5	26.8
W-W contiguity	0.22	0.17
W-W neck size, µm	10.5	9.3

Is the dihedral different for these two situations?

8.8. Would you expect the action of a wetting agent to be effective if it is simply added to the composite formulation? Why or why not?

8.9. Use the micrograph in Fig. 8.7 to estimate the dihedral angle distribution.

8.10. In measuring connectivity in using the number of contacting grains of the same phase, larger grains have more contacts and smaller grains have fewer contacts. Why is this expected?

References

1. R.T. DeHoff, F.N. Rhines (eds.), *Quantitative Microscopy* (McGraw-Hill, New York, 1968)
2. W. Rasband, Research Services Branch, National Institute of Mental Health, Bethesda, MD, file download at http://rsb.info.nih.gov/ij/download.htm
3. M. Sahimi, *Applications of Percolation Theory* (Taylor and Francis, London, 1994)
4. C.S. Smith, Some elementary principles of polycrystalline microstructures. Metall. Rev. **9**, 1–48 (1964)
5. D.A. Aboav, T.G. Langdon, The distribution of grain diameters in polycrystalline magnesium oxide. Metallography **1**, 333–340 (1969)
6. J.N. Boland, X.S. Li, Microstructural characterisation and wear behaviour of diamond composite materials. Materials **3**, 1390–1419 (2010)
7. S. Li, D. Xiong, M. Liu, S. Bai, X. Zhao, Thermophysical properties of SiC/Al composites with three dimensional interpenetrating network structure. Ceram. Int. **40**, 7539–7544 (2014)
8. J. Zeschky, J. Lo, T. Hofner, P. Greil, Mg alloy infiltrated Si-O-C ceramic foams. Mater. Sci. Eng. **A403**, 215–221 (2005)
9. R.M. German, P. Suri, S.J. Park, Review: liquid phase sintering. J. Mater. Sci. **44**, 1–39 (2009)
10. K.S. Churn, R.M. German, Fracture behavior of W-Ni-Fe heavy alloys. Metall. Trans. **15A**, 331–338 (1984)
11. R.A. Krock, L.A. Shepard, Mechanical behavior of the two-phase composite, tungsten-nickel-iron. Trans. Metall. Soc. AIME **227**, 1127–1134 (1963)
12. S.Y. Fu, X.Q. Feng, B. Lauke, Y.W. Mai, Effects of particle size, particle/matrix interface adhesion and particle loading on mechanical properties of particulate-polymer composites. Compos. Part B **39**, 933–961 (2008)
13. I.A. Ibrahim, F.A. Mohamed, E.J. Lavernia, Particulate reinforced metal matrix composites - a review. J. Mater. Sci. **26**, 1137–1156 (1991)
14. D. Zhang, K. Sugio, K. Sakai, H. Fukushima, O. Yanagisawa, Effect of spatial distribution of SiC particles on the tensile deformation behavior of Al-10 vol% SiC composites. Mater. Trans. **48**, 171–177 (2007)
15. A. Contreras, E. Bedolla, R. Perez, Interfacial phenomena in wettability of TiC by Al-Mg alloys. Acta Mater. **52**, 985–994 (2004)

16. F. Danusso, G. Tieghi, Strength versus composition of rigid matrix particulate composites. Polymer **27**, 1385–1390 (1986)
17. K.E. Easterling, H.F. Fischmeister, E. Navara, The particle to matrix bond in dispersion hardened austenitic and ferritic iron alloys. Powder Metall. **16**, 128–145 (1973)
18. M. Ruhle, A.G. Evans, High toughness ceramics and ceramic composites. Prog. Mater. Sci. **33**, 85–167 (1989)
19. X. Liang, C. Jia, K. Chu, H. Chen, J. Nie, W. Gao, Thermal conductivity and microstructure of Al/diamond composites with Ti coated diamond particles consolidated by spark plasma sintering. J. Compos. Mater. **46**, 1127–1136 (2011)
20. M.G. Nicholas, T.M. Valentine, M.J. Waite, The wetting of alumina by copper alloyed with titanium and other elements. J. Mater. Sci. **15**, 2197–2206 (1980)
21. N. Eustathopoulos, M.G. Nicholas, B. Drevet, *Wettability at High Temperatures* (Pergamon, Oxford, 1999)
22. T.P.D. Rajan, R.M. Pillai, B.C. Pai, Review: reinforcement coatings and interfaces in aluminum metal matrix composites. J. Mater. Sci. **33**, 3491–3503 (1998)
23. K.K. Chawla, *Composite Materials: Science and Engineering* (Springer, New York, 2012)
24. K. Kondoh, Titanium metal matrix composites by powder metallurgy (PM) routes, in *Titanium Powder Metallurgy*, ed. by M.A. Qian, F.H. Froes (Elsevier, Oxford, 2015), pp. 277–297
25. A.P. Sutton, R.W. Balluffi, *Interfaces in Crystalline Materials* (Clarendon, Oxford, 1995)
26. S. Suresh, A. Needleman, *Interfacial Phenomena in Composites: Processing, Characterization and Mechanical Properties* (Elsevier Applied Science, New York, 1989)
27. E. Saiz, R.M. Cannon, A.P. Tomsia, High-temperature wetting and the work of adhesion in metal/oxide systems. Annu. Rev. Mater. Res. **38**, 197–226 (2008)
28. S.J. Monte, G. Sugerman, Processing of composites with titanate coupling agents – a review. Polym. Eng. Sci. **24**, 1369–1382 (1984)
29. F. Delanny, L. Froyen, A. Deruyttere, Review: the wetting of solids by molten metals and its relation to the preparation of metal matrix composites. J. Mater. Sci. **22**, 1–16 (1987)
30. D.A. Mortimer, M. Nicholas, The wetting of carbon and carbides by copper alloys. J. Mater. Sci. **8**, 640–648 (1973)
31. J.M. Yang, S.M. Jeng, Interfacial reactions in titanium-matrix composites. J. Met. **41**(11), 56–59 (1989)

Chapter 9
Design

Designs employing particulate composites consider the materials, forming process, and target properties to select the component size, shape, and features. In spite of much focus on properties, often design decisions include cost as a top priority. Particulate composites excel with low lifetime costs. The selection of a fabrication process includes assessment of geometric features and dimensional tolerances.

Introduction

To satisfy the needs of an application, the engineering design concept encompasses decisions on the appropriate combinations of geometry, composition, and processing. Design specifications point to acceptable materials and possibly even which powders to employ. In some situations the fabrication route is specified to ensure delivery of a properly functioning component.

The geometric aspects of a component design are captured in the engineering "definition" that includes geometric factors such as dimensions, tolerances, and surface finish. It is common to see the definition specify applicable standards and even inspection criteria. Geometric attributes include specifications for the dimensions, size and location of holes as well as slots, protrusion, indents, and related features. Further, critical functional aspects are included, such as flatness or intended fit. Final attributes included in the specification might include color and handling factors such as removal or all burrs, or packaging specifics. Whenever possible, the specification is quantitative, for example an allowed size range or allowed hardness minimum. Likewise, the specification might detail the inspection criteria as part of the product qualification.

A desire in many particulate composites is to combine hardness with toughness. Three factors are important;

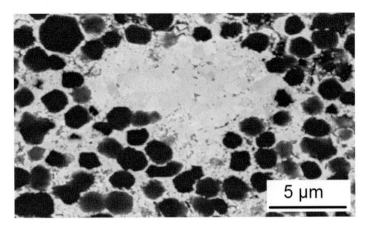

Fig. 9.1 Scanning electron micrograph of a three phase composite, WC-25Al₂O₃-7Co (the alumina is *dark* and the tungsten carbide is *white*) with an inhomogeneity in the microstructure that is termed a "lake"

- ductile constituents (composition)
- high interface strength (fracture strength)
- homogeneity (microstructures).

Ignoring these factors causes significant difficulties. Thus, the specification needs to specify the ratio of phases; for example, Mo-20Cu implying 20 wt% of ductile copper. Property tests specify the minimum strength or hardness. Homogeneity is defined in terms of the microstructure sale. In some industries this is achieved using comparative micrographs as standards, making it easy for production technicians to accept or reject. Figure 9.1 is a micrograph of a composite with a large phase cluster. It is appropriate to describe the size of the inhomogeneity and the examination magnification. For example, no single phase region should be larger than a specified size, say 20 μm, to avoid features such as shown in this figure. Pores larger than the grain size, as evident in Fig. 9.2, are significantly detrimental to properties. Accordingly, microstructure specifications might include grain size, porosity, or pore size. For WC-Co the specifications identify the maximum allowed porosity in vol% for three sizes designated A, B, and C. The A pores are smaller than 10 μm, B pores are larger than 10 μm, and C pores are defects noted by comparison to industry standard pictures.

Another difficulty arises from highly scattered properties. As an example, Fig. 9.3 plots the fracture toughness for Al-SiC composites with varying silicon carbide contents based on a variety of studies. The shaded region embraces the nominal upper and lower values for composites with as much as 66 vol% SiC. Note for example the 20 vol% SiC composition delivers fracture toughness from 5 to 20 MPa√m. If a design is based on the lower value, it will be bulky and wasteful. On the other hand, if a design assumes the higher value, then significant service failure is possible.

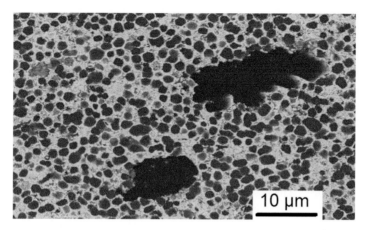

Fig. 9.2 Example large pore defects formed in a hard composite by improper mixing and consolidation; this is the same material as imaged in Fig. 9.1

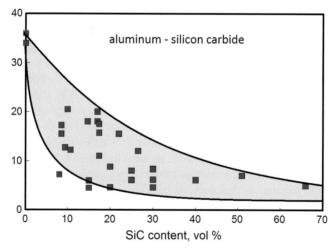

Fig. 9.3 Fracture toughness plotted against silicon carbide content for Al-SiC composites. The data were taken from commercial sources, ranging up to 66 vol% SiC. The property scatter corresponds to differences in interface quality and microstructure homogeneity

Since fracture toughness varies over a considerable range, conservative designs rely on the lower value. This is a reflection an aversion to the risk of failure during use. Thus, design specification need to state the testing procedures and minimum allowed property as part of vendor qualification. Further, periodic random testing is advised to ensure production is meeting the planned property level. In a litigation over a failed component, one company lost the case because they were unable to verify test results; the component specification required two random fracture tests every 2 h. Note, it was never proven the component was weak, but the fact the firm failed to perform the required periodic strength test was sufficient to indicate disregard for the specification.

Case History

Design involves a sequence of decisions. A search for the optimal design starts with property data and cost information for comparison with the performance objectives. An iterative loop is typical, where the application dictates lead to a first conceptual design. That first design becomes a basis for analysis such as finite element analysis, geometric refinements, and first specifications. The specification is refined by further analysis, especially focused on properties and component geometry. As the design advances, alternative materials are considered in light of potential cost or fabrication ease. As the design reaches maturity more attention is given to cost issues.

Wear materials are an area where much effort is required to optimize the design [1]. Oil well drilling tips are a good example. Deep drilling requires tools that lasts days under extremely harsh conditions. If the drill fails prematurely, then the tip must be extracted and replaced, at a high cost. Accordingly, the cost of failure skews any evaluation toward high performance with secondary regard for costs. Three properties dominate service life—wear resistance, hardness, and fracture toughness. The solution is usually a drill tip based on WC-Co, possibly with cutting edges coated with diamond. Actually the diamond is a composite called polycrystalline diamond, reflecting high diamond content at 80 vol% of 2–50 μm grains.

The geometric design is simple, typically a cylinder shape with multiple protruding cylinders forming cutting teeth, as seen in Fig. 9.4. This particular tip has arrays of polycrystalline diamond tips brazed into a carbide body with an inner steel core. The polycrystalline diamond has hardness of 5000 HV, compressive strength of 7 GPa, and friction coefficient near 0.05. The fracture toughness is 10 MPa√m. Such a combination of properties provides durability in the harsh drilling conditions. The tips survive peak temperatures up to 1200 °C (1473 K) caused by friction, while delivering the extended component life required to avoid costly drilling rig downtime.

Design optimization requires sufficient polycrystalline diamond thickness to reach the predefined drilling depth without loss of hardness. Normally this results in individual teeth 10–75 mm in diameter with a 3 mm thick layer of polycrystalline diamond. The assembly is bonded into layers as pictured in Fig. 9.5. The top beveled disk is diamond and the lower segment is cemented carbide. One new variant relies on silicon liquid infiltrated into diamond to form a reactive composite of diamond and silicon carbide.

A mixture of particle sizes is employed to initiate sinter bonding between the diamond grains. The small grains dissolve and reprecipitate to bond the large

Fig. 9.4 A photograph of
an oil well drilling tip with
six arrays of cylindrical
cutters consisting of
polycrystalline diamond
faces on cemented carbide
backing blanks. This shape
is formed by slurry casting
the carbide body, sintering,
infiltration with a copper
alloy, and brazing of the
diamond faces

Fig. 9.5 An oil well
drilling cylinder cutter face
consisting of a WC-Co base
and polycrystalline
diamond face

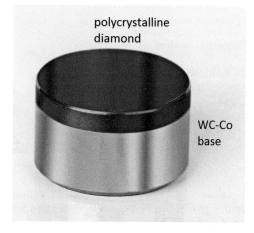

grains. To prevent diamond decomposition into graphite, consolidation is by spark
sintering at pressures of 6 GPa and temperatures near 1500 °C (1773 K). The final
microstructure consists of angular diamond grains, varying in grain size, bonded
together with interspersed WC-Co. Figure 9.6 is a scanning electron micrograph of
the structure, with the diamond appearing as the dark phase and cemented carbide
appearing as the light phase. This material is not stable at high temperatures. Only a
few minutes exposure are possible at temperatures over 1200 °C (1473 K) before
the diamond decomposes. Longer times are possible at lower temperatures.

Fig. 9.6 Microstructure of polycrystalline diamond where the majority of the structure consists of bonded diamond grains with the remainder consisting of WC-Co cemented carbide [courtesy A. Griffo]

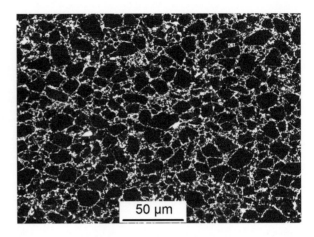

Fig. 9.7 A wire saw for cutting hard stones, such as marble or granite tiles. The bead consists of a diamond composite with tungsten and tungsten carbide shaped by injection molding and infiltrated during sintering [courtesy N. Williams]

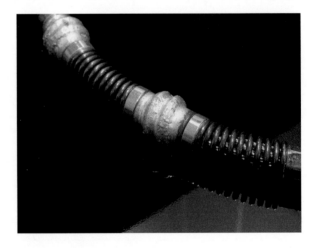

Diamond composites are an expensive option. To improve toughness, an array of matrix phases are employed in different industries. The bonds are based on resin, metal, cemented carbide, and silicon carbide [2]. At the upper performance range are metal bonded diamonds, such as imaged in Fig. 9.7. This is a picture of a wire saw with injection molded diamond composite beads. This wire saw is pulled through stone, like a cheese cutter, to slice tiles from marble, granite, and similar minerals. If diamond composites are too expensive, then lower cost and lower performance systems are employed. One example is steel containing alumina or glass containing alumina, but with substantially lower strength, hardness, and wear resistance.

Objective Based Design

Design of particulate composites involves simultaneous satisfaction of multiple objectives [3–15]:

- geometric attributes of size, shape, features, tolerances,
- constituent selection, materials and particles,
- microstructure features, grain size, connectivity,
- fabrication process to satisfy geometric requirements,
- production cost, tooling cost, and life cycle cost, including recycle considerations.

Life cycle costs differ from the purchase price, especially for long-lasting components. Additional requirements arising in design these days include consideration of sustainability, reuse, recycle, and greenhouse gas release during production. When performed in depth, design analysis can lead to surprising findings. For example, paper grocery bags are perceived as being friendlier to the environment versus plastic grocery bags, but when audited the paper bag production consumes 2.7-fold more energy, 17-fold more water, and releases 1.6-fold more greenhouse gases. Of course the plastic bag decomposes slowly and requires more effort in disposal.

Future engineers need to include life cycle and environmental impact as part of the design cycle. Technically we recognize the requirements for a taxicab differ considerably from the requirements for a sports car. While fuel economy, purchase price, and reliability dominate taxicab design, for a sports car the design emphasizes performance, handling, and appearance. The life cycle cost would be very high if we elected to use sports cars as taxicabs.

Successful designs depend on embracing the various challenges and constraints [14, 15]. Some of the common issues relate to the component geometry and features that cause additional manufacturing steps. Complexity is expensive both in manufacturing and in inspection. One common issue relates to surface finish, where a polished surfaces might easily double the cost. Finally, joining is often overlooked, but is often a weakness in the system.

Some component features add considerably to the cost. For example, perpendicular holes are more expensive to form versus parallel holes, and blind holes are more difficult to form compared to through holes. Likewise, material selection leads to added expense. Materials such as polycrystalline diamond are difficult to fabricate, so they are only available in simple shapes dictated by the high-pressure, high-temperature processing [16]. Of all aspects of engineering, component design is the most challenging. Unfortunately, it is mastered only through years of practice.

For some property-geometry combinations, the selection process is rather straightforward, especially if there is prior history of success. However, for other property-geometry combinations, especially in new designs, there is a need to synthesize evaluation criteria. This means the dominant factors are identified, ranked, and combined to provide an index of merit. Usually a few criteria dominate—such as strength, wear resistance, or corrosion resistance. A helpful technique is to build a design index that combines the dominant properties into a single parameter. In aerospace component design an early version was the specific modulus, calculated by dividing the elastic modulus by the material density.

As an example of the design index concept, consider the selection of a material for precision instrumentation. Let us assume the device needs to be dimensionally stable at the micrometer level, in spite of ambient temperature and load fluctuations. A high elastic modulus is desirable to reduce size fluctuations due to loading. At the same time a low thermal expansion coefficient minimizes temperature induced dimensional change. Additionally, a high thermal conductivity ensures thermal equilibrium during heating and cooling cycles [17, 18]. The search for materials and the identification of potential composites is enabled by material property databases. The first step is to map the general properties—for example thermal conductivity and elastic modulus. Such a map is given in Fig. 9.8. Note the scaling is logarithmic to handle the vast differences between natural rubber at one extreme and copper, aluminum, and diamond at the other extreme. Composites represent an opportunity to form materials to fill in areas where there are missing entries.

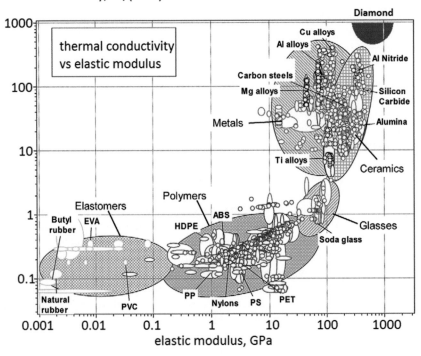

Fig. 9.8 A logarithmic scatter plot of thermal conductivity versus elastic modulus, representing about 3000 engineering materials. The shape of the plot shows the high thermal conductivity materials are also stiff. Note the unique position for diamond

Building from the two-dimensional map, it is appropriate to create a numerical design index [19]. This index allows for evaluation of several attributes, including cost. Further, the attributes can be weighted to give emphasis to some parameters. For the situation of a high thermal conductivity K, low thermal expansion coefficient α, and high stiffness as measured by the elastic modulus E, the resulting index I is as follows:

$$I = \frac{K E}{\alpha} \qquad (9.1)$$

In this form, each of the three design factors are given the same weighting. Accordingly, the index I leads to the conclusion that diamond would be an ideal material. This ignores cost. If cost is added to the denominator, then a choice would be silicon, and as more constraints are added the design becomes better focused.

As guidelines, the search needs to start broadly and gain focus as information is accumulated,

- start broadly and set hierarchical evaluation criteria,
- set priorities as a basis for conflict resolution,
- generally geometric attributes are most important, then engineering material, and finally manufacturing process, so keep priorities straight,
- examine system questions, such as inspection, assembly, environmental degradation, repair, and life cycle cost,
- objectively compare to evaluation criteria and boundary conditions, without emotional involvement.

Likewise, geometric attributes are grouped and ranked, with the production process selected after the geometric, compositional, and performance aspects are completed [5–9, 15, 20–22]. It is important to remain objective and not select the production process or vendor until after the best design is attained. This avoids sacrifice of design integrity. As design progresses, cost becomes a concern. However, as knowledge is gained on different composites and production routes, design modifications or property compromises are possible to lower cost [23]. Design accounts for 5 % of the component cost, but decisions made during design determine 90 % of the component cost. Skimping on the design work might make for costly problems during production.

An area of detailed study are bone implants consisting of porous titanium scaffolds with infused hydroxyapatite. Applications include fixation pins, facial reconstructions, replacement bones, hearing implants, and dental implant anchors [21]. One application is for an implant pin shroud, such as shown in Fig. 9.9. This pin is used in surgical attachment of tendons to bone.

Implants are focused on tissue ingrowth in pores of 100–200 μm. An important option is to match the elastic modulus and strength of bone using distended titanium [24, 25]. The high porosity titanium scaffold is combined with hydroxyapatite (calcium phosphate of the approximate composition $Ca_{10}(PO_4)_6(OH)_2$) filling the pores. The hydroxyapatite induces bone ingrowth while the porous titanium

Fig. 9.9 A small implant device used in surgical repair for securing a tendon to bone where biocompatibility is a dominant concern, the strength is 200 MPa

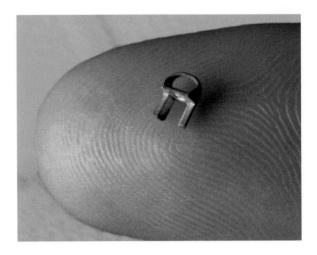

provides strength and toughness to match bone. The suitable compositions are nearly 60 % porous. Sintered strengths are close to bone, typically near 100 MPa.

Hidden Difficulties

Difficulties are often hidden in the design arena and turn up as surprises only when physical models are built. Engineers learn to anticipate the unanticipated, if that really is possible. For example, some materials decompose, melt, or react on heating. What starts as an exciting design turns into a disaster unless broad testing is employed prior to production. Thermodynamic calculations help avoid the formation of unanticipated new phases during use. Indeed, it is appropriate to apply "run-away tests" to simulate the worst case situations. It is possible to consult handbooks to rank the relative stabilities of candidate materials. Interfacial barrier coatings are available to avoid decomposition reactions during use. Good examples are titanium, zirconium, or chromium. In the same manner, dissimilar materials induce galvanic reactions, so consultation with electrochemical tables provides a means to assess situations that accelerate corrosion.

Weak bonding of phases typically results in low toughness, ductility, and fatigue strength. Signs of weak bonding are evident on fracture surfaces. Figure 9.10 is a scanning electron micrograph of a high toughness surgical stainless steel reinforced with silica particles. The fracture surface evidences failure nucleation at the weak interface with the reinforcing phase. Although the silica adds strength and hardness, the poor bonding between the two phases results in premature fracture and low ductility.

Mechanical stability is sensitive to stress and temperature oscillations. Both induce differential strains in the two phases. A difference in elastic modulus or

Fig. 9.10 Fracture surface of a surgical stainless steel reinforced with SiO_2 silica. The silica particles were not well bonded to the stainless steel, so under tension failure nucleated at the interface, leading to low ductility and low toughness

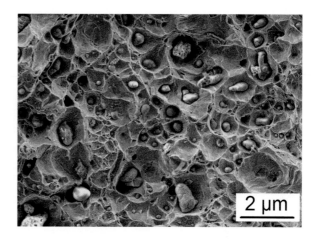

thermal expansion coefficient produces stress at the interface between the two phases. The resulting differential strains induce delayed failures. Indeed thermal fatigue is a problem in electronics due to on-off heating and cooling cycles. Differential strain within a particulate composite might induce fracture [26].

Another pervasive difficulty is with declining toughness as strength increases. Ideally, any cracks that form encounter an interface to blunt the crack and interrupt continued propagation, assuming a strong interface. However, in many high strength composites there is insufficient interface strength for this role. As a consequence, once a propagating crack reaches an interface, continued propagation is easier since the interface provides a short-cut path. Fracture surfaces showing preferential exposure of the interface indicate insufficient attention to interface strength. This difficulty is corrected using treatments to preferential improve interface strength. A thin interface layer is often successful in improving bonding. When properly manipulated, fracture paths avoid or minimize interface propagation.

Design for Processing

As a component design emerges, it is important to further consider how it might be fabricated. Independent of the process, the design should follow a few general rules [7, 19, 27–29]:

1. keep the component as simple as possible,
2. select the process based on the component features,
3. ensure the component ejects from the tooling without difficulty,
4. design for long tool life,
5. keep thickness as uniform as reasonable.

Fig. 9.11 Microelectronic packages formed by injection molding represent a high level of component complexity with over 200 dimensional specifications

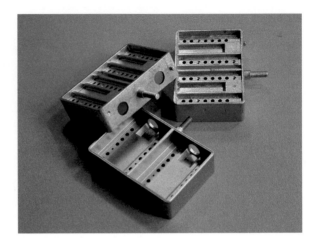

A few considerations dominate the process selection. A key factor is the component size, possibly categorized by the component mass. This is followed by considering the general shape. Certain shapes are naturally suited to different fabrication routes. Finally, tool cost is high for some options, so the production quantity is a factor to decide if the tool cost is justified [28]. Generally, high tooling costs are best justified by high production quantities, where the amortized tooling cost is small per component.

The popular production routes are introduced in the following sections. Process design information is organized around key attributes. Besides mass and overall size, secondary sorting factors include the shape complexity, best assessed by the number of geometric specifications. A cylinder is specified by just two dimensions (length and diameter), but an electronic circuit mount as shown in Fig. 9.11 requires hundreds of dimensional specifications. Nominally, a few parameters are useful in identification of a best candidate for production.

Die Compaction

Component production by uniaxial compaction is applied to flat, squat shapes, such as the magnetic pulse generator shown in Fig. 9.12. Table 9.1 lists various design features that align with the die compaction process. Listed are desirable features, restrictions, and some options. The typical production quantity is 150,000 per year, but some components range up to 20 million or more per year. Smaller productions quantities, say 2000 per year, are discouraged by the high tool cost. The typical mass is 40 g, with 40 dimensional specifications. Tolerances are generally about $\pm 0.3\%$ in terms of standard deviation divided by mean size.

Designs envisioning die compaction make sure that features are rationalized to the pressing direction and long tool life. For example, holes are only possible

Fig. 9.12 Magnetic rotational sensors called tone wheels, formed by die compaction of a mixture of iron and iron phosphide particles. These illustrate the flat, squat character of ideal die compaction shapes

Table 9.1 Outline of component design for die compaction

Desirable features
Squat structure, low height
Slight taper for ejection
Largest dimension below 200 mm
Weight less than 100 g
Wall thickness less than 30 mm
Allowed features
Multiple through holes in pressing direction
Hexagonal, square, round holes
Indents, marking, lettering
Tapered side walls
Restricted features
No closed cavities
No undercuts perpendicular to pressing direction
Hole diameter larger than 0.1 mm
Wall thickness greater than 0.2 mm
Weight under 15 kg

parallel to the pressing axis. Early in the design cycle it is common to redesign the component to avoid problems. A few redesign situations are illustrated in Fig. 9.13. Note the shapes can be complicated when viewed from the top, but from the side view the shape is relatively simple. The sketches in this figure show how features that are difficult to compact can be included in a redesign to better match with the process. For example, sharp corners are problematic, so rounded corners are suggested as an alternative.

Multiple level components are possible, with concomitant increases in the demands on the tooling and compaction press. Holes are created by core rods running inside the punches, and steps are created using segmented punches.

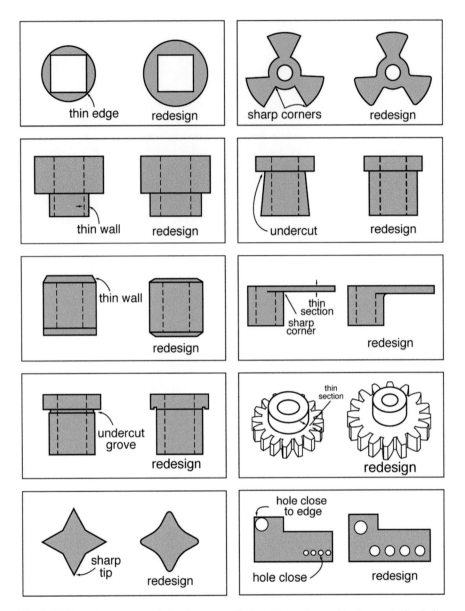

Fig. 9.13 Several component design issues are illustrated here along with changes to the design intended to ease production, mainly by avoiding situations where tool damage would occur

Round holes are common, but other shapes are possible, such as keyways, half-circles, and squares. Blind holes are possible, but not common.

Flat components are easiest to press, but there are means to add flanges and steps. Wall thickness is more than 1.5 mm to avoid damage to the tooling and to the

Fig. 9.14 An example of an everyday component fabricated by die compaction and infiltration. This automotive ehaust flange is a composite of steel and infiltrated copper alloy

pressed compact. Long and thin components are avoided because of both short tool life and significant density and property gradients.

Typical tolerances for die compaction range are about 0.3 %. Tighter tolerances require grinding or machining as secondary manufacturing steps. Dimensions perpendicular to the pressing direction vary less when compared to dimensions aligned with the pressing direction.

Infiltration of liquid into a die compacted shape is a means to form a composites. In this approach, the major phase powder is die compacted to form a blank, nominally about 65–85 % dense. Subsequently during heating to the sintering temperature, a second phase is melted and infiltrated into the pores. The infiltrated component exhibits about twice the dimensional variation of the pressed component. One example application is in the fabrication of automotive exhaust flanges as pictured in Fig. 9.14. The steel matrix is die compacted and then infiltrated with a copper alloy, providing the desired resistance to gas leakage as well as high strength and toughness. Similar ideas are used in forming electrical circuit breaker contacts from W-Ag, where first tungsten powder is compacted and then heated in contact with a silver foil that melts and infiltrates the pores.

To summarize, die compaction, with subsequent sintering or sintering and infiltration, is best suited for components averaging 40 g with squat shapes, where the complexity is primarily visible along the pressing direction.

Injection Molding and Extrusion

Injection molding is applied to smaller three-dimensional components, typically around 10 g in mass and 25 mm in maximum size, with about 75 specified features. An example of the shape complexity is illustrated in Fig. 9.15. This cobalt-chromium carbide housing has holes and support fins formed as part of the molding

Fig. 9.15 An example of
an injection molded wear
component consisting of
chromium carbides
dispersed in cobalt,
illustrating the significant
complexity possible with
this fabrication route

Table 9.2 Injection molding
component design guidelines

Desirable features
Gradual section thickness changes
Largest dimension below 100 mm
Weight less than 100 g
Wall thickness less than 10 mm
At least one flat surface for support
Allowed design features
Holes at angles to one another
Hexagonal, square, blind, and flat bottom holes
Stiffening ribs
Knurled and waffle surfaces
Protrusions and studs
Threads
Part number or identification
Restrictions
No closed cavities
No internal bore undercuts
Corner radius greater than 0.075 mm
2° draft on long parts for ejection
Smallest hole diameter 0.4 mm
Minimum thickness 0.2 mm

process. Table 9.2 provides qualitative design guidelines, including desirable,
optional, and undesirable features. The minimum section thickness relates to the
particle size, where the thickness must be more than ten times the particle diameter.
Particle size also limits the sharpness of corners, since several particles are needed
to fill out the dimension. A practical lower limit is 0.05 mm edge radius and 0.1 mm

Fig. 9.16 A WC-10Co
spindle formed by injection
molding. It would be very
difficult to machine these
features

corner radius. Some allowed design features include square holes, flat bottomed holes, knurled surfaces, pins, fins, slots, grooves, and features such as threads. Although these features are not required, they are possible and add to the design flexibility. Figure 9.16 is a photograph of a hard WC-10Co spindle component fabricated with holes, slots, and undercuts using injection molding. Due to the high hardness, adding these features by machining would be difficult and expensive. Such complexity is much less expensive using injection molding, assuming the production quantity is large enough to justify the tool fabrication.

The wall thickness is typically below 3 mm. Thicker walls are possible, but add to cost due to slow production. As outlined in Fig. 9.17, it is most desirable to have uniform and thin walls to avoid stresses that might distort the component. Thick walls slow processes such as debinding, so redesigns are appropriate. Tolerances for injection molded components are generally in the ±0.3 % range, and in a few instances less than ±0.1 % variation has been achieved in production. A smaller dimensional scatter usually requires machining or grinding of the component.

Component design needs to allow for placement of tool blemishes that include parting line, ejector pin blemishes, vent, and gate, hopefully where these features do not detract from the performance or aesthetics. A slight draft angle 0.5–2° eases ejection from the mold.

Similar aspects apply to extrusion of powder-binder feedstock. The same binder-powder mixture as used in injection molding is forced through a die to form long, constant cross-section shapes. Some common uses are in forming cemented carbide rods, tubes, or helical twist drills. Length is not a limitation with extrusion, but - cross-sectional area is limited by the press capacity and is usually less than 100 mm^2.

In summary, injection molding is ideally suited to 10 g components, but can be applied to larger mass ranges to even 100 g. The maximum dimension for the component is usually less than 100 mm, but long-thin components up to 200 mm or more are in production. The shape is three-dimensional, with typically

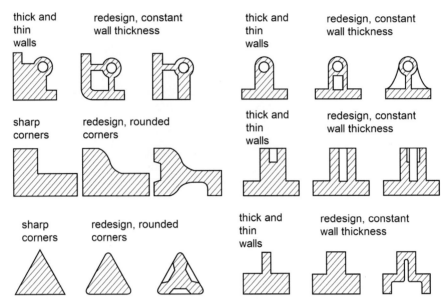

thick and thin walls | redesign, constant wall thickness

sharp corners | redesign, rounded corners

thick and thin walls | redesign, constant wall thickness

sharp corners | redesign, rounded corners

thick and thin walls | redesign, constant wall thickness

thick and thin walls | redesign, constant wall thickness

Fig. 9.17 Some component redesign suggestions for injection molding, where sharp corners and thick-thin sections are removed where possible

75 dimensional specifications. Usually, die compaction is a lower cost forming process compared to injection molding, so injection molding is better applied to more complicated designs that cannot be die compacted.

Forging and Hot Pressing

Two powder consolidation options apply stress and heat simultaneously. At high densification rates are forging processes. Hot pressing is at a lower strain rate, lower stress so it delivers slower densification. Compact shape is limited with these approaches, typically producing flat components. Heat is used to soften the particles and densification occurs by pressure-induced deformation. For hot pressing the first step is filling the die with loose powder, but for forging the starting point is a previously sintered preform. That preform is designed to drop into the forging cavity where it flows while densifying. The temperature and pressure combination for both approaches requires special forming tooling and forming devices.

A typical hot pressed component is flat and rather simple in shape, such as the diamond cutting segments pictured in Fig. 9.18. Forging starts with a preform that is crafted to allow more powder flow occurs during densification, giving more complicated shapes. When properly design, the final shape conforms to that set by the

Fig. 9.18 Diamond
segments formed by hot
pressing, illustrating the
relative shape simplicity
dictated by the compaction
conditions and weak
graphite tooling

Fig. 9.19 A truck engine
connecting rod formed by
forging a mixture of iron,
copper, and graphite
powders to form a full
density steel-copper
structure. Note the rounded
corners associated with
forged shapes

tooling while reaching full density at the end of the forging stroke. The material
flow provides for more final component shape complexity, as illustrated in
Fig. 9.19. Neither approach is precise; dimensional scatter tends to range up to
$\pm 0.4\,\%$. Production quantities are potentially large, and a few applications reach
10 million components per year. However, tool wear requires reworking the dies
every few thousand compacts due to the high temperature and high stress. The mass
and size range is limited by the press capacity, where 200 mm and 2 kg would be
considered large components.

After densification the component is machined to achieve desired final dimensions, especially the forged components. Hot pressing and forging are best suited to
squat, rounded components that lack sharp corners.

Slurry and Isostatic Processes

Slurry approaches rely on filling a rigid mold with a viscous particle-liquid mixture. Options range from casting liquid metals with entrained ceramic particles to casting polymer-particle slurries, such as in slip casting. The dimensional tolerances are low, about $\pm 3\%$, so these forming approaches are best applied to freestanding components that do not need to mate with other parts. This is because the low gravity forming pressure is not able to induce uniform mold filling. Further, gravity causes settling over time, leading to gradients and nonuniform dimensions. This reflects the need for a low viscosity slurry to enable pouring into the mold, but that low viscosity allows solid separation after mold filling.

Slurry approaches tend to succeed for limited production (100 per year) of components in the 100 g to 10 kg range, with maximum dimensions up to 0.5 m. Wall thickness is at least 1 mm. For stand along objects, slurry techniques are ideal. Tooling can be very inexpensive, consisting of rubber or Plaster of Paris, so slurry approaches are useful for limited production situations, in spite of the dimensional variation. Figure 9.20 is an example of a plastic mold fabricated by slurry casting tool steel and then infiltrating the pores with copper. After fabrication, the mold is machined to reach close tolerances.

Cold isostatic compaction is applied to components from 10 g to 1000 kg. The smaller components tend to be abrasive tips, such as seen in the dental office, and the larger components are for mining or excavation tools, such as seen in tunnel drills. The length is limited by the pressure chamber size, up to 2 m maximum size. These components are fabricated with considerable dimensional variation due to the soft tooling. Tolerances for cold isostatic pressing are scattered, typically showing $\pm 5\%$ variation within a single component. For this reason the compacts are machined to final sizes. Improved dimensional control is possible using rigid inserts in the mold. A few abrasive composites, used for grinding and drilling, are produced at high production rates approaching 6 per minute, but more typical are production rates of 4 per hour. The shapes are simple so tooling is not expensive.

Fig. 9.20 A mold formed by slurry casting tool steel powder with subsequent bronze infiltration to form a composite structure that is then used for plastic injection molding

Hot isostatic pressing is similar to cold isostatic pressing in dimensional capabilities, with similar limitations due to pressure chamber size. Dimensional scatter is typical, unless special precautions are taken to ensure uniform powder filling and heating. Rigid containers need to anticipate the anisotropic shape change, especially near corners, to reduce final compact scatter. Variations in shrinkage is as much as 6% between locations, giving a dimensional scatter of 1% within a component. There is less scatter over repeat fabrication runs. The general protocol is to machine the compact after consolidation. When hot isostatic pressing is applied to densify previously sintered components, in the containerless mode, dimensional scatter is not as large. Because the hot isostatic pressing cycle time is long, measured in hours, the technique is applied to lower production quantities, as low as one compact per day. The mass range is up to 1000 kg; one recent component reached 16,000 kg, but this is not typical. Since tooling is used only once, tool costs are high.

Specifications

To summarize points made in this chapter, component specification might include some of the following, depending on the situation:

1. composition, phases, and ratio of phases,
2. geometric design, captured in the engineering definition or drawing,
3. information on the inspection plan, critical features, and expected properties,
4. tests and required property ranges, such as minimum strength,
5. statistical aspects of acceptance; maximum percent allowed out of specification,
6. required density, maximum allowed pore size,
7. qualification tests and criteria,
8. applicable industry standards,
9. microstructure features, such as grain size,
10. surface finish, packaging, marking, and other delivery details.

A high level of detail is required to ensure the delivered product is suitable for the intended application. It is appropriate to reference industry standards or tests to ensure conformance to the design conceptualization.

Study Questions

9.1. Small rockets employ guidance vanes to steer during flight. For a short duration rocket, with flight lasting about 60 s, the specifications require 500 MPa room temperature strength and melting temperature in excess of

2000 °C (2273 K). Additionally, a ductile material is desired to survive handling. What are some possible composite materials?

9.2. For the rocket guidance vane in Study Question 9.1, one suggestion is W-20Cu where the copper melts and evaporates during use. The evaporation enthalpy provides cooling, similar to how sweat cools an athlete. There is concern the structure will be too weak at a typical operating temperature of 1200 °C (1473 K). What is the estimated strength for this composite at 1200 °C (1473 K)?

9.3. The diamond beads used in wire saws for cutting architectural stone rely on a high drag force for the cut. Is it possible frictional heating might decompose the diamond in this application? How might such heating be reduced?

9.4. Faced with the wide scatter in properties for Al-SiC composites, as shown in Fig. 9.3, suggest steps you might specify to remove the scatter from purchased components made from this material.

9.5. In the production of an automotive component, a new facility is commissioned. To qualify the component, 100 % inspection is required to ensure a defect rate less than 1 part per million (a million parts are tested and only 1 is defective = 1 ppm). If the facility is producing 8640 parts per hour in a 24-7 operation (around the clock), what is the minimum time to demonstrate compliance? The customer requires 3 months of such testing—what is the intent of this long qualification period?

9.6. Sometimes a component fails third-party testing, such as that performed by Underwriters Laboratories. Many applications require compliance to these specifications. As a design engineer, responsible for a component that fails initial testing, outline your response to assure management the problem will be permanently corrected.

9.7. A high voltage electrical contact is required for turning power generation equipment on and off. Nominally the contact operates at 440 V AC and 20,000 A. How would you approach the design of such a contact, including materials selection and specification of minimum properties?

9.8. For the electrical contact described in the previous Study Question, outline qualification tests that should be performed to ensure proper function.

9.9. A knife blade is designed with the goal of never needing sharpening. Additionally, resistance to corrosion is required. A composite rim is envisioned that would be located on the cutting edge of the knife. What composite composition might be considered for this application?

9.10. Machines used to knit and weave wire meshes, such as screen doors, exhibit rapid wear. Outline a program to evaluate composites that might be used to form the weaving equipment. Specifically what tests are required?

References

1. K. Friedrich (ed.), *Composite Tribology* (Elsevier Science, Amsterdam, 1993)
2. J. Konstanty, *Powder Metallurgy Diamond Tools* (Elsevier, Amsterdam, 2005)
3. M.F. Ashby, Y.J.M. Brechet, Designing hybrid materials. Acta Mater. **51**, 5801–5821 (2003)

4. E.J. Barbero, *Introduction to Composite Materials Design* (CRC Press, Boca Raton, 2011)
5. D.M. Bryce, *Plastic Injection Molding: Manufacturing Process Fundamentals*, vol. 1 (SME, Dearborn, 1996)
6. G.E. Dieter (ed.), *Materials Selection and Design*. ASM Handbook, vol. 20 (ASM International, Materials Park, 1997)
7. P. Dewhurst, C.C. Reynolds, A novel procedure for selection of materials in concept design. J. Mater. Eng. Perform. **6**, 359–364 (1997)
8. A. Fujiki, Present state and future prospects of powder metallurgy parts for automotive applications. Mater. Chem. Phys. **67**, 298–306 (2001)
9. R.M. German, A. Bose, *Injection Molding of Metals and Ceramics* (Metal Powder Industries Federation, Princeton, 1997)
10. A. Mortensen, J. Llorca, Metal matrix composites. Annu. Rev. Mater. Res. **40**, 243–270 (2010)
11. D. Shetty, *Design for Product Success* (SME, Dearborn, 2002)
12. A.I. Taub, Automotive materials: technology trends and challenges in the 21st century. Mater. Res. Soc. Bull. **31**, 336–343 (2006)
13. S. Torquato, Optimal design of heterogeneous materials. Annu. Rev. Mater. Res. **40**, 101–129 (2010)
14. M.P. Groover, *Fundamentals of Modern Manufacturing*, 4th edn. (Wiley, Hoboken, 2010)
15. G. Boothroyd, P. Dewhurst, W. Knight, *Product Design for Manufacture and Assembly*, 3rd edn. (CRC Press, Boca Raton, 2010)
16. D. Belnap, A. Griffo, Homogeneous and structured PCD.WC-Co materials for drilling. Diam. Relat. Mater. **13**, 1914–1922 (2004)
17. B.S. Zlatkov, H. Danninger, O.S. Aleksic, Cooling performance of Tube X-Cooler shaped by MIM Technology. Powder Injection Moulding Int. **2**(1), 51–54 (2007)
18. S.J. Park, Y.S. Kwon, S. Lee, J.L. Johnson, R.M. German, Thermal management application of nano tungsten - copper composite powder, in *Proceedings PM 2010 World Congress on Powder Metallurgy*, Florence Italy October, European Powder Metallurgy Association, Florence, Italy, 2010
19. M.F. Ashby, *Materials Selection in Mechanical Design*, 4th edn. (Butterworth-Heinemann, Burlington, 2011)
20. T.W. Clyne, P.J. Withers, *An Introduction to Metal Matrix Composites* (Cambridge University Press, Cambridge, 1993)
21. J.R. Davis (ed.), *Metals Handbook Desk Edition*, 2nd edn. (ASM International, Materials Park, 1998)
22. R. Warren (ed.), *Ceramic-Matrix Composites* (Blackie, Glasgow, 1992)
23. J.L. Johnson, Opportunities for PM processing of metal matrix composites. Int. J. Powder Metall. **47**(2), 19–28 (2011)
24. M.A. Meyers, P.Y. Chen, *Biological Materials Science: Biological Materials, Bioinspired Materials, Biomaterials* (Cambridge University Press, Cambridge, 2014)
25. K. Kondoh, Titanium metal matrix composites by powder metallurgy (PM) routes, in *Titanium Powder Metallurgy*, ed. by M.A. Qian, F.H. Froes (Elsevier, Oxford, 2015), pp. 277–297
26. C. Zweben, Advanced composites and other advanced materials for electronic packaging thermal management, in *Proceedings International Symposium on Advanced Packaging Materials* (IEEE, New York, 2001), pp. 360–365
27. J. Hafner, C. Wolverton, G. Ceder, Toward computational materials design: the impact of density functional theory on materials research. Mater. Res. Soc. Bull. **31**, 659–665 (2006)
28. R.M. German, *Powder Metallurgy and Particulate Materials Processing* (Metal Powder Industries Federation, Princeton, 2005)
29. J.D. Meadows, *Geometric Dimensioning and Tolerancing* (Marcel Dekker, New York, 1995)

Chapter 10
Optimization

Design optimization involves maximization or minimization of selected attributes with simultaneous satisfaction of secondary goals. The mapping of properties versus adjustable parameters, such as composition, helps identify an optimal solution. Unfortunately, optimization of properties often conflicts with lower cost. A multiple-parameter approach is reduced to two-dimensional plots to guide parameter selection in reaching an optimal design.

Conceptualization

Specifications for particulate composites focus on the property combination required to deliver the required performance. Engineers focus on features (such as tolerances, strength, hardness, and toughness), while users focus on benefits (long lasting, scratch resistant, resists corrosion). For example, a coffee cup should have a low thermal conductivity as a feature. The benefit is an ability to hold hot coffee without burning your hand. Design transforms consumer perceived benefits into specific engineering features against which design optimization occurs. Sales and marketing appropriately focus on the benefits derived from included features, while engineering focuses on the properties and specifications:

benefits ↔ features ↔ properties ↔ specifications.

Optimization is frequently mentioned in advertising. Automotive companies tout "optimal cornering" or "optimum ride comfort". These messages are telling us about design compromises. Although not explicitly stated, a trade-off is made to settle on a perceived balance. Hidden in these statements are many assumptions and constraints. Unfortunately optimization against multiple objectives is difficult. What is optimal for one application is not necessarily optimal for another. Two-wheel drive transmissions are better for fuel economy, but four-wheel drive is better for winter driving. This is a case of different situations with different

© Springer International Publishing Switzerland 2016
R.M. German, *Particulate Composites*, DOI 10.1007/978-3-319-29917-4_10

"optimal" designs—one for fuel economy and one for inclement weather. A compromise of three-wheel drive is rejected since a more realistic solution is part time four-wheel drive, switching from two-wheel drive in dry conditions to four-wheel drive in snow.

Engineering optimization requires a response model that includes the adjustable variables linked to the parameters being optimized [1]. This requires a mathematical equation, computer simulation, spreadsheet, or related platform for analysis. Usually there are constraints on the range for the adjustable variable search. In addition, experience from earlier designs identifies prior successes and failures. To handle interactions between multiple variables requires computer models. The solution quality rests on the accuracy of the response model. Novel programs generate mathematical fits to raw data without regard to the underlying physics. Such models are helpful, but limited in scope, especially if projected outside the data range used in their generation.

In short, optimization requires

- performance objectives,
- constraints statement,
- adjustable variables,
- parametric relations linking variables to performance attributes.

For multiple objectives it is possible to give relative weighting to emphasize parameters. For example, in a design considering cost, strength, and toughness, the parameter weighting might be cost 50 %, strength 35 %, and toughness 15 %. Compromise is required to strike a balance between the objectives. Otherwise, if a luxury automobile is truly optimized for riding comfort, then it would be large, sluggish, heavy, and rely on underinflated tires, resulting in terrible fuel economy, poor parking ease, and an enormous cost. On the other hand, a hybrid vehicle with overinflated tires is nimble, fuel efficient, easy to park, and inexpensive due to government subsidies. Optimization in inherent to system design, but the optimal solution depends on the boundary conditions.

The optimal solution produces acceptable values for the object functions within the range deemed practical. Effectively, an optimized design is based on acceptable trade-offs of parameters [2]. At least 30 commercial computer-based solution techniques are available to handle design optimization. They require boundary conditions, parametric relations, and weighting factors. The merits of the different solution techniques are the subject of whole books and review articles.

In particulate composites, the phase morphology and component dimensions are optimized using numerical techniques. Often there are surprising solutions, such as the discovery of enhanced fatigue endurance strength in a heterogeneous steel composite [3]. The properties depend on several design factors, including the amount and placement of each phase. Computer optimization calculations apply intentional changes to identify improved configurations. Some efforts rely on random variations coupled to finite element analysis. After a few iterations,

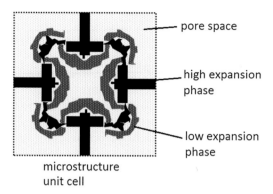

microstructure
unit cell

Fig. 10.1 An example of computer optimized three phase composite structure. This case optimizes the layout for a negative thermal expansion coefficient, formed using pores, low thermal expansion phase, and high thermal expansion phase. This unscaled unit cell is repeated to make a three-dimensional body. Optimization routines identify such structures using random search protocols

gradients are calculated. Those gradients are used to set the direction for the next iteration. The process continues until either a maximum or minimum is found.

The children's game "Marco Polo" relies on a pulse-echo idea. It is played in a swimming pool with one player blindfolded. The other players scatter and when the blindfolded player yells "Marco" the other players respond with "Polo". The echo helps the blindfolded player decide on direction to tag someone else. With repeated pulses the person that is "it" eventually makes sufficient course corrections to tag one of the other players. In a similar manner, an initial design provides a pulse, and analysis of the structure gives the echo. Repeated design modifications help identify beneficial changes. Repeated cycles lead to an optimized design.

The gradient approach to optimization results in novel structures. For example, Fig. 10.1 is an idealized cell for a low thermal expansion composite. It is based on an optimized mixture of pore, low thermal expansion solid, and high thermal expansion second phase; the thermal expansion of the two solid phases differ by a factor of 10. This design evolved from the concept outlined above where changes are made based on property gradients identified through progressive random changes. The result is a novel design; the structure allows for local expansion while the bulk structure undergoes contraction.

The optimization function is based on partial derivatives. To outline the approach, a functional relation, say between composition and performance, is expressed by $F_i(X_j)$. Here X_j is one of k adjustable parameters (such as composition, particle size, grain spacing, or particle hardness) and $F_i(X_j)$ is a functional relation between the adjustable parameters and performance. The weighting of different attributes is by w_i. Optimization occurs by seeking the minimum or maximum for a functional relation;

$$\min/\max \sum_{i=1}^{k} w_i F_i(X_j) \qquad (10.1)$$

This is assessed using the partial derivative with respect to an adjustable parameter, finding the gradients on how each attribute F_i changes with small variations in each of the adjustable parameters X_j. Experimental data are needed to feed these models, so a combination of calculation and experimentation is required [4].

Solution routines guide the selection to the best overall response based on the governing relations. A favorite procedure examines property gradients. Small incremental changes determine the gradient. Just as water flows downhill, the computational solution seeks to find minimum or maximum. Similar ideas are seen in using a topographic map. Elevation is plotted versus latitude and longitude. As an example, Fig. 10.2 is topological map near Lake Hodges in San Diego, California. The contour lines indicate equal elevations. Regions where the contour lines are closely spaced are steep, while regions of large contour line spacing are

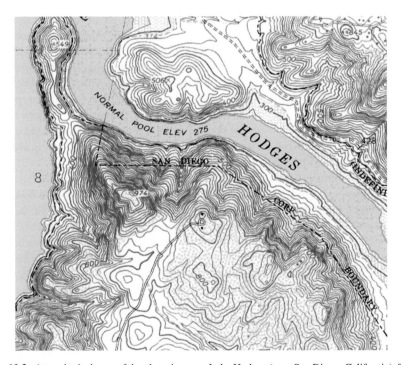

Fig. 10.2 A topological map of the elevation near Lake Hodges (near San Diego, California). The wavy topographic lines indicate uniform elevations. Steep gradients occur where the lines are close and indicate favored directions for rainfall drainage. In the same sense, an optimization map is used to calculate parametric changes as the independent parameters are tested to sense gradients, leading to either maximization or minimization based on parametric changes

nearly flat. The lake is surrounded by steep terrain. If rain falls on the peak at 974 ft, gravity causes it to flow along the downhill gradient into the lake. At any point on the map, it is possible to calculate the steepest gradient to estimate the water flow direction. In the same manner, optimization routines test for the local gradients, using repeated steps to determine the gradients pointing to the lowest point.

Optimization response is complicated if a large number of adjustable parameters exist. Further, the optimum design changes depending on what factors are included and the relative weighting of those factors. For example, the optimization of a Ti-TiB automotive component is driven by cost, while medical applications are more focused on properties [5, 6]. Depending on the weighting function for cost, then different optimizations result. While one automobile is optimized for fuel efficiency (with minimum sound insulation), another is optimized for ride comfort and includes copious sound insulation with a sacrifice in fuel efficiency.

It is important to set boundary conditions on the solutions. This eliminates unreasonable situations, outside the predefined range of conditions. Further, solutions are constrained by practical issues such as material availability, as well as toxicity and equipment capacity. Otherwise the optimization might suggest ridiculously expensive ingredients. Robust solutions require the system be immune to normal fluctuations in raw materials, fabrication steps, or use conditions. A robust solution implies a lack of sensitivity to such variations and a high quality solution.

A wide variety of simulation routines are applied to engineering optimization. It is not the intent to treat the approaches here. However, the overall sense on how design optimization occurs is illustrated in this chapter for a few particulate composite examples.

Optimization Protocol

The optimization approach advocated here starts with construction of response models. Each of the important properties is linked to composition and microstructure using constitutive relations trained with existing data. In some cases the parameters are interlinked. Solution then comes from iterative approximations that eventually converge. That approach requires boundary conditions and assessment parameters to capture overall merit. For example, exceeding a minimum strength say 700 MPa, but at the lowest cost. What is an ideal solution this year might fall out of favor next year due to change. For example, the catalytic converter in an automobile varies in formulation between platinum, palladium, or rhodium depending on which material has the lowest market price.

In this chapter, three examples are used to illustrate optimization. The first example deals with heat dissipation, and a trade-off between cost, thermal conductivity, and thermal fatigue failure. It focuses on copper-molybdenum composites. The second example relies on the concept of a composite within a composite. In this case agglomerates of WC-Co are enmeshed in cobalt as a matrix to form a wear resistance composite. At low magnification the agglomerate phase appear as reinforcement grains, but at high magnification each of the agglomerates is itself a composite. Here optimization involves consideration of wear resistance, fracture toughness, and cost. A third example is a composite of alumina-zirconia for artificial hip implants, where the cost of installation far outstrips the device cost. Accordingly, optimization is focused on defect-free structures for long service life with the intent of avoiding repeat surgery. In this situation there is less concern with material and fabrication costs.

Optimized Heat Dissipation Considering Thermal Fatigue

Silicon, the basis for microelectronics, has a low thermal expansion coefficient. During operation silicon loses conductivity as it heats, and it tends to heat since it is a poor electrical conductor. To preserve high response rates and circuit efficiency, it is desirable to rapidly remove heat from the semiconductor [7]. The problem is especially acute with high current levels, such as encountered in radar, hybrid vehicle, and inverter systems. High currents densities cause resistive heating, resistive heating then lowers conductivity—effectively initiating a spiral of declining performance. To circumvent this problem, heat spreaders are bonded to the underside of the silicon semiconductor [8]. Ideally the heat spreader is high in thermal conductivity. Thus, diamond, gold, copper, or silver would be good choices. Aluminum is often selected since due to lower cost, in spite of lower performance. Flowing air removes the heat, often requiring a small fan in the device. During use thermal loads can reach the 100–200 W/cm^2 range; for example, in personal computers 125 W is distributed over 1.5 cm^2. For the higher loads an alternative is used wherein a pumped water radiator is installed, similar to how an engine is cooled.

However, during on-off cycles the silicon heats and cools, expanding and contracting. If the heat spreader is different in thermal expansion coefficient, elastic modulus, and other properties with respect to silicon, then differential strains arise. Cyclic strain causes thermal fatigue at the interface, similar to how cyclic stress or cyclic strain cause mechanical fatigue [9]. To minimize separation of the silicon from the heat spreader, a close match of thermal expansion coefficients is desired. Since silicon is fixed, then the heat spreader is adjusted to minimize thermal fatigue.

Depending on weight restrictions, the heat spreader might be a composite of W-Cu, Mo-Cu, Al-SiC, Al-AlN, SiC-Si, or SiC-Cu. For the case illustrated here the materials and solution approach trace to data for silicon power transistors bonded to Mo-Cu. It is possible to repeat these calculations for other candidate materials. If mass is a limitation, such as in notebook computers, then material density is added to the problem. For this solution a particulate composite sheet is envisioned as sketched in Fig. 10.3. Performance is deemed unsatisfactory when partial delamination occurs, defined by a 25 % loss in interface thermal conductivity. Table 10.1 gives measured thermal fatigue results for 6 mm thick heat spreaders, recording the number of cycles (2 min heating, 3 min cooling) until loss of integrity for three

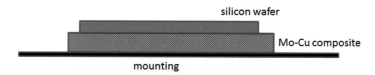

Fig. 10.3 The idea of a heat dissipation device acting to cool a semiconducting circuit to reduce loss of performance due to self-heating

Table 10.1 Thermal fatigue results for power semiconductors

Temperature change, °C	Interface stress, MPa	Cycles to failure
47	85	4000
65	117	1800
110	198	800

temperature changes. For high reliability systems the stipulated temperature change is 110 °C, corresponding to cooling to −20 °C and heating to 90 °C, a test reflective of the potential cycle anticipated in power electronics.

Density

For copper-molybdenum, a first consideration is the linkage of composition to density. The variables and constants are as follows:

ρ_{Cu} = density of copper = 8.96 g/cm³
ρ_{Mo} = density of molybdenum = 10.2 g/cm³
W_{Cu} = weight fraction of copper
V_{Cu} = volume fraction of copper.

As dependent variables, the molybdenum weight fraction W_{Mo} and volume fraction V_{Mo} are $W_{Mo} = 1 - W_{Cu}$ and $V_{Mo} = 1 - V_{Cu}$, respectively. The relation between weight fraction and volume fraction is,

$$V_{Cu} = \frac{\frac{W_{Cu}}{\rho_{Cu}}}{\frac{W_{Cu}}{\rho_{Cu}} + \frac{(1-W_{Cu})}{\rho_{Mo}}} \tag{10.2}$$

Based on composition, the particulate composite density is determined from the volume fraction and densities, as follows:

$$\rho_C = \rho_{Cu-Mo} = V_{Cu}\,\rho_{Cu} + (1 - V_{Cu})\rho_{Mo} \tag{10.3}$$

The composite density range is limited because of the relatively small difference between copper and molybdenum densities, assuming no porosity. Since porosity is detrimental to thermal conductivity and strength, full density is assumed.

Cost

The composite cost is a function of composition. Let C_{Cu} represent the cost of high purity copper particles, nominally $48 per kg, and C_{Mo} represent the cost of deagglomerated, high purity molybdenum particles, nominally $60 per kg. The cost for Cu-Mo per mass C_{Cu-Mo} depends on the composition,

$$C_{Cu-Mo} = W_{Cu} C_{Cu} + (1 - W_{Cu}) C_{Mo} \qquad (10.4)$$

As with many designs, the heat spreader is fixed in volume, so cost must be s rationalized to a volume basis using the composite density. Thus, the cost calculation involves determining C_V the cost per volume as,

$$C_V = \rho_C C_{Cu-Mo} \qquad (10.5)$$

These relations give a material cost per device, recognizing the processing cost is an additional factor, assumed to be essentially the same for each composition. Processing costs are about $10 per kg are overshadowed by material cost. A lower copper cost is possible, using copper infiltration into a porous molybdenum skeleton, lowering the copper cost for an equivalent conductivity to about $26 per kg. However, infiltration with sheet material adds additional fabrication steps, including final machining to remove surface residue. Essentially, infiltration reduces material cost, but proportionately increases processing cost, so that alternative is ignored. Also, infiltration is limited to a narrow composition range, so should be considered only after an ideal composition is identified.

Thermal Conductivity

Thermal conductivity needs to be as high as reasonable. Typically the design goal is to exceed 200 W/(m °C) to be in the range of extruded or die cast aluminum products, which suffer a disadvantage of a high thermal expansion coefficient. Since 200 W/(m °C) is a minimum thermal conductivity, detailed calculations involving microstructure are not required and a volumetric rule of mixtures is adequate. Accordingly, the composite thermal conductivity K_C is estimated from the constituents as follows:

$$K_C = K_{Cu-Mo} = V_{Cu} K_{Cu} + (1 - V_{Cu}) K_{Mo} \qquad (10.6)$$

where K_{Cu} is the thermal conductivity of pure, dense copper at 395 W/(m °C) and K_{Mo} is the corresponding value of 138 W/(m °C) for pure molybdenum. Both values change with impurities, residual strain, and temperature. More details models are available, including elastic interactions and phase connectivity, changing the calculated composite conductivity by about 10–20 %.

Thermal Expansion Coefficient

Any difference between thermal expansion coefficient for silicon and the copper-molybdenum causes thermal fatigue. This is why low-cost aluminum is not useful. To estimate the number of cycles to failure requires knowledge of the thermal expansion coefficient α. Detailed calculations involve the elastic modulus, Poisson's ratio, and thermal expansion coefficient of both phases to take into account the interaction strains, especially with large temperature fluctuations. For first analysis the elastic interaction strains are ignored so a volumetric rule of mixtures approach is assumed for the composite,

$$\alpha_C = \alpha_{Cu-Mo} = V_{Cu} \alpha_{Cu} + (1 - V_{Cu}) \alpha_{Mo} \qquad (10.7)$$

The two input values are $\alpha_{Cu} = 17 \times 10^{-6}$ (1/°C) and $\alpha_{Mo} = 5 \times 10^{-6}$ (1/°C). High copper contents deliver more thermal conductivity with a negative impact on thermal expansion coefficient, implying a shorter life due to thermal fatigue.

Thermal Fatigue

For silicon mounted on copper-molybdenum a linkage is required between design features and cycles to failure N. Experimental data show that for any given situation the failure probability follows a Weibull distribution, dependent on three parameters;

$\Delta\alpha$ = difference in thermal expansion coefficient between the composite and silicon

$$\Delta\alpha = \alpha_c - \alpha_{si}$$

assume $\alpha_{Si} = 3 \times 10^{-6}$ (1/°C) and α_C is calculated from the composition.
ΔT = the temperature range, assumed here at 110 K or 110 °C to reflect a worst case.
L = heat spreader thickness, set at 6 mm.

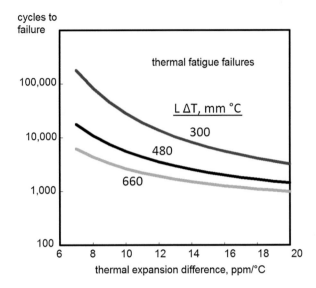

Fig. 10.4 The thermal fatigue behavior for different heat dissipation designs. The plots give the logarithm of cycles to failure against the difference in thermal expansion coefficient between silicon and the heat spreader. The three curves are for different combinations of ($L \, \Delta T$), the heat spreader thickness L (in mm) and the temperature fluctuation ΔT (in °C)

Figure 10.4 is a plot of the thermal fatigue behavior measured using different heat spreaders, thicknesses, temperature changes, and materials. The scaling is for the mean failure behavior, meaning about half failed by this number of cycles. The findings show lower thermal expansion differences, lower temperature changes, and thin structures are most helpful in minimizing thermal fatigue. The relation between these parameters is represented as follows:

$$N = A \, exp\left(\frac{\beta}{\Delta\alpha \, \Delta T \, L}\right) \qquad (10.8)$$

A key factor is the difference in thermal expansion coefficient, which depends on composition. Trial data with several materials, thermal cycles, and thickness find $A = 370$ and $\beta = 1.3 \times 10^{-2}$ mm.

Optimal Solution

Solution to the composite design problem is contingent on setting a boundary condition for the minimum life assuming thermal fatigue failure. Early personal computers operated with about 1500 on-off cycles prior to failure. As the technology evolved quickly the computer often was obsolete prior to thermal fatigue failure. Even so, folklore said to leave computers on all the time to avoid thermal fatigue, usually resulting in failure of the power supply instead of the motherboard. Now personal computers are cycled a few times per day, so for an extended warranty of 36 months this implies about 3000 cycles. Assuming 3000 cycles as current thermal fatigue goal, then Eqs. (10.8) and (10.9) are used to solve for the maximum allowed thermal expansion coefficient for the composite,

$$\alpha_{Cu-Mo} = \alpha_{Si} + \frac{\beta}{L \, \Delta T \ln(N/A)} \tag{10.9}$$

Substituting $N = 3000$, $A = 370$, $\beta = 1.3 \times 10^{-2}$ mm, $L = 6$ mm, $\Delta T = 110°C$, and $\alpha_{Si} = 3 \times 10^{-6}$ (1/°C) gives a value of α_{Cu-Mo} of 12.4×10^{-6} (1/°C) as the solution.

Returning to the thermal expansion relation in Eq. (10.7), with the composite goal set to 12.4×10^{-6} (1/°C), enables calculation of the optimal composition, giving 0.62 copper volume fraction (62 vol% Cu). The predicted Cu-Mo properties are as follows:

volume fraction copper $= 0.62$
weight fraction copper $= 0.59$
composite theoretical density $= 9.4$ g/cm^3
material cost $= \$53$ per kg
volumetric material cost $= \$0.50$ per cm^3
thermal conductivity $= 297$ W/(m °C)
thermal expansion coefficient $= 12.4 \times 10^{-6}$ (1/°C)
thermal fatigue life for 6 mm thick, 110 °C temperature change $= 3000$ cycles.

Besides these features, calculations are possible on the unit fabrication cost, suggesting about \$2 each including materials. Curiously, firms fabricating consumer microelectronic circuits initially set an upper limit of \$1 per device. This restricted the choice to aluminum, resulting in a poor thermal expansion match and accelerated thermal fatigue failure. However, as performance demands increase, the tolerable cost is now a more realistic \$2 per device. For microwave and cell phone relay stations, internet servers, radar power inverters, and military electronic devices, thermal fatigue failure is not acceptable. As a consequence, more expensive systems such as copper-tungsten, copper-diamond, aluminum-aluminum nitride, silicon carbide-silicon, and aluminum-silicon carbide used. Not only are the ingredients expensive, but often hot pressing is accepted as the fabrication option.

Wear Optimization Using Cemented Carbide Agglomerates

A novel multiple-level composite structure arises from ideas first used to form durable swords and knives. Within the structure one phase is a composite, and that composite is in turn dispersed in another phase to form a composite. For ancient swords the microstructure consisted of high carbon bands (steel and cementite) laced with low carbon steel bands. High carbon regions deliver hardness, relying on ferrite (body-centered cubic iron) with interstitial carbon and cementite (iron carbide, Fe_3C). The cementite forms micrometer sizes dispersed grains. But at the

Fig. 10.5 An illustration of how a composite phase is used as one constituent in a multiple-level composite. At a low magnification the individual agglomerate balls form a composite with the matrix. At a high magnification the agglomerate balls are evident as composites themselves. Two levels of composite are evident depending on the magnification

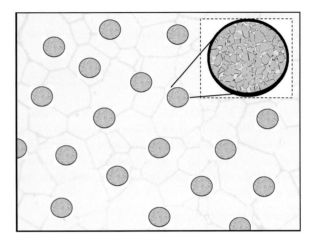

millimeter scale, the composite is laminated with surrounding regions of softer, tougher low carbon steel.

Ferrite with a low carbon content forms steel with good strength and ductility, while cementite provides hardness. The overall structure provides toughness and wear resistance. From a chemical standpoint, the ferrite and cementite phases are both iron-based, but differ in carbon and vanadium content. As a trace element, vanadium and the hard VC phase was critical to success.

The concept of composite phases used to form a composite is a means to deliver another level of properties [10, 11]. A sketch of the idea is given in Fig. 10.5. One example structure relies on a matrix of hardened aluminum alloy reinforced with clusters of Al-SiC as a second phase. At high magnification the reinforcement phase is a mixture of hard (SiC) and soft (Al) phases, but at low magnification that reinforcement appears as a single phase. Several variants are known, largely using tough phases reinforced with hard metal-ceramic clusters. Such structures excel in applications where hardness or wear resistance is needed in a strong, tough, or high temperature composite with tailored properties. For example, stainless steel reinforced with a composite phase of NiAl-B_4C gives high hardness and strength with good fracture resistance.

Sometimes called double composites, or composites within a composite, the viability lies in the expanded range of options possible when composites are used as reinforcement phases. One variant is illustrated in Fig. 10.6 where WC-10Co composite balls are dispersed in a matrix itself consisting of a composite of stainless steel, chromium boride, and bronze. The WC-Co balls provides hardness and wear resistance, the stainless steel and bronze provide toughness and corrosion resistance, and the chromium boride hardens the stainless steel. The composite is customized for use in hard tooling for forming, molding, or extrusion [12].

Such composites add to the property combinations. This evident in fracture toughness and hardness combinations. Although hardness is dominated by

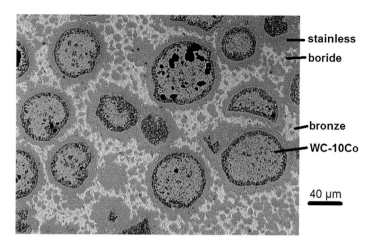

stainless
boride

bronze
WC-10Co

40 µm

Fig. 10.6 An example microstructure for a multiple level composite. This tooling material consists of four phases, including composite WC-10Co balls (*gray*) surrounded by stainless steel, with precipitated chromium boride particles, infiltrated with liquid bronze to give the final microstructure. The cemented carbide balls have some porosity (*black*)

composition, the new microstructure options allow for high toughness. The hard grains are dispersed such that cracks are arrested by the intervening ligaments of tough phase.

To illustrate design optimization using a composite within a composite, the reinforcing phase is WC-Co agglomerates [13, 14]. These agglomerates range in size from 55 to 133 µm, each having from 6 to 16 wt% cobalt. For example, spherical agglomerates near 106 µm in diameter are formed by sintering spray dried powders at 1200 °C (1473 K) for 8 h, resulting in discrete 3 µm WC grains within the 106 µm agglomerates. Effectively, the spheres are like snowballs of many small grains. In turn, the sintered agglomerate balls are mixed with additional cobalt powder to up to 30 vol%. Consolidation of the mixture is by hot pressing for 1 h at 1250 °C (1523 K) under 35 MPa pressure. All of the composites are dense and the hot pressing temperature was sufficiently low to avoid significant grain growth.

To optimize the composite, data were generated for density, hardness, fracture toughness, and wear resistance. The latter was measured in terms of the 1000s of revolutions on an abrasive test required to generate a 1 cm^3 volume loss. A high number indicates greater wear resistance. Tabulated data for the experimental composites is in Table 10.2. The question then is what is the optimal composite for high wear resistance and low cost? Life-cycle cost is considered more important versus the initial cost; a more expensive item might last considerably longer, so its life-cycle cost could be lower.

Building a response model for optimization is a multiple step process involving calculations on density, hardness, wear resistance, and cost.

Table 10.2 Cemented carbide in cobalt composite design testing

Carbide agglomerate size, µm	Agglomerate cobalt content, wt%	Agglomerate carbide content, wt%	Agglomerate cobalt content, vol%	Added cobalt content, vol%	Total cobalt content, vol%
110	6	94	10	0.0	10.0
110	6	94	10	10.0	19.0
110	6	94	10	20.0	28.0
110	6	94	10	30.0	37.0
110	11	89	18	0.0	17.7
110	11	89	18	10.0	26.0
110	11	89	18	20.0	34.2
110	11	89	18	30.0	42.4
110	16	84	25	0.0	24.9
110	16	84	25	10.0	32.4
110	16	84	25	20.0	40.0
110	16	84	25	30.0	47.5
110	6	94	10	0.0	10.0
110	11	89	18	0.0	17.7
110	16	84	25	0.0	24.9
55	11	89	18	30.0	42.4
71	11	89	18	30.0	42.4
91	11	89	18	30.0	42.4
106	11	89	18	30.0	42.4
133	11	89	18	30.0	42.4

Density

A first consideration is calculation of density versus composition. The composite has cobalt within the agglomerates and cobalt between the agglomerates. For the density calculation the variables and constants are as follows:

ρ_{CO} = density of cobalt = 8.96 g/cm^3

ρ_{WC} = density of tungsten carbide = 15.96 g/cm^3

W_{CC-Co} = cobalt weight fraction in cemented carbide agglomerate = 0.06, 0.11, 0.16

W_{CC-WC} = tungsten carbide weight fraction in cemented carbide agglomerate = $1 - W_{CC-Co}$

V_{Co} = volume fraction of added cobalt = 0.00, 0.10, 0.20, 0.30.

The theoretical density requires determination of the total cobalt volume fraction. The volume fraction inside the cemented carbide agglomerates V_{CC-Co} is as follows:

$$V_{CC-Co} = \frac{\frac{W_{CC-Co}}{\rho_{Co}}}{\frac{W_{CC-WC}}{\rho_{WC}} + \frac{W_{CC-Co}}{\rho_{Co}}} \qquad (10.10)$$

This results in 10, 18, and 25 vol% cobalt inside the agglomerates.

The total cobalt content as a volume fraction V_T, is the sum of the fraction inside the carbide agglomerates plus the added cobalt,

$$V_T = V_{CC-Co} (1 - V_{Co}) + V_{Co} \qquad (10.11)$$

This gives from 10 to 47.5 vol% (0.10 to 0.475 fraction) total cobalt for the various test compositions.

Based on the composition, density for the composite is derived from the volume fraction of each phase (Co or WC) and the constituent densities, as follows:

$$\rho_C = V_T \rho_{Co} + (1 - V_T)\rho_{WC} \qquad (10.12)$$

The result is composite densities from 12.5 to 15.0 g/cm^3 assuming no porosity.

Hardness

Hardness data show a small role from the cemented carbide agglomerate size. At 30 vol% added cobalt for 11 wt% Co agglomerates, the hardness varies from 743 HV for 55 μm agglomerates to 728 HV for 133 μm agglomerates. On the other hand, the cobalt content has a large influence on hardness, peaking at 1660 HV reported for 110 μm agglomerates with 6 wt% Co and no added cobalt, corresponding to a traditional cemented carbide.

Hardness data reveal the cobalt content is dominant. A regression between hardness HV and the two sources of cobalt (agglomerate cobalt volume fraction V_{CC-Co} and added cobalt volume fraction V_{Co}), results in a highly significant relation as follows (correlation coefficient of 0.963):

$$HV = 1769 - 2360 \, V_{CC-Co} - 2040 \, V_{Co} \qquad (10.13)$$

For hardness the standard error of the prediction is ± 62 HV. At zero cobalt Eq. (10.13) gives pure WC at 1769 HV, in agreement with the 1768 HV literature value for 3 μm WC without cobalt [15]. Other estimates for the hardness of pure tungsten carbide range to 2200 HV for aligned crystal structures [16]. For composites, the value of 1769 HV is accepted to best represent random crystals structure arrays.

The cobalt within the agglomerates degrades hardness slightly more than the cobalt between the agglomerates. The relation between measured and predicted hardness is evident by the correlation plot of Fig. 10.7, with hardness in units

Fig. 10.7 Hardness variation in terms of Vickers hardness number (HV) in cemented carbide reinforced cobalt, plotting the agreement between predicted and experimental behavior

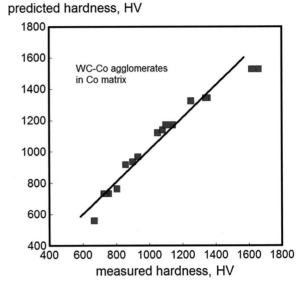

of HV. The disagreement at the very high hardness corresponds to two cases of no added cobalt between the agglomerates.

The carbide agglomerate diameter is intentionally varied at 55, 71, 91, 106, and 133 μm. Over this range the cemented carbide agglomerate size has a small effect on hardness; smaller carbide agglomerates lead to higher hardness. For example, at 30 vol% added cobalt using 18 wt% Co agglomerates (total 42.4 vol% Co), a 55 μm cemented carbide agglomerate produces 743 HV while 133 μm agglomerate gives 728 HV. However, no statistically significant relation exists between hardness and agglomerate size.

Thus, density and hardness are linearly dependent on composition [17]. The double composite delivers hardness similar to that attainable from straight carbides. For hardness, the total cobalt content is dominant, independent of its placement in the carbide agglomerates or between the agglomerates. Composite hardness has a 0.975 correlation with the total cobalt content, independent of where it is added.

Fracture Toughness

Properties such as fracture toughness depend on the microstructure details, especially grain size, cobalt content, porosity, and to a lesser degree cooling rate since it impacts residual stress [18, 19]. The data in Table 10.3 indicates four parameters dominate fracture toughness—added cobalt content, total cobalt content, density, and hardness. These last three attributes are interrelated. For the composite the following relation gives an accurate representation for fracture toughness [20, 21],

Table 10.3 Properties of cemented carbide in cobalt composites (from Table 10.2)

Total cobalt content, vol%	Composite density, g/cm^3	Composite hardness, HV	Fracture toughness, MPa√m	Wear resistance, 1000 revolutions per cm^3
10.0	14.96	1620	11	14.8
19.0	14.36	1250	18	7.4
28.0	13.76	1050	24	4.7
37.0	13.16	856	35	4.3
17.7	14.45	1350	13	4.6
26.0	13.90	1080	20	3.5
34.2	13.35	899	28	2.7
42.4	12.80	732	36	2.0
24.9	13.97	1100	17	2.5
32.4	13.47	930	23	1.8
40.0	12.97	804	32	1.5
47.5	12.46	667	38	1.4
10.0	14.96	1660	11	15.0
17.7	14.45	1340	13	4.6
24.9	13.97	1140	17	2.5
42.4	12.80	743	29	1.5
42.4	12.80	754	32	1,7
42.4	12.80	736	33	1.9
42.4	12.80	725	35	2.0
42.4	12.80	728	38	2.3

$$K_{Ic} = 23 + 37\, V_{CC-Co} + 74\, V_{Co} - 165/\sqrt{G} \qquad (10.14)$$

where K_{Ic} is the measured fracture toughness in MPa√m, V_{CC-Co} is the cobalt volume fraction in the agglomerates, V_{Co} is the added cobalt volume fraction, and G is the cemented carbide agglomerate size in μm. This equation fits the data with a correlation of 0.992 and a standard error of 0.9 MPa√m, about the scatter typical to fracture toughness measurements. To illustrate the response model, Fig. 10.8 compares the predicted values with the measured fracture toughness, both expressed in units of MPa√m.

The added cobalt located between the cemented carbide agglomerates is twice as potent in improving fracture toughness versus the cobalt within the agglomerates. The WC-Co agglomerates provide for improved fracture toughness, beyond that typical with homogeneous cemented carbide microstructures.

Wear Resistance

As widely noted for WC-Co, wear resistance improves with hardness [22, 23]. The average wear resistance is given by the number of abrasive wheel revolutions required to remove 1 cm^3 volume in a standardized test. Wear data are scattered,

Fig. 10.8 Fracture
toughness correlation
showing how the predicted
behavior matches with the
measured behavior.
Reliable models are
important to optimization
analysis. The plots are in
units of MPa√m

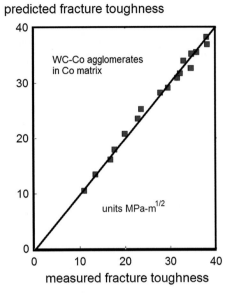

Fig. 10.8 Fracture toughness correlation showing how the predicted behavior matches with the measured behavior. Reliable models are important to optimization analysis. The plots are in units of MPa√m

with a coefficient of variation of 0.36 (standard deviation from repeat tests divided by the mean). Hence, with so much inherent variation any postulated response model will not be accurate, at least no more accurate than the input data, which are scattered.

Statistical correlation identified the wear resistance R as measured by 1000s of revolutions for 1 cm^3 removal as a function of hardness HV, fracture toughness K_{Ic}, agglomerated volume fraction granule cobalt content V_{CC-Co}, and total cobalt V_T volume fraction as follows:

$$\log(R) = -6.7 + 1.4\log(HV\,K_{Ic}) - 0.57\log(V_{CC-Co}) - 1.5\log(V_T) \quad (10.15)$$

This relation is recast in the following form,

$$R = 2\ 10^{-7}\ \frac{(HV\,K_{Ic})^{1.4}}{(V_T)^{3/2}\ \sqrt{V_{CC-Co}}} \quad (10.16)$$

This relation predicts wear resistance R with a correlation of 0.998, with a standard error of 0.15. Wear resistance is driven by the combination of a high hardness and high fracture toughness, but degraded by a high total cobalt content.

Figure 10.9 compares the predictions and measurements for wear resistance. Prior reports of wear resistance and microhardness suggest a slightly nonlinear behavior. But for the current purposes, the nonlinear behavior is only apparent at the very high values. High cobalt levels degrade hardness, but improve fracture toughness.

Fig. 10.9 Wear resistance
of cemented carbide
agglomerate reinforced
cobalt where the wear
resistance is determined by
the 1000s of revolutions
required to remove 1 cm³ of
material. This plot
compares the measured
experimental wear
resistance with the
predicted behavior

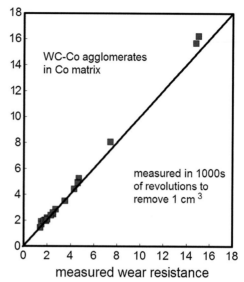

Cost

Cost is the final parameter involved in this optimization problem. Since there are only two ingredients and all processing is essentially the same, then cost is a dominated by the raw ingredients, following a linear composition function relation as follows:

$$C_M = (1 - V_{Co}) C_{CC} + V_{Co} C_{Co} \qquad (10.17)$$

where C_M is the mass based cost in \$/kg for the mixture of cobalt and cemented carbide agglomerates, based on the volume fraction of added cobalt V_{Co}. Here C_{CC} is the cemented carbide agglomerate cost (\$/kg) and C_{Co} is cobalt cost (\$/kg). The cost for sintered spray dried cemented carbide agglomerates depends slightly on the cobalt content V_{CC-Co} as calculated from the composition,

$$C_{CC} = 30 + V_{CC-Co} C_{Co} + (1 - V_{CC-Co}) C_{WC} \qquad (10.18)$$

The cost of cobalt C_{Co} is \$33/kg and for the cost of tungsten carbide C_{WC} is \$45/kg. The value of \$30 in Eq. (10.18) reflects the processing cost to mill, spray dry, sinter, and screen the agglomerates on a unit kg basis.

Most wear components are fixed in volume, so cost needs to be rationalized to a volume basis by multiplying by the composite density. Thus, the final cost calculation involves determining C_V the cost per volume as,

$$C_V = \rho_C \, C_M \tag{10.19}$$

Once these relations are isolated, then calculation is possible for any optimization goals.

Optimal Solution

Combinations using the above relations provide a means to design around target properties, especially in a case where composition effects are dominant. Plotted in Fig. 10.10 are composite hardness (HV), fracture toughness (MPa√m), and wear resistance (1000s revolutions) versus the added cobalt volume fraction based on 110 μm cemented carbide agglomerates, each containing 10 vol% cobalt (WC-6Co). The wear behavior and hardness depends on the added cobalt volume fraction and decline as the higher cobalt content reduces hardness. On the other hand, toughness increases with added cobalt.

   In terms of wear performance, a minimum fracture toughness is required to avoid field difficulties and based on work with ceramics, that minimum is set to 15 MPa√m. Thus, one criterion in optimization is to avoid solutions below this threshold. Then a solution is sought with the highest merit index, I, that being the largest wear resistance per volumetric cost, expressed as;

Fig. 10.10 Three performance parameters are plotted versus the added cobalt content for composites relying on cemented carbide balls to reinforce the cobalt structure. Illustrated in this plot are the nearly linear variations in hardness, fracture toughness, and wear resistance versus the volume fraction of added cobalt. These plots are for cemented carbide agglomerate balls containing 10 vol% cobalt

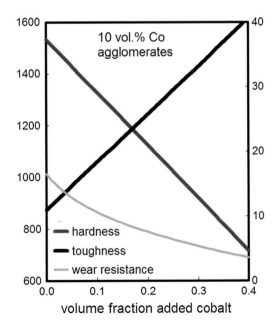

$$I = \frac{R}{C_V} \; for \; K_{Ic} > 15 \; MPa\sqrt{m} \qquad (10.20)$$

Emerging from this construct is an array of solutions, with the best fit occurring using 110 μm cemented carbide WC-6Co agglomerates with an additional 6 vol% added cobalt. The anticipated properties are,

density $= 14.6$ g/cm^3
hardness $= 1500$ HV
fracture toughness $= 15$ MPa√m
wear resistance $= 12{,}100$ revolutions per cm^3
cost per kg $= \$71.35$
cost per cm$^3 = \$1.04$.

This is the optimal solution on the basis of wear resistance for a fixed volume at the lowest cost, realizing the constraint of at least 15 MPa√m fracture toughness. If the criteria are modified, then a slightly different optimization arises.

The use of a composite to form one phase within a composite has much appeal and has resulted in patents. Clearly overall composition effects are dominant, independent of the double composite construct. Properties are predominantly functions of the cobalt content, with less dependence on its location in the carbide agglomerates or between the carbide agglomerates. In the end, the added manufacturing cost is not easily justified by performance gains. Even so, constitutive relations, including cost, allow for the selection of a best composition to satisfy target performance criteria. In this case the highest wear resistance is with a low cobalt content, limited by a minimum fracture toughness, leading to a composite of WC-6Co agglomerates intermixed with an additional 6 vol% cobalt.

Biomedical Implant Optimized for Long Life

Replacement knees, hips, shoulders, and other implant components are increasingly encountered as the population ages. Orthopedic implants are installed with the intent of a one-time replacement. Success depends on a high resistance to failure during long-duration immersion in the body, where combinations of wear, stress, and corrosion can induce premature failure [24]. Early human implants relied on stainless steels, cobalt-chromium alloys, titanium alloys, and tantalum, and these were sometimes coated with high density polymers. More recently a calcium phosphate phase known as hydroxyapatite has become the mainstay treatment to improve biocompatibility.

The concerns for biocompatibility and fracture resistance balance oxide ceramics, such as alumina and zirconia, as particulate composites for hip replacement structures. Several considerations drive the solution search, but understanding failure mechanisms are a dominant focus. The goal is to design away from failure. If a hip replacement is made at age 65 with a design life of 20 years, then the patient must undergo another surgery at age 85. Complications with surgery at this age are severe and should be avoided. Thus, it is preferred that hip replacement surgery happens only once in a lifetime, typically after the patient reaches 55. The implant life must exceed the remaining life expectancy of the patient, quantified as 99.99 % implant survival for 30 years.

Failures for hip replacements arise from wear, fatigue, or corrosion. The resistance to surface abrasion due to wear particle abrasion drives the designs to high hardness designs. The resistance to fatigue fracture is quantified by crack propagation velocity versus stress intensity acting on the crack. Most useful are designs that do not nucleate cracks under normal stress cycles. Implant deterioration by corrosion usually includes simultaneous wear, a form of stress corrosion. Data on material loss over 10 years for hip implants gives a significant difference in the surface penetration of the design materials as follows:

ultrahigh molecular weight polyethylene 1300 μm
austenitic stainless steel 270 μm
cobalt-chromium-molybdenum 60 μm
alumina 10 μm.

The dense alumina, with high hardness and good resistance to corrosion in the body is attractive. Zirconia is lower in hardness versus alumina, but when 3 % yttria is added to the zirconia the fracture toughness is improved. Thus, a composite of alumina and yttria stabilized zirconia provides a desirable combination of hardness, corrosion and wear resistance, biocompatibility, and fatigue crack growth resistance.

Crack growth is a severe concern. In a brittle ceramic, two factors are important. The first factor is the threshold stress intensity, below which there is no measurable crack growth, designated K_O. The second factor is the critical fracture toughness for rapid crack propagation, K_{Ic}. Like friction, where there is a different friction associated with the initiation of flow versus sustaining flow, fracture toughness has a different stress intensity to initiate cracking versus rapid crack propagation. In this regard, test data for mechanical properties versus composition for alumina-zirconia composites provide a basis for optimization.

The key parameters in composite design for corrosion resistant biomedical implants are hardness, fracture toughness, and threshold stress intensity for no crack growth. Material and component fabrication costs are additional considerations, but are secondary to service life considerations since the cost of replacement surgery is extraordinarily high.

Density

The alumina-zirconia composite density is a linear function of composition:

ρ_{Al2O3} = density of alumina = 4.0 g/cm^3

ρ_{ZrO2} = density of yttria stabilized zirconia = 6.0 g/cm^3
W_{Al2O3} = weight fraction of alumina
V_{Al2O3} = volume fraction of alumina.

The zirconia weight and volume fractions W_{ZrO2} and V_{ZrO2} are $W_{ZrO2} = 1 - W_{Al2O3}$ and $V_{ZrO2} = 1 - V_{Al2O3}$, respectively. The relation between weight fraction and volume fraction is,

$$V_{Al2O3} = \frac{\frac{W_{Al2O3}}{\rho_{Al2O3}}}{\frac{W_{Al2O3}}{\rho_{Al2O3}} + \frac{(1 - W_{Al2O3})}{\rho_{ZrO2}}} \tag{10.21}$$

The composite density ρ_C depends on the volume fraction of each phase, resulting in a linear function:

$$\rho_C = V_{Al2O3}\, \rho_{Al2O3} + (1 - V_{Al2O3})\rho_{ZrO2} \tag{10.22}$$

The composite density range is from 4 to 6 g/cm^3, assuming all pores are eliminated (most likely by hot isostatic pressing).

Hardness, Elastic Modulus, and Fracture Strength

Both the alumina and zirconia phases are relatively hard, with alumina being harder. A high hardness is desirable; it depends on composition, porosity, grain size, and purity. A full density is assumed. Likewise, ignoring the grain size effect finds the composite hardness varies approximately by the rule of mixtures, as plotted in Fig. 10.11; representative values for 2 μm alumina (H_{Al2O3}) and 0.5 μm zirconia (H_{ZrO2}) are 1600 and 1290 HV, allowing estimation of composite hardness H_C as a function of composition as follows:

$$H_C = V_{Al2O3}\, H_{Al2O3} + (1 - V_{Al2O3})\, H_{ZrO2} \tag{10.23}$$

Likewise, strength and elastic modulus vary with composition. Table 10.4 assembles representative end values for alumina and zirconia (with 3 % yttria).

Cost

The composite cost depends on the composition, with the constituents at C_{Al2O3} = $35/kg for high purity alumina and C_{ZrO2} = $100/kg for yttria stabilized zirconia. The cost for the composite C_C is calculated as,

Fig. 10.11 The Vickers hardness number (HV) as a function of zirconia content for nearly full density alumina-zirconia composites. In this case the zirconia contains 2 mol % yttria. The behavior is linear with composition on a volume fraction basis. Factors such as porosity, grain size, and impurity content further influence the hardness

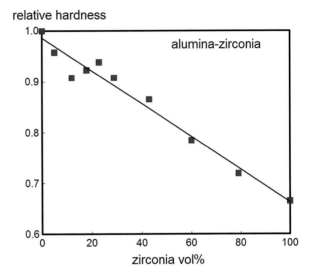

Table 10.4 Representative properties for alumina and zirconia (3 % yttria)

	Alumina	Zirconia
Density, g/cm^3	4.0	6.0
Elastic modulus, GPa	390	210
Fracture strength, MPa	600	240
Hardness, HV	1600	1290
Fracture toughness, MPa√m	4.2	5.5

$$C_C = W_{Al2O3}\, C_{Al2O3} + (1 - W_{Al3O3})\, C_{ZrO3} \qquad (10.24)$$

As with many designs, the volume is fixed, so cost is rationalized to a volume basis C_V given as,

$$C_V = \rho_C\, C_C \qquad (10.25)$$

These relations allow for estimation of the normalized material cost per implant, recognizing a processing cost is an additional factor. The final finishing (polishing) and inspection costs are high in this situation, far exceeding the material and processing costs. Industry reports a manufacturing costs for a ceramic replacement hip at about $5000 each, so material cost represents only a few percent of the total.

Fracture Resistance

One factor driving optimization in this application is reduced fracture in service [25]. The surgical procedure passed $40,000 per hip in 2014, making repeat of the procedure much more expensive than any possible savings based on materials.

Fig. 10.12 Data on the
fracture toughness for
alumina-zirconia
composites as a function of
composition, illustrating a
maximum near 10 vol%
zirconia. The placement of
the toughness maximum
depends on several factors,
including porosity, grain
size, impurity content, and
size ratio of the two phases

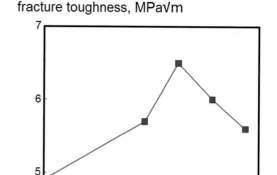

Thus, design optimization is driven by fracture resistance, with less attention to cost as long as the hardness, corrosion, and wear life are acceptably long.

Zirconia stabilized with yttria provides a gain in the toughness and crack resistance for alumina [26]. A plot of fracture toughness versus zirconia content is given in Fig. 10.12 for alumina-zirconia consolidated by hot pressing. The fracture toughness, K_{Ic}, is highest for 10 vol% zirconia. Grain size is an additional factor, with preference given to smaller grain sizes. For this analysis the properties are associated with 0.3 μm zirconia and 1.2 μm alumina.

Near the fracture toughness composite with 10 vol% zirconia, the alumina-zirconia composite is hard, corrosion and wear resistant, biocompatible, and resists fracture. That latter attribute is evident in Fig. 10.13, plotting crack velocity versus stress intensity for three compositions—pure alumina, pure zirconia, and alumina-10 vol% zirconia (toughened with yttria). The threshold stress intensity for no measurable crack growth, denoted K_O, is highest for the composite. Likewise the point of rapid fracture, corresponding to K_{Ic} is also highest for the composite.

Optimal Solution

The dominant factor for a long service life requires a high resistance to crack growth. The high cost for a replacement surgery and the patient danger from an operation at an advanced age mandate long service life as the optimization

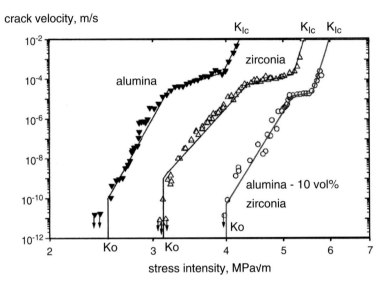

Fig. 10.13 Crack growth velocity for alumina, zirconia, and alumina-zirconia versus the stress intensity acting on a crack. The higher threshold measured for the crack growth in the composite ensures the longest life in service

parameter. Thus, alumina with 10 vol% or 14 wt% zirconia is the material of choice, with the following properties:

density $= 4.2$ g/cm^3
material cost $= \$44.10$ per kg
volumetric material cost $= \$0.19$ per cm^3
elastic modulus $= 370$ GPa
hardness $= 1570$ HV
fracture strength $= 560$ MPa
fracture toughness $= 5.9$ MPa$\sqrt{m}$
threshold crack growth toughness $= 4.0$ MPa$\sqrt{m}$.

One important advance is the addition of a titanium sleeve on the high stress region to achieve twice the service life. A key point is the high value in this case for longevity in service, due to the cost of replacement surgery. Accordingly, material cost is a secondary consideration. Similar optimization calculations arise in the design of jet engines, rocket motors, and military armaments, as well as life-critical medical devices.

Note the solutions shown above involve straightforward models with reduced sets of variables. For the student seeking the rigorous formalism a few references are recommended [1, 2, 4, 27, 28].

Study Questions

10.1. If a heat dissipation structure is desired without regard to cost, would copper-diamond be an attractive option? Why or why not?

10.2. The procedure for accelerated thermal fatigue testing relies on large temperature swings from high and low temperature extremes to shorten the time to failure. Estimate for your laptop computer the range ΔT and number of thermal cycles per day. From this, estimate how many years would be required to induce failure, assuming the failure tracks Eq. (10.8)?

10.3. Generally wear resistance improves with hardness, but at the same time the opposing material suffers more wear. In the design of industrial equipment what are some design steps useful in minimizing wear of the opposing materials?

10.4. Investigate five different automobile designs (possibly a small economy car, sports car, sport utility vehicle, and so on), collecting data on the vehicle weight, fuel tank capacity, and driving range. Determine if there is a correlation between these factors. Now add fuel economy to the analysis (miles per gallon or liters for 100 km). Which of the five selected vehicles is optimized for fuel economy? From the database, what features are giving the fuel economy?

10.5. For metal-ceramic particulate composites the change in ductility as ceramic is added to the metal often results in nonlinear behavior. Find data for ductility in a metal-ceramic system where trials are reported over a range of compositions. Plot this behavior and suggest why ductility typically behaves in this manner.

10.6. Automotive tires typically last 40,000 miles or 65,000 km, but still fail prematurely due to punctures. Tires designed to be puncture proof, or to "run flat", cost up to ten times as much, especially those used for politicians. On the other hand, tires designed to run 50 miles (80 km) after a puncture cost a premium of 1.2 times a standard tire. After 50 miles it must be repaired. This is a problem in the cost of replacement. Do an economic analysis of the three options—standard, run flat, and 50 mile puncture. Which is the best option for you personally? What are the assumptions and boundary conditions? Describe situations where you can envision each option being optional?

10.7. Hydrogen-oxygen rocket engines are proposed using two design options. The first is to rely on a lower melting temperature copper alloy with high thermal conductivity, that is actively cooled on the outside by the flow of liquid hydrogen on its way to combustion. The second option is to rely on a high melting temperature a hafnium carbide composite with silicon carbide to resists melting during use. Analyze if there is merit to a composite of copper alloy bonded hafnium carbide, giving conceptual strengths and weaknesses. What factors need to be considered in optimizing the HfC-Cu composite with respect to retention of shape during operation?

10.8. A composite of aluminum and graphite is proposed for a heat spreader material. Estimate the density, thermal conductivity, and thermal expansion coefficient for a mixture of 5 wt% graphite.

10.9. Repeat the above calculations for a composite of copper-diamond with 20 wt% diamond.

10.10. The electrical contacts in medical computer tomography imaging devices are a mixture of high conductivity phase, arc erosion resistant phase, and lubricating phase. Outline the considerations that would help in optimization this composite, including cost, fabricability, and the necessity of uninterrupted signal over a long service life.

10.11. Several governments have banned lead in shooting ammunition, but the common substitutes of steel or copper are lower in density so they suffer with a lower accuracy. Due to firearm design, the bullet is constrained by volume independent of the material. How might you synthesize a particulate composite to match the 11.2 g/cm^3 density of lead bullets? What are three composites that would offer a density match?

10.12. For the ammunition problem in Study Question 10.11, which of those three is the best candidate from the standpoint of fabrication and cost?

10.13. A material widely used for kitchen counter tops consists of aluminum trihydrate ($Al(OH)_3$) bonded by acrylic polymer (polymethyl methacrylate with repeat units of $C_5O_2H_8$). What engineering test factors need to be considered in optimizing the material for widespread kitchen use?

10.14. For the composite kitchen counter top in Study Question 10.13 what might be the practical limits on composition?

References

1. S. Torquato, Optimal design of heterogeneous materials. Annu. Rev. Mater. Res. **40**, 101–129 (2010)
2. S.S. Rao, S.S. Rao, *Engineering Optimization Theory and Practice* (Wiley, Hoboken, 2009)
3. T.Y. Baba, Y. Yamanishi, T. Honda, H. Miura, Fatigue failure test for high strength MIM sintered alloy steels. J Jpn. Soc. Powder Powder Metall. **43**, 863–867 (1996)
4. A. R. Parkingson, R. Balling, J.D. Hedengren, *Optimization Methods of Engineering Design* (Brigham Young University, Salt Lake City, 2013)
5. T. Saito, The automotive application of discontinuously reinforced TiB-Ti composites. J. Met. **56**, 33–36 (2004)
6. S.C. Tjong, Y.W. Mai, Processing-structure-property aspects of particulate and whisker-reinforced titanium matrix composites. Compos. Sci. Technol. **68**, 583–601 (2008)
7. C.J.M. Lasance, Thermal management of air cooled electronic systems: new challenges for research, in *Thermal Management of Electronic Systems*, ed. by C.J. Hoogendoorn, R.A.W.M. Henkes, C.J.M. Lasance (Kluwer Academic, Amsterdam, 1994), pp. 3–24
8. R.M. German, K.F. Hens, J.L. Johnson, Powder metallurgy processing of thermal management materials for microelectronic applications. Int J. Powder Metall. **30**, 205–215 (1994)
9. G.A. Lang, B.J. Fehder, W.D. Williams, Thermal fatigue in silicon power transistors. IEEE Trans. Electron Devices **ED17**, 787–793 (1970)

10. K.M. Prewo, V.C. Nardone, J.R. Strife, *Microstructurally Toughened Metal Matrix Composite Article*, U. S. Patent 4,999,256, issued 12 March 1991

11. K.M. Prewo, V.C. Nardone, J.R. Strife, *Microstructurally Toughened Metal Matrix Composite Article and Method of Making Same*, U. S. Patent 5,079,099, issued 7 January 1992

12. T.J. Weaver, J.A. Thomas, S.V. Atre, R.M. German, Time compression - rapid steel tooling for an ever changing world. Mater. Des. **21**, 409–415 (2000)

13. X. Deng, B.R. Patterson, K.K. Chawla, M.C. Koopman, C. Mackin, Z. Fang, G. Lockwood, A. Griffo, Microstructure/hardness relationship in a dual composite. J. Mater. Sci. Lett. **21**, 707–709 (2002)

14. X. Deng, B.R. Patterson, K.K. Chawala, M.C. Koopman, Z. Fang, G. Lockwood, A. Griffo, Mechanical properties of hybrid cemented carbide composite. Int. J. Refract. Met. Hard Mater. **19**, 547–552 (2001)

15. K.J.A. Brookes, *Hardmetals and Other Hard Materials*, 3rd edn. (International Carbide Data, Hertsfordshire, 1998)

16. H.O. Pierson, *Handbook of Refractory Carbides and Nitrides: Properties, Characteristics, Processing, and Applications* (Noyes, Westwood, 1996)

17. S. Luckx, The hardness of tungsten carbide - cobalt hardmetal, in *Handbook of Ceramic Hard Materials*, ed. by R. Riedel, vol. 2 (Wiley-VCH, Weinheim, 2000), pp. 946–964

18. J.L. Chermant, F. Osterstock, Fracture toughness and fracture of WC-Co composites. J. Mater. Sci. **11**, 1939–1951 (1976)

19. R.A. Cutler, A.K. Virkar, The effect of binder thickness and residual stresses on the fracture toughness of cemented carbides. J. Mater. Sci. **20**, 3557–3573 (1985)

20. F. Osterstock, Model describing the fracture toughness of cemented carbide, in *Fracture Mechanics of Ceramics*, ed. by R.C. Bradt, A.G. Evans, D.P.H. Hasselman, F.F. Lange (Plenum Press, New York, 1983), pp. 243–253

21. L.S. Sigl, H.F. Fischmeister, On the fracture toughness of cemented carbides. Acta Metall. **36**, 887–897 (1988)

22. K. Jain, T.E. Fischer, Sliding wear of conventional and nanostructured cemented carbides. Wear **203**, 316–318 (1997)

23. J. Pirso, S. Letunovits, M. Viljus, Friction and wear behaviour of cemented carbides. Wear **257**, 257–265 (2004)

24. J.K. Davis (ed.), *Handbook of Materials for Medical Devices* (ASM International, Materials Park, 2003)

25. A.H. Deaza, J. Chevalier, G. Fantozzi, M. Schehl, R. Torrecillas, Crack growth resistance of alumina, zirconia and zirconia toughened alumina ceramics for joint prostheses. Biomaterials **23**, 837–945 (2002)

26. J. Wang, R. Stevens, Review: zirconia-toughened alumina (ZTA) ceramics. J. Mater. Sci. **24**, 3421–3440 (1989)

27. K.A. Laurie, A.V. Cherkaev, Effective characterization of composite materials and the optimal design of structural elements, in *Topics in the Mathematical Modelling of Composite Materials*, ed. by A. Cherkaev, R. Kohn (Springer, New York, 1997), pp. 175–258

28. R. de Borst, T. Sadowski (eds.), *Lecture Notes on Composite Materials* (Springer, New York, 2008)

Chapter 11
Applications

Applications for particulate composites are illustrated by several examples. The cases span a range of materials, fabrication routes, and target properties, with information included here on the design philosophy and microstructure. Properties are rationalized to show the convergence to optimal combinations, with example uses in a variety of situations, such as dental restorations, automotive brakes, and stereo loudspeakers.

Overview

Particulate composites span a tremendous range of applications. Some easily identified composites are used every day in the form of concrete, electrical switches, automotive tires, refrigerator magnets, patio wood decking, dental amalgams, and kitchen counter tops. The more technical applications extend the array to include artificial bones and metal cutting tools. Detailing all of these is beyond the space available in this book and beyond the interest of most readers. Hence, this chapter highlights a few examples.

As shown in Fig. 11.1, the design behind particulate composites is focused on performance. Decisions arise around this focus with attention to microstructure, homogeneity, porosity, and interface character. This chapter is arranged to illustrate a diversity of materials and applications.

Aluminum-Silicon Carbide

Aluminum is a lightweight, tough metal and is easily formed via most standard metal forming technologies. Because aluminum lacks exceptional strength, ceramic additives are used to boost strength and stiffness to better compete with steels.

© Springer International Publishing Switzerland 2016
R.M. German, *Particulate Composites*, DOI 10.1007/978-3-319-29917-4_11

Fig. 11.1 Applications for
particulate composites rely
on attaining novel
performance levels. Several
factors provide the platform
for determining
performance, ranging from
the phases to the fabrication
details, with an overarching
economic concern

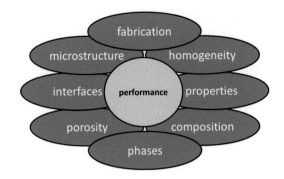

Table 11.1 Property comparison of aluminum and steel with an Al-20SiC composite [1, 8]

material	Al 2014—T6 treatment	8640—heat treated	Al 2014—20 vol% SiC
Density, g/cm^3	2.8	7.9	2.9
Elastic modulus, GPa	75	207	104
Thermal expansion, 10^{-6}/°C	23	13	17
Thermal conductivity, W/(m °C)	160	45	180
Yield strength, MPa	440	1430	410
Tensile strength, MPa	490	1520	600
Fracture elongation, %	12	15	6
Fatigue strength, MPa	133	570	300
Fracture toughness, MPa√m	38	56	19
Specific modulus, GPa/(g/cm^3)	27	26	36
Specific strength, MPa/(g/cm^3)	175	192	207

The additives for aluminum include various oxides, nitrides, carbides, borides,
as well as intermetallics, graphite, glass, and diamond. The most popular of these
composites rely on silicon carbide, termed AlSiC [1–7]. The intent is to combine
the low density, oxidation and corrosion resistance of aluminum with the strength
and stiffness of silicon carbide to increase the "specific" properties. Specific
strength is strength divided by density. It is an important parameter in the evaluation of materials used for transportation systems—bicycles to airplanes. Besides
mechanical properties, ceramic phases are used to reduce thermal expansion coefficient while preserving thermal conductivity. Table 11.1 compares properties for a
standard aircraft alloy (2014-T6), a quench and temper steel (8640), and aluminum
containing silicon carbide (2014 with 20 vol% SiC) [1, 6–8]. The specific strength
and specific elastic modulus are highest for the composite, but with a decrease in
ductility and toughness.

Fig. 11.2 Comparative property changes for Al-SiC composites versus composition, illustrating density, elastic modulus, and thermal expansion variations

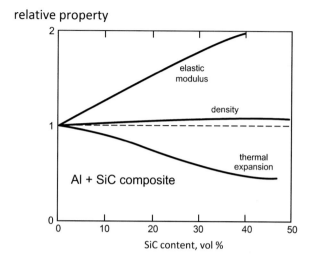

Fig. 11.3 This Al-SiC horseshoe is intended for endurance races, where the low density aluminum is favorably combined with the high wear resistance of silicon carbide

   Silicon carbide as the added phase is a favorite because of hardness, stiffness, and the possibility of good bonding to aluminum. As plotted in Fig. 11.2 the density is little changed by silicon carbide, but thermal expansion decreases and elastic modulus increases. Various aluminum alloys and silicon carbide contents result in many compositions, ranging up to 70 vol% SiC. The loss of ductility and toughness at high SiC contents occurs with increases in strength and stiffness. A common compromise is a composition near 2–25 vol% SiC.

   This was employed in sporting equipment and automotive components. Some of these applications have been displaced by carbon composites (graphite with epoxy). AlSiC is particularly valuable in rotating components, such as automotive air conditioner rotors. Figure 11.3 illustrates the use for endurance racing horseshoes where high wear resistance and low mass is of value. Another application is in heat spreaders for electronics. Silicon carbide lowers the thermal expansion coefficient of aluminum to better match the thermal expansion of common semiconductors. Low thermal expansion composites are required for wide bandgap semiconductors used in hybrid vehicles and military computers.

In AlSiC, the silicon carbide is usually an angular particle. Rounded SiC particles provide higher strength and ductility, since they do not induce stress concentrations associated with angular particles, but rounded SiC particles are more costly. Early SiC was produced as a byproduct from rice hulls. Today, direct fabrication reacts carbon with silica (SiO_2) to produce SiC whiskers about 5–20 μm long. Longer whiskers provide better strengthening because of better load transfer is between the aluminum and whisker. Improved interface bonding between the two phases is possible with a thin titanium or magnesium coatings on the carbide prior to consolidation.

Matrix phases range from pure aluminum to various aluminum alloys, such as 6061 (Al with Mg, Cu, Si, Cr) or 2014 (Al with Cu, Mg, Mn, Si). Both alloys contain magnesium to assist in bonding to the SiC. Alloying aluminum degrades thermal conductivity, but improves hardness, fatigue life, and fracture toughness, so alloy selection is rationalized to the application and the appropriate balance of performance characteristics. Pure aluminum as the matrix phase is reserved for thermal management applications, but not structural applications.

Properties such as hardness depend on alloying, composition, particle size, heat treatment, and homogeneity. A few direct comparisons illustrate the difference in response with alloying:

Al 6061-T6 alloy with no SiC has a hardness of 100 HV
Al 6061 alloy with 25 vol% SiC has a hardness of 140 HV
Al 2014-T6 alloy with no SiC has a hardness of 150 HV
Al 2014 alloy with 25 vol% SiC has a hardness of 118 HV.

In one case the SiC improves hardness, but in the other case it degrades hardness. Several composites are identified in Table 11.2, organized by increasing silicon carbide content. Elastic modulus increases with SiC content up to 70 vol%. Interface bonding dominates fracture toughness and ductility. By implication, a low ductility also results in reduced tensile strength. Thermal conductivity peaks near 25 vol% SiC. Thus, depending on the desired properties and application, different compositions are employed, with optimal compositions near 20–25 vol% SiC.

Fabrication of the AlSiC composite is by several approaches:

- mix SiC and Al powders and consolidate the mixture by pressing and sintering,
- infiltrate molten Al into a SiC preform,
- cast a slurry of SiC mixed into molten Al,
- co-deposit the two phases as sprays.

For the mixed powder route, prealloyed aluminum powder is created using gas atomization. Rapid cooling using helium gas minimizes oxidation. However, to prevent fires and explosions, that powder is intentionally oxidized to passivate against spontaneous reactions. The oxidation adds 0.1 wt% oxygen to the powder.

Table 11.2 Example AlSiC composite properties [1, 6]

vol% SiC	0	10	20	25	25	30	30	55	70
Aluminum alloy	2014	a	6061	2014	b	Al	2014	6061	6061
Density, g/cm³	2.8	3.0	2.8	2.8	2.9	2.9	2.9	3.0	3.1
Hardness, HV	150	130	115	118	210	122	145	145	90
Elastic modulus, GPa	75	97	100	120	167	125	125	180	260
Yield strength, MPa	390	440	355	425	440	300	440	488	255
Fracture elongation, %	6	12	5	2	4	4	5	0	0
Fracture toughness, MPa√m	36	13	17	14	15	15	12	7	5
Thermal expansion, $10^{-6}/°$C	23.2	21.0	15.2	15.5	15.5	13.7	13.1	10.5	6.2
Thermal conductivity, W/(m °C)	160	95	125	170	165	158	125	139	170

[a]Al-8.5Fe-1.7Si-1.2V
[b]Al-4.5Cu-1.5Mg-1Mn

The resulting particles are spherical with a typical size of 25–100 μm. Cold isostatic pressing is an alternative used to form long, thin shapes.

For infiltration, a SiC preform is generated by die compaction for simple shapes or injection molding for complex shapes. The SiC preform is sintered at a low temperature to build strength without densification. Molten aluminum alloy is infiltrated into that preform. In some cases, pressure is required to force the melt into the silicon carbide pores. An alternative is to add a wetting agent such as titanium to the aluminum.

The casting option stirs SiC particles into molten aluminum. That semisolid mixture is slurry cast or die cast into the desired shape. On larger components the density difference between the Al and SiC lead to phase separation during cooling. Smaller components chill quickly and are less susceptible to phase separation.

In spray forming, mixed powders are fed into a plasma torch for simultaneous melting and deposition, or more commonly molten aluminum alloy is sprayed onto a mandrel and solid SiC particles are simultaneously injected into the deposit.

The least expensive component fabrication route is the press and sinter approach. The resulting microstructure is shown in Fig. 11.4, consisting of SiC grains dispersed in the aluminum alloy matrix. To increase strength, deformation by high pressure extrusion, rolling, or forging is used to form rods, tubes, or other simple shapes. Ductility declines and strength increases due to deformation, making it difficult to bend or reshape the product.

Each fabrication option has inherent difficulties. For the mixed powder approach, powder handling induces separation. Polymer additions are used to agglomerate the constituents prior to consolidation. Otherwise SiC clustering in the final microstructure provides an easy failure path with degraded properties. Polymers processing aides need to be extracted to avoid contamination. This is accomplished by heating to temperatures near 450 °C (723 K) in vacuum. Sintering is at temperatures in the 600–630 °C range (873–903 K) for times up 120 min. Nitrogen is an effective sintering atmosphere, although some aluminum nitride

Fig. 11.4 Cross-section micrograph of the Al-SiC composite. The angular silicon carbide grains are well dispersed in the aluminum alloy matrix [courtesy of H. Shen]

forms. The sintered product is nominally 92–98 % dense, so a final forging densification treatment is required to reach full density. Other options include repressing and resintering, high pressure extrusion, hot isostatic pressing, and hot rolling. Extrusion of tubes and rods uses a 25:1 extrusion ratio at temperatures near 500 °C (773 K).

In liquid metal infiltration, after the aluminum solidifies any excess alloy is removed by machining. These structures also might be treated using hot isostatic pressing to remove residual pores. Efforts to mix the SiC into the molten aluminum and to cast the slurry encounter phase separation and are generally avoided since the SiC contents are low and subject to stratification within the component.

Cemented Carbides (Hard Metals)

Cemented carbides consist of grains of tungsten carbide (WC) and low concentrations of other carbides (TaC, TiC, VC, Mo_2C, and Cr_3C_2) in a metallic matrix [9–14]. A key characteristic of these composites is captured by the nickname "hard metals." The matrix is usually cobalt, but other transition metals such as nickel are employed. The many possible combinations lead to a large array of options for what are called cemented carbides. These composites are the workhorse for machining, mining, drilling, sawing, tunneling, and other taxing operations.

Tungsten carbide was discovered in the 1890s by French chemist Henri Moissan, winner of the 1906 Nobel Prize. He used an electric arc furnace to create new compounds and inadvertently discovered tungsten carbide while trying to made artificial diamonds. The compound is very hard (HV 2100–2400), but brittle. It was adopted as a need arose for drawing dies to produce tungsten wire for light bulb

Table 11.3 Evolution of cemented carbides tool materials [9]

Approximate year	Main ingredients
1900	Tool steels
1909	Cobalt-chromium alloys
1914	Cast tungsten carbides
1922	WC – Ni, Co, Fe
1929	WC + TiC – Co
1930	WC + TaC – Co WC + VC – Co WC + NbC – Co
1938	WC + Cr$_3$C$_2$ – Co
1956	WC + TiC + TaC + NbC + Cr$_3$C$_2$ – Co
1959	WC + TiC + HfC – Co

Fig. 11.5 The microstructure of WC-Co. The hard and angular WC grains are bonded together by pockets of solidified cobalt, which was liquid during fabrication [courtesy of J. T. Strauss]

20 µm

filaments. Tungsten is a hard metal, so it quickly wore drawing dies. Early data on WC alluded to a hardness near diamond, making WC an attractive substitute for diamond drawing dies.

The first carbide drawing dies were formed by casting WC, but they were brittle and weak. By 1914 mixtures of WC and Mo$_2$C were mixed and sintered at 2200 °C (2473 K), but these were brittle. To find alternatives the search shifted to infiltration of porous carbide bodies with molten iron. In 1922, powder mixtures of WC and Co were sintered to produce cemented carbide composites. The WC-Co formulation was the one adopted to form drawing dies for tungsten wire. A succession of improvements arose around the WC-Co composition, as outlined in Table 11.3 [9]. The discoveries were driven by a desire to lower cost and improve properties, leading to 6–12 wt% transition metal (Co, Ni, or Fe) as the usual added phase.

A cemented carbide microstructure is given in Fig. 11.5. It consists of angular WC grains bonded with a solidified alloy. The grains are angular since the surface energy for WC is anisotropic, resulting in a prismatic grain shape. Random metallographic cross-sections capture a mixture of two-dimensional triangles, squares,

Fig. 11.6 A wedding ring with an added gold flash in the recess. These are sold as "tungsten" wedding rings, but are actually tungsten carbide composites that hold an excellent surface finish due to the high hardness

Table 11.4 Properties of WC-Co cemented carbides ranked by cobalt content [10, 13, 14]

Cobalt, wt%	Hardness, HRA	Strength, MPa[a]
3	92	1220
6	92	1585
10	91	1930
12	89	2140
15	87	2500
25	84	3500

[a]Transverse rupture strength; tensile strength is typically 60 % of this value

and rectangles, reflecting the underlying random prismatic grain orientation. The WC grains bond into a continuous skeleton with an interpenetrating network of solidified alloy. Chemical additives are used to change grain size, grain shape, and grain hardness, leading to a large number of compositions and property combinations.

The global commercial value of cemented carbide products is in excess of $20 billion in 2015. Applications include metal cutting inserts, spray nozzles, mining tools, ball point pen tips, compaction tools, metal shears, oil and gas drilling bits, and wire drawing dies. Drawing dies are still needed for filament wires used in lamps. Additionally, wire drawing is used for the production of electrical cables, radial tires, surgical sutures, champagne cork retainers, orthodontic wires, and spiral notebooks. In recent years, cemented carbides have moved into consumer luxury products, such as the wedding band pictured in Fig. 11.6. However, the mainstay application is in metalworking tools that require wear resistance.

Table 11.4 assesses the trade-off between hardness and strength for WC-Co composites of increasing cobalt content [10, 13, 14]. The appropriate choice of properties depends on the application. Trade-offs are required since a high value of one attribute is usually accompanied by a low value of another attribute. For

Table 11.5 Example properties of mixed carbide composites [12]

wt% WC	94	85.3	78.5	60
wt% Co	6	12	10	9
wt% other carbides (TaC, TiC, VC, …)	0	2.7	11.5	31
Density, g/cm^3	14.9	14.2	13.0	10.6
Hardness, HV	1580	1290	1380	1560
Strength, MPa	2000	2450	2250	1700
Elastic modulus, GPa	630	580	560	520
Fracture toughness, MPa√m	10	13	11	8
Thermal conductivity, W/(m °C)	80	65	60	25
Thermal expansion, 10^{-6}/°C	5.5	5.9	6.4	7.2

example, a high wear resistance implies a low fracture toughness. These composites are difficult to machine, usually requiring diamond tools. However, once created the cemented carbides provide long service life versus the alternatives, namely tool steels and cobalt-chromium alloys.

During sintering, the average grain size enlarges by coalescence and coarsening. This means the number of grains declines as the grain size grows. Grain growth is slowed by inhibitors such as vanadium carbide (VC). Small grain sizes, in the 0.5–1 μm size range, are good for metal cutting. Coarser grain sizes are used for mining applications. Nanoscale grain sizes, below 0.1 μm, are used for precision tools such as printed circuit board drills.

Cobalt is the typical matrix phase. Cobalt has solubility for WC at the sintering temperature, enabling liquid phase sintering. At 1250 °C (1523 K), the solubility of WC in Co is 22 wt%, but for Ni the solubility is 12 wt%, and for Fe it is 7 wt%. Although early carbides involved iron and nickel, the properties are best using cobalt. Little success has come from lower cost alternatives. Ruthenium is one exception. It is costly, but at low concentrations provides a high performance WC-Co formulation. Similar property gains are possible using controlled cooling rates from the sintering temperature. Other additions strengthen the sintered composite by slowing grain growth. Accordingly, a range of strength and hardness properties are possible, as evident in Table 11.5 [12]. As hardness increases fracture toughness falls, as plotted in Fig. 11.7 [10, 12, 13]. A typical fracture toughness is near 15 MPa√m, about three times that of diamond, but the hardness is significantly less than diamond.

The starting point for cemented carbide production combines the powders by milling. A goal is to smear the Co onto the WC particles. Sintering to full density is improved by a homogeneous cobalt dispersion. Additives, such as VC, TiC, Cr_3C_2, MoC_2, or TaC, are introduced during milling to ensure they are likewise homogeneously distributed throughout the mixture.

After milling the agglomerated powder is mixed with wax and solvent to form a slurry. These ingredients are agglomerated into granules to assist in handling. Spray drying is used, where the powder-wax-solvent slurry is forced through an atomization nozzle in a heated chamber. The spray forms droplets from which the solvent

Fig. 11.7 Scatter plot of fracture toughness versus hardness for several cemented carbide composites, showing evidence of an inverse correlation between these two parameters [10, 12, 13]

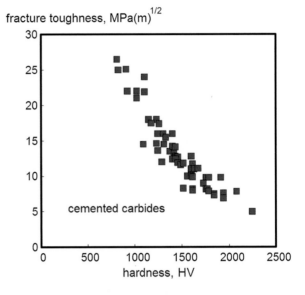

Fig. 11.8 This scanning electron micrograph shows agglomerated tungsten carbide and cobalt (WC-11Co) powders after spray drying. The individual grains are submicrometer sized, but the agglomerates are large balls held together by wax

evaporates in freefall inside a hot chamber. The resulting granules are spherical clusters of the particles as pictured in Fig. 11.8, a scanning electron micrograph of spray dried WC-11Co powder. Each agglomerate consists of thousands of individual particles. Improved powder flow and easier handling without the dusting are the main reasons for spray drying. The improved flow allows the agglomerates to uniformly and rapidly fill compaction tooling.

Die compaction is the main shaping approach. Tool motions are controlled to ensure homogeneous compacts to avoid warpage during densification. Modern compaction presses are computer numerically controlled to ensure delivery of desired shapes with tight tolerances. Tolerances for metal cutting tools are often in the ± 10 μm range on dimensions of 20 mm.

Powder injection molding of WC-Co is accomplished using about 50 vol% wax-polymer binder mixed with the milled powder. It is applied to three-dimensional shapes that would be difficult to die compact. Long shapes, such as rods, tubes, or twist drills, are extruded using powder-binder formulations similar to those employed in injection molding. Cold isostatic pressing is a slower process, but it is employed for simple, low tolerance, and larger bodies.

After shaping, the next step is to slowly drive off the wax, either by heating in hydrogen or vacuum. Hydrogen requires precautions, but provides better heat transfer and wax removal. Final heating to the sintering temperature is performed in a graphite vacuum furnace. At a temperature between 1280 and 1300 °C (1553–1573 K), cobalt forms a eutectic liquid with the tungsten carbide. If the cobalt is homogeneously distributed in the compact, then the liquid immediately pulls the carbide grains together and densifies the structure by liquid phase sintering. Sintering shrinkages ranges from 15 to 25 % for each dimension. The higher the pressed compact density, the lower the sintering shrinkage. A peak temperature near 1400 °C (1673 K) ensures uniform liquid formation. This temperature is held for up to 120 min under vacuum to ensure full densification. Grain growth occurs during this final densification step. Any oxygen present at the high temperature causes carbon loss and loss of stoichiometry. Thus, carbon and oxygen control are controlled, requiring attention to the carbon potential during sintering.

One option is to apply pressure during sintering to force final densification. Pressure is applied late in the sintering cycle to squeeze residual pores closed. This is usually applied during sintering, but in some cases is performed in a separate hot isostatic pressing (HIP) cycle. When incorporated into the sintering cycle it is termed sinter-HIP or pressure-assisted sintering. In this approach, the initial part of the cycle involves liquid phase sintering in vacuum. Late in the same cycle, when the compact is about 95 % dense, the furnace is pressurized.

Sintering rounds sharp edges and corners, so cutting edges are added by grinding after sintering. If close final dimensions are required, they too are attained after sintering using laser ablation or diamond grinding. Further performance advances are derived from hard vapor deposited coatings on the sintered carbide. Coatings are alumina, titanium carbo-nitride, diamond, or similar hard materials. In metal cutting the tool heats, so a high conductivity coating helps avoid high temperature corrosion. Coatings such as the gold colored titanium nitride are common.

Cemented carbides are difficult to machine, so for less demanding situations other hard materials are selected. These include ferrotic, a composite of 50 vol% titanium carbide (TiC) dispersed in tool steel. It is produced by liquid phase sintering milled powder with final densification by hot isostatic pressing to give a density of 6.5 g/cm^3 with strength of 1200 MPa.

During manufacturing, the key production tests are density, magnetic behavior, and hardness. Residual pores are detrimental to properties and component life during service. Pores are categorized by size and amount using microscopy; "A" pores are below 10 μm and "B" pores are larger than 10 μm. Loss of the WC stoichiometry degrades performance and is categorized as "C" defects. These letter designations are followed by the amount of each. For example, A-02 indicates

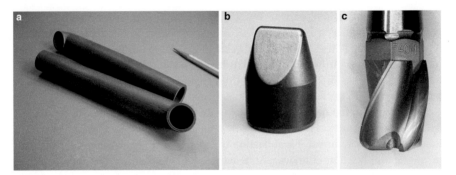

Fig. 11.9 Pictures of cemented carbide components; (**a**) bent tubes for use in centrifugal separation of oil and sand, (**b**) tip used for rotary drilling, and (**c**) end mill used for machining flat surfaces

0.02 % porosity with a size below 10 μm. Microscopy is used to measure carbide grain size from the average linear intercept size. Magnetic testing is useful since cobalt is a magnetic phase. Coercivity is the resistance to demagnetization and it measures loss of WC stoichiometry. Also, coercivity depends on the cobalt grain size; a high coercivity indicates small cobalt pockets that go hand in hand with small WC grains. Magnetic saturation provides a measure of the free carbon and hard eta phase (W_2C). Both phases are detrimental to performance. Magnetic properties provide quick, nondestructive quality checks. Microhardness is another confirmation of performance. It is possible to also estimate the fracture toughness from crack propagation in the microhardness test.

Because cemented carbides excel in wear resistance, they are widely used in applications where other materials would fail. Some examples are depicted in Fig. 11.9. The first (a) shows hollow bent tubes used in centrifuges to separate sand from oil prior to refining. The second (b) is a carbide insert used for drilling. It is appropriate to rely on a high hardness to attain the longest service. The end mill with the spiral cutting edge pictured in (c) is an example of a titanium nitride coated cemented carbide.

Mechanical properties depend on stoichiometry, porosity, grain size, and alloying. The highest strength is near the equiatomic WC composition. Porosity severely reduces fracture strength and toughness since pores act to nucleate premature failure. Accordingly, full density via sinter-HIP is a means to ensure a high fracture strength. One of the most common formulations, WC-10Co, has the following properties:

density = 14.6 g/cm^3
heat capacity = 209 J/(kg °C)
thermal expansion coefficient = 4.3 10^{-6}/°C
thermal conductivity = 95 W/(m °C)
electrical resistivity = 16 μΩ-cm
elastic modulus = 608 GPa
Poisson's ratio = 0.24

Fig. 11.10 A plot of
transverse rupture strength
for WC-12Co compositions
with varying carbide grain
sizes, giving a peak fracture
strength near a 4 μm grain
size [13]

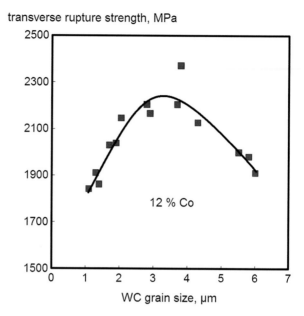

hardness $= 1310$ HV
transverse strength $= 3040$ MPa
compressive strength $= 4620$ MPa
fracture toughness $= 14.5$ MPa$\sqrt{m}$.

Hardness and strength follow an inverse square-root of grain size behavior, known as the Hall–Petch relation. Strength tests are performed using three-point transverse rupture tests. For a constant cobalt content, the highest strength occurs with an intermediate grain size. As an example, Fig. 11.10 plots transverse rupture strength versus carbide grain size for the WC-12Co composition [13]. For this composition, the maximum strength is with 4 μm grains. A similar peak occurs if the plot is based on the cobalt ligament size between grains, as is evident in Fig. 11.5. As cobalt content increases there is improved strength because the material is tougher but hardness decreases, leading to a compromise on strength, hardness, and toughness in terms of composition and grain size.

In high stress applications, such as metal cutting, considerable heat arises during use. Accordingly, thermal softening occurs with a loss of strength, especially at temperatures over 1000 °C (1273 K). Temperatures in this range are reached in several applications, leading to short life. Simultaneous oxidation further degrades properties.

High wear resistance is a recognized benefit with cemented carbides. A common measure of wear resistance is mass loss or volume loss. For example, the number of revolutions of a rubbing abrasive wheel required to generate a 1 cm^3 volume loss. An alternative is to report the mass loss using rubbing with entrained abrasive grit. As illustrated in Fig. 11.11, abrasive wear resistance is dominated by hardness. A

Fig. 11.11 Abrasive wear resistance (large values are more resistant) on a logarithmic scale versus Vickers microhardness for various WC-Co composites. Hardness is the dominant parameter in setting abrasive wear resistance

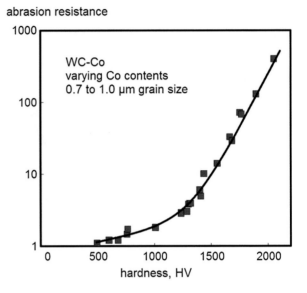

variety of compositions are used for this plot, all with mean grain sizes in the 0.7–1.0 μm range.

Efforts to improve the cemented carbides explore both composition and microstructure variations. One approach is based on a composite of WC-Co balls dispersed in a tough matrix. For example, sintered WC-Co balls are mixed with stainless steel, sintered, and residual porosity is filled by infiltration. This microstructure is shown in Fig. 10.6. This multiple phase composite offers wear resistance in a tough composite. The image in Fig. 11.12 corresponds to the fracture surface. The predominant fracture is across the carbide grains, as desired for high fracture resistance.

Related ideas examine WC-Co embedded in a cobalt matrix to form a heterogeneous composite. The WC containing regions provide reinforcement to the otherwise softer matrix. Hardness decreases as the total cobalt content increases, but fracture toughness and wear resistance benefit from the heterogeneous microstructure.

The long history of research on cemented carbides provides a rich benchmark of property-composition-microstructure relations for particulate composites. Property models developed for cemented carbides are starting points for evaluating other composites. The direction of current research is toward harder and higher toughness compositions. One approach is to form components with hard surfaces and tough cores in functional gradients.

Fig. 11.12 Fracture surface
from a multiple-scale
composite, consisting of
WC-Co balls in a matrix of
stainless steel and bronze.
The large hard balls provide
wear resistance and the
mixture of phases between
the balls provides toughness

Fig. 11.13 Dental crown
and bridge structures rely on
porcelain bonded to a metal
crown. This picture is of a
three unit bridge sitting on a
mirror, showing the metal
substrate onto which the
porcelain is bonded

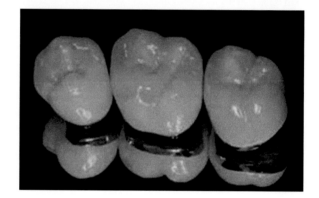

Dental Porcelain

Dental porcelain is one of the unnoticed but widely used composites. The success of the system is because of the difficulty in differentiating normal dentition from a porcelain restoration. Dental porcelain is formulated for semi-translucency, color, and fluorescence to match natural teeth [15, 16]. Further the composite is hard, strong, wear resistant, corrosion/staining resistant, biocompatible, and easily processed.

Restorative dentistry moved from the gold crown to a thin metal coping with fired porcelain veneer to improve aesthetics. When properly constructed, the restoration has the appearance of natural teeth as evident in the photograph in Fig. 11.13. The dentist prepares an injured or damaged tooth by removing decayed and defective regions. To restore chewing surfaces, a strong conformal metal cap is fabricated to fit the remaining tooth. The metal is created using investment casting or laser additive manufacturing. Overlaying the metal are layers of fired dental porcelain, starting with an opaque inner layer and building to a semi-transparent

Fig. 11.14 A sketch of the
dental crown consisting of
dental porcelain as the outer
visible layer. The porcelain
restores chewing function
while providing an aesthetic
appearance

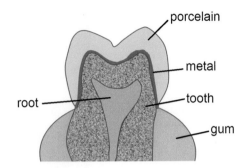

outer layer. The color is matched to neighboring dentition. A cross-section of the
final structure is sketched in Fig. 11.14. The finished restoration is cemented onto
the tooth. Dental porcelain consists of about 50 vol% crystalline ceramic in a glass
matrix.

There are two routes to forming dental porcelain. One is based on natural
minerals, mixed to form a paste that starts as glass, but transforms into a ceramic-
glass composite during firing. The second approach is similar with reactions
induced by firing, but synthetic chemicals are employed. A common formulation
relies on a mixture of compounds such as silica (SiO_2), feldspar ($KAlSi_3O_8$),
kaolinite ($Al_2Si_2O_5(OH)_4$), and sodium or potassium oxide (Na_2O or K_2O).

Porcelain was first used in dentistry in 1723, and metal jacketed crowns arose in
the 1885–1900 period using pure platinum or palladium as the inner core. Porcelain
dentures gained traction with the invention of vacuum firing furnaces. Finally the
durable porcelain bonded to metal coping arose in the 1960s. Early applications for
glass-ceramic composites were in dinnerware, and from that arose formulations for
dental.

For a dental appliance, two critical issues are fit and strength. The metal coping
is adhesive bonded to the remaining tooth structure. Distortion during fabrication is
unacceptable, since a poor fit provides an avenue for decay. Strength increases
during cooling from the firing temperature due to residual strains traced to differential thermal expansion coefficients between the porcelain and metal coping.

The porcelain restoration requires a limited thermal expansion coefficient to
avoid cracking during cooling. The first layer on the metal is opaque, and layered
over this are translucent and transparent layers to give an appearance similar to
natural dentition. It is possible to eliminate the metal coping by forming the entire
restoration from porcelain. The viscous glass is formed into the desired shape, then
heat treated to precipitate the strengthening crystal phase. This system listed in
Table 11.6 based on precipitation of platelet crystals of $K_{0.8}Mg_{2.6}Si_4O_{10}F_2$ in a
silicate glass at 1075 °C (1348 K) [16, 17]. The crystals are about 2 μm in diameter
and occupy 70 vol% of the structure. Thus, both phases, glass and crystal, are

Table 11.6 Properties of dental porcelains [16, 27]

	Pressure molding	Sintered laminates
Approach	Pressure mold pellet by viscous flow into cavity, crystallize after shaping	Layer powder onto metal coping, sinter layers, crystallize during firing
Starting form	Shaped blank	Powder slurry
Chemistry	SiO_2-MgO-K_2O-F-ZrO_2	SiO_2-Al_2O_3-K_2O-Na_2O-CaO-P_2O_5-F
Crystal phase	$K_{0.8}Mg_{2.6}Si_4O_{10}F_2$	$Ca_5(PO_4)_3$—Cl/F/OH and $KAlSi_2O_6$
Density, g/cm^3	2.7	2.6
Elastic modulus, GPa	68	43
Bend strength, MPa	153	80
Compress strength, MPa	828	340
Thermal expansion coefficient, 10^{-6}/°C	7	13
Relative optical transmission	0.6	0.4

percolated in the structure. Unfortunately fracture toughness, is low, near 2 MPa√m, restricting the use of pure porcelain crowns.

The starting powder is in the 20–40 μm size range. A dental technician dispenses the powder into water to form a slurry that is layered to build up the tooth structure. Firing induces densification and crystallization at peak temperatures near to 1100 °C (1373 K). The result is a two phase structure consisting of apatite and leucite. The former is a transparent calcium phosphate akin to tooth enamel and bone, nominally $Ca_5(PO_4)_3$ with Cl, F, or OH in the repeating unit. Leucite is a potassium aluminum silicate, $KAlSi_2O_6$, that is harder than apatite with less transparency. Initially crystals nucleate near 500–600 °C (773–873 K), and grow during firing. The firing is performed in vacuum at a peak temperature of 980 °C (1253 K) or higher for 10–20 min. After firing, the structure consists of 20 vol% precipitated leucite about 2 μm in diameter with 5–10 vol% apatite crystals as 1–2 μm long whiskers, giving the properties listed in Table 11.6.

Glass-ceramic porcelain composites are very successful in dental applications because of aesthetics and biocompatibility. Several variants are under development. For example, new restorations are machined from colored zirconia, avoiding the firing cycle and metal coping.

Diamond Impregnated Metals

Composites consisting of diamond dispersed in a metallic matrix are known as diamond impregnated metals. They are widely employed in asphalt cutters, marble tile slicers, glass lens grinders, concrete cores saws, granite tile saws, as well as ceramic machining bits [18–21]. As composites they combine the hardness of diamond in a metal matrix that provides toughness. In contrast with the early variants consisting of natural diamond bonded using resins, today's diamond tools rely on synthetic diamond bonded with metals. Several variants exist such as the following:

• polishing pads with surface mounted diamonds bonded to metal,
• diamonds entrained in electroplating deposits,
• liquid phase sintered diamonds also known as polycrystalline diamonds
• diamond dispersed in a metal matrix.

An example of the structure for a metal bonded diamond tool is shown in Fig. 11.15. The design optimizes the diamond size and content for optimal life.

To form diamond tools, first the diamonds are milled and classified into size groups, such as 0.9–1.5 μm, 1–5 μm, 4–8 μm, on up to 180–220 μm, almost always using synthetic diamonds. The particles are polygonal in shape. Smaller diamonds are used for glass applications and larger particles are used for concrete sawing and drilling. One means to improve performance is to coat each diamond with a hard phase, such as titanium, silicon carbide, or chromium alloy. The coating increases interface strength to improve diamond grain retention in the tool during use. The ideal situation is shown in cross-section in Fig. 11.16. The diamond protrudes out of the softer matrix, yet is retained by the matrix phase. As the wear occurs, eventually the diamond is pulled from the matrix, allowing another diamond grain to take over the wear burden.

Fig. 11.15 An example of a cement coring tool that consists of a steel tube with diamond composites laser bonded to the steel tube to form the cutting surface. Some segments are shown with the tool

Fig. 11.16 Cross-section micrograph of diamond protruding out of a metal matrix as desired for best cutting performance. After the diamond becomes dull, it is pulled out of the softer matrix and a subsequently lower diamond then is exposed to continue cutting [courtesy L. G. Campbell]

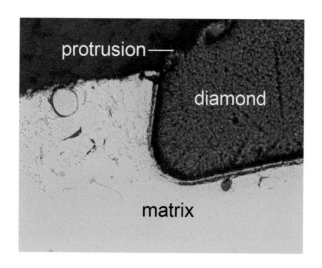

For bonding, the matrix metal must wet the diamond. The most common matrix is cobalt or cobalt alloys, because of wetting and high strength. However, matrix variants include Fe-20Cu-10Co, Cu-35W-10Sn, and Fe-27Co-7Cu. However, contamination by sulfur or oxygen degrades properties. A typical matrix has a hardness of 300–350 HV, yield strength of 400–600 MPa, and fracture elongation below 10 %.

Most diamond-metal tools rely on 15–25 vol% diamond to minimize easy fracture at diamond-diamond contacts. Lower diamond contents result in faster wear and pull-out of the diamond grain (the primary failure mechanism) and higher diamond contents result in a higher machine loads, loss of toughness, premature tool fracture, and higher cost.

Fabrication by hot pressing relies on a rapid heating cycle to avoid diamond decomposition into graphite. Decomposition is slow at 700 °C (973 K), but progressively increases in rate at higher temperatures. Typical hot pressing is at 35 MPa and 800 °C (1073 K) for 3 min. Infiltration of a porous diamond body with a copper alloy and even liquid phase sintering are options for some shapes.

Figure 11.17 is a top view of a diamond polishing tool. The diamond grains intentionally protrude from the stainless steel surface. In use, the diamond carries the load by acting as a plow while the softer matrix phase wears away. Eventually, the diamond dulls and heats, resulting in failure of the interface with the matrix. The matrix wears until a new diamond is exposed. Thus, a homogeneous dispersion of diamonds is critical to uniform wear.

At high diamond contents, the diamond grains are connected. The product is a mixture of diamonds with a matrix, termed polycrystalline diamond. Sintering diamond requires a pressure of 6 GPa and temperature of 1500 °C (1773 K). Polycrystalline diamond reaches to 94 vol% diamond, with cobalt as the matrix. These composites are restricted to compressive applications, such as in wire drawing dies and oil well drilling teeth. A variant combines diamond with tungsten

Fig. 11.17 Optical
micrograph of diamond
particles protruding from
the top surface in a diamond
composite tool used for
semiconductor polishing

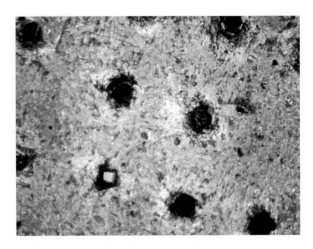

carbide to attain significant wear resistance. One composite consisting of 40 vol% diamond, 22 vol% titanium nitride, and 38 vol% alumina reached a hardness of 4200 HV. Consolidation required 800 MPa pressure at 1100 °C (1373 K). Another variant combines polycrystalline diamond (94 vol% diamond and 6 vol% cobalt) with 40 vol% WC-14Co to produce a high toughness composite with 3150 HV hardness. In relative terms, the fracture toughness is low, ranging from 5 MPa√m for low cobalt contents (6 vol% Co) to 12–15 MPa√m for the toughest structures.

Efforts with diamond dispersed in ceramic matrices generally are not successful. For example, alumina with 10 vol% of 1.6 μm diamond results in a strength of 250 MPa and fracture toughness of 5 MPa√m. Alumina without the diamond reached twice the strength at the same fracture toughness, so the only merit from adding diamond is missing. The diamond grains acts to initiate fracture at a lower stress. Cutting and drilling tools based on diamond require a high toughness matrix.

Electrical Contacts

Composites are widely used for electrical contacts, current interrupters, circuit breakers, and electron emission devices [22–25]. Make-break electrical contacts provide a means to turn power on and off. Sliding contacts transfer current to motors without the arcing seen in make-break contacts. When electrons or plasma flow in a gap between two conductors in the form of a spark, the situation is termed arcing. It arises from vapor generated at one of the contact surfaces. For an on-off contact, the greatest arcing difficulty arises when the contact surfaces are in proximity to one another. The arc forms to cause local heating, erosion, welding, or other failures. Most upsetting is the situation where current continues to flow through the contacts in spite of switching to the off position. In the worst situation, the contacts weld together, refusing to allow switching into the off position.

Fig. 11.18 An arc erosion resistant high voltage switching electrical contact that is a composite of tungsten and silver, in this case with excess silver on the exterior

The composites used to form electric contacts consist of two phases; a high melting temperature refractory phase for arc erosion resistance and a high conductivity phase for heat and current dissipation. The most popular compositions are tungsten-copper and tungsten-silver composites. They are complimented by WC-Ag, Mo-Cu, and Ag-CdO composites. Cadmium oxide is toxic, so its use is limited and alternatives now include cerium oxide (CeO_2) and antimony oxide (Sb_2O_3). Accordingly a range of options exist, but W-Ag and W-Cu are most common.

A high voltage electrical contact, such as the one pictured in Fig. 11.18, carries considerable current density (thousands of A/m^2), making it easy for the contact to weld closed. Accordingly, a time delay occurs from when mechanical switching occurs and the halt to current flow. In high current density applications, contact opening is not necessarily sufficient to stop current flow. Indeed, turning off current is quite difficult in large buildings, machines, and utilities.

Proper function of an electrical contact depends on the following influences:

- environment—atmosphere and contaminants,
- design—contact area, current density, magnetic fields, and contact force,
- operation—current type, power cycle, peak voltage and current, frequency of use,
- composition—conductive phase and arc erosion phase,
- properties—strength, hardness, thermal conductivity, arc erosion resistance,
- processing—defects, homogeneity, voids,
- microstructure—grain size, homogeneity, porosity, pore size.

High current density electrical contacts are formed from mix powders. The composition is customized to the use cycle. For example, low voltage and low current contacts are copper or silver with no refractory phase. At higher current densities, when arc erosion resistance is required, conductivity is sacrificed to add a refractory ingredient such as tungsten. Spot welding contacts are copper-alumina. Copper is high in thermal and electrical conductivity, is easily formed, but oxidizes over time and welds during use. Silver is similar with more resistance to oxidation and arcing, but has a higher cost. Usually, the high melting temperature phase

Fig. 11.19 Cross-sectional micrograph of the Ag-CdO composite used in low current density electrical contacts. Silver is the continuous phase and cadmium oxide is the dark, angular phase

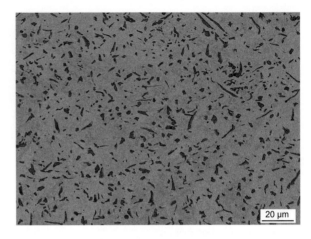

Fig. 11.20 The microstructure of a W-Ag composite designed for use in high current density electrical contacts. The silver is the continuous matrix phase and tungsten is the cubic phase. Some residual porosity (*black*) is evident at the W-W bonds

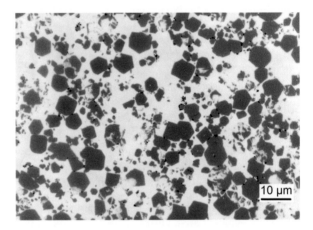

provides arcing resistance. Thus, refractory metal additions to either silver or copper prove most effective. An option is to add graphite to the copper or silver to improved welding resistance.

One route to forming contacts is to die compact mixed mix powders to give a button shape. This is common for composites with more than 75 vol% tungsten or molybdenum. Another option for 40–75 vol% of refractory phase is to press just that powder and then infiltrate liquid copper or silver into the voids during sintering. Below 40 vol% refractory phase, there is insufficient refractory solid to hold shape during sintering, so the contact is formed from mixed powders. Sintering temperatures are lower to avoid melting the copper or silver, almost always in hydrogen.

Figure 11.19 is the microstructure for an Ag-CdO contact and Fig. 11.20 is a cross-section microstructure through a W-Ag composite, showing the individual tungsten grains in a silver matrix. The dark spots located at the tungsten grain-grain contacts are pores. The tungsten phase is interconnected, as is the silver phase.

Some tungsten-tungsten sintering occurs during the heating cycle to provide rigidity to the compact and wear resistance during use. Below 20 vol% tungsten the refractor phase is discontinuous. The preferred microstructure is essentially full density (less than 2 % porosity) with silver and tungsten particles of 1–5 μm in a homogeneous dispersion. Arc erosion improves as the tungsten grain size increases.

Porosity is detrimental to performance, so residual pores are eliminated by repressing the contact after sintering. When followed by a second sintering treatment, the process is called double-press and double-sinter. Residual pores larger than the tungsten grain size indicate problems with mixing or milling. Pores smaller than the tungsten grain size indicate under-sintering. On occasion a gas reaction occurs during sintering, resulting in surface blisters. For example, the reaction of an oxide with hydrogen during sintering produces a steam blister. The cure is to remove the oxygen contaminant by slower heating.

Figure 11.21 is a low magnification view of a W-Cu electrical contact in cross section. The contact brazed onto a copper substrate at the bottom. A silver coating is on the outside. The serrated lower surface is filled with electroplated silver at the copper interface. This contact is applied to switching situations carrying currents of 800 A at voltages up to 600 V, specifically for use in portable power generation facilities requiring backup electrical service.

Contact shapes range from simple buttons and squares to complicated rocker arm geometries. The higher the current density, the more exotic the design. Flat designs are usually required for high force contacts to ensure the load is distributed over a large area. Rockers are used where arcing and welding are problems. Arc erosion resistance is improved by immersion in oil. Tungsten-silver contacts are used up to 800,000 V with currents up to 100,000 A. The arc temperature reaches 10,000 K, exceeding the melting temperature of all materials. In these applications, silver evaporation is used for adiabatic cooling to help extinguish any arcs. At the very high current levels, over 10,000 A, infiltrated contacts are superior because of the uniform microstructure.

Fig. 11.21 Cross-sectional view of a W-Ag electrical contact at low magnification. The contact has serrations for bonding and silver coating between the contact and the copper substrate visible at the bottom

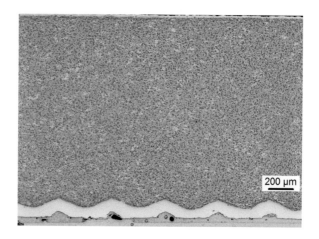

Table 11.7 Properties of W-Ag electrical contacts [1, 22, 25]

Composition wt% W	Composition vol% W	Density g/cm^3	Hardness HV	Strength MPa	Conductivity % IACS	Relative contact resistance
90	83	17.5	240	1330	35	1000
80	69	16.3	210	1050	45	500
70	56	15.2	150	800	55	250
60	45	14.4	110	700	60	180
50	35	13.5	90	570	70	150
30	10	12.0	50	385	85	120

IACS international annealed copper standard

Table 11.7 gives property variation with composition for tungsten-silver compositions [1, 22, 25]. A high arc resistance comes with the higher tungsten content,
but with higher contact resistance. Additives provide a means to selectively adjust
attributes. A high thermal conductivity dissipates heat to reduce arcing. For tungsten-silver at 30 wt% tungsten the thermal conductivity is 327 W/(m °C) and this
falls to 222 W/(m °C) for 90 wt% W. As an example of the durability, W-40Ag
contacts reach 150,000 switching cycles at 10 kW loads.

Filled Polymers

Filler polymers arose early in the development of plastics. First examples date from
the 1950s when minerals such as calcite (limestone) were added to lower cost.
Besides minerals, additions include glass and metals (silver, nickel, copper, iron,
and aluminum) for property adjustment [26–30]. Early success was with thermosetting compounds, such as Bakelite, where the added phase had some benefit on
properties. More property gain occurred with thermoplastics, so filled polymers
grew as polyethylene, polypropylene, and related polymers emerged. The combination of cost reduction and improved properties resulted in the following
combinations:

- common polymers as the matrix phase such as nylon, polycarbonate, polypropylene, polyacetal, polyphenylene sulfide, acrylonitrile butadiene styrene, and
  polyphenylene oxide,
- common fillers as the filler such as silica (SiO_2), glass, talc ($Mg_3Si_4O_{10}$), titania
  (TiO_2), antimony trioxide (SbO_3), cellulose (($C_6H_{10}O_5)_n$), mica, limestone
  ($CaCO_3$), clay, hydrated alumina ($Al(OH)_3$), boronic acid ($R\text{-}B(OH)_2$), aluminum (Al), copper (Cu), and graphite (C).

Justifications for filled polymers include both technical and economic factors:

- Reduce oxygen permeation, such as in tires and tennis balls with graphite or talc additions.
- Reduce cost since minerals such as talc, mica, and calcite are less expensive compared with polymers, especially higher performance polymers.
- Increase stiffness where the filler is high in elastic modulus compared to the polymer, as seen in glass filled nylon.
- Lower thermal expansion coefficient since the additives reduce the polymer thermal expansion.
- Antimicrobial additives to retard adhesion and breeding of bacteria, often relying on copper.
- Improve heat stability by increasing dimensional rigidity of the polymer to avoid distortion during excursions to warm temperatures, especially in automotive engine compartment components.
- Add color or texture, such as the role of titania or graphite to make white or black plastics, or wood flour to form composite lumber; coloration helps prevent discoloration during exposure to sunlight.
- Retard flames and to resist burning in fires, such as antimony trioxide or hydrated alumina dispersions in polymers for passenger compartment components in aircraft.
- Improve wear resistance, such as by adding hard minerals (clay, talc, limestone) to polymers for street-side garbage containers where dragging on sidewalk and road surfaces induces wear.
- Provide electrical conduction to devices where static electricity will do harm, such as in semiconductor processing facilities, typically metallic particles or graphite, most recently including carbon nanotubes. The same features appear in electronic shielding where a conductive phase is added to the computer or instrument plastic case to prevent the device from broadcasting radio waves.

Based on tonnage, the addition of particles to polymers amounts to 90 % or all composite production. Automotive tires are an especially visible application. Carbon black is used to provide ultraviolet protection. This application consumes 90 % of all carbon black. These graphite particles are small with sizes near 0.2 μm that agglomerate into 10 μm clusters. Carbon black is formed by partial combustion of hydrocarbons. Besides tires, carbon black is used in inks, photocopy printers, and paints.

Electronic circuit packaging relies heavily on conductive plastics that contain added metallic particles. Flame retardants are another additive for polymers, but those additives are toxic once ignited. Synthetic wood is widely used outdoors where weather, insect, bacteria, and sunlight resistance are concerns. Toys are an application for wood filled polymers. The formability of the composite is a major attraction versus traditional wood cutting.

A classic illustration of particle modified plastics involves low-cost mica additives. Mica corresponds to plate-like potassium aluminum silicates that contain water, iron, and fluorine. Because mica has an easy fracture crystal plane, the formation of small flakes is possible via milling or ultrasonic treatment. Mica

Table 11.8 Properties of mica filler [27, 31]

Property	Typical range
Theoretical density, g/cm^3	2.76–3.20
Elastic modulus, GPa	172
Strength, MPa	250–300
Thermal conductivity, W/(m °C)	0.7
Decomposition temperature, °C	1300
Thermal expansion, 10–6/°C	8–25
Heat capacity, J/(kg °C)	850–900

Table 11.9 Properties of polypropylene filled with mica

vol% mica	wt% mica	Density, g/cm^3	Strength, MPa	Elastic modulus, GPa	Elongation, %	Thermal expansion coefficient, 10^{-6}/°C
0	0	0.9	35	1.5	100	91
3	10	1.0	43	1.9	20	81
7	20	1.1	54	2.4	10	68
17	40	1.3	59	13	1	–
23	50	1.4	75	18	0	–
31	60	1.5	61	24	0	–
50	77	2.0	173	38	0	–

particles used for filled polymers are from 2 to 50 µm diameter, averaging about 20 µm, with a thickness about 1 % of the diameter. The plate-like particle are effective in reducing oxygen permeability, so are used in sporting and automotive structures. Properties of mica are listed in Table 11.8 [27, 31]; however, these properties change with water content. Improved interfaces with the polymers require silane surface treatments.

The case of polypropylene containing various mica is outlined in Table 11.9. The strength increases with mica additions, but results in a ductility loss. At high filler contents, the mica particles are damaged during mixing, so properties decline with too much additive.

Similar results are obtained with graphite additions, where polypropylene with 35 vol% graphite reaches 90 MPa strength, but ductility falls to 3 %. As expected, the elastic modulus is a linear function of volume fraction. Thermal characteristics are linear with composition on a volume fraction basis. Strength is less certain, mostly because interfaces play a large role. Fracture toughness very much depends on interface quality, and in cases of good particle-polymer bonding the composite is significantly improved over the pure polymers. A key is the introduction of a wetting agent on the particle surface prior to mixing the particles into the polymer.

As the electronics industry grew, plastic enclosures for computers were required to meet radio interference requirements. This led to significant efforts to understand how to preserve plastic moldability (low additive content) while forming a conductive network. One solution was to add carbon black (particle size near 0.2 µm) at 10–25 vol%. These high concentrations add cost, make forming difficult, and lower strength. Thus, carbon nanotubes are now proving to be an alternative, since concentrations of just 3 vol% are successful. Unfortunately, the cost of the carbon nanotubes remains high.

Friction Products

Durable braking materials are required to stop vehicles, such as trains, bicycles, automobiles, trucks, motorcycles, or airplanes. The brakes are simple geometries, such as pads. Clutches rely on similar friction material. Other friction material uses are in hand tools, such as electric screwdrivers with torque limiting features. These are broadly termed friction composites [32, 33].

A high friction coefficient with low wear rate is desired from a brake. It is important for the brake to resist degradation due to heating in service, especially in sever situations such as panic stopping. Operating pressures are a few MPa and peak temperatures reach to 600 °C (873 K). Accordingly, brakes are a mixture of polymer, metal, and ceramic phases, and low cost mineral phases are typically included. Formerly, asbestos was used in brakes, but that was discontinued in US automobiles manufactured after 1985. The combination of heat resistance and high friction coefficient (0.8) cast iron rotors is most desirable.

Brake linings were invented in 1897 using cloth impregnated with resin. Early transportation devices, such as the *Spirit of Saint Louis,* saved weight by not including brakes.

The three main friction composite groups are metallic (sintered), semi-metallic (bonded), and organic (polymer bonds). The metallic variants rely on iron and copper as the primary ingredients. The organic formulations are mixtures of resins, minerals, graphite, fiberglass, and even recycled rubber. Table 11.10 outlines the typical phases used in each of these products. There is nothing standard about the formulations. For example, one automobile brake formulation consists of alumina, glass, phenolic, organic filler, scrap rubber, and oxides. A metallic formulation consists of iron-copper, mullite, silica, graphite, and other oxides. There is confusion in comparing materials, since some formulations term abrasive phases as fillers, lubricant phases as binders, and fillers as abrasives. A brief attempt to clarify the terminology follows.

Abrasive phases, such as alumina, are added to clean the brake surface during use. Reinforcement phases provide strength, such as possible from sintered iron-

Table 11.10 Example automotive friction product formulations

Formulation	Example	#1	#2	#3	#4
Binder, resin	Phenolic resin	9	6	0	0
Abrasive	Aluminum oxide	20	40	19	20
Filler	Rubber	34	15	70	40
Reinforcement	Fiberglass	27	10	6	35
Lubricant	Graphite	10	29	5	5

Fig. 11.22 Microstructure of a friction product consisting of several phases, with best performance arising from this heterogeneous structure, with copper, carbon (graphite), tin, and iron as the main ingredients [courtesy P. Cuzzo]

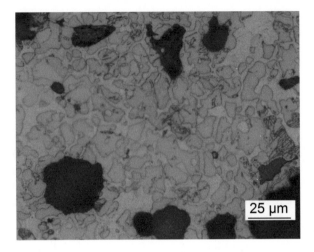

25 µm

copper-tin particles. Fillers plug the spaces in the structure. These are low cost ingredients such as limestone, mica, cashew shell dust, recycled rubber, or barium sulphate. Binder phases are coke, pitch, or phenolic phases. Lubricants such as zircon sand are added to clean the opposing material rubbing against the brake pad, typically cast iron. In some formulations, graphite and brass are added to modify performance by improving heat dissipation.

A friction material is heterogeneous and not fully dense. Resin bonding relies on compression molding, similar to hot pressing, to cure phenolic resin to provide strength to the mixture. Resins bonding is being replaced by sintering as a means to reach higher strengths. Figure 11.22 is a micrograph from a sintered metallic brake, showing iron-rich and copper-rich regions, with large pores where tin previously melted. This structure is based on a composition of Fe-20Cu-15C-4Sn with added carbon in the form of coke, synthetic graphite, and natural graphite. Sintering leads to bronze bonds around the steel grains.

The sintered brake pad pictured in Fig. 11.22 is formed using mixed 60–70 µm iron and copper powders with a smaller 18 µm tin powder. The graphite powder is 100 µm. The mixture is first pressed at 414 MPa, with 0.1 wt% oil added as a binder, giving a green density of 73 %. The trapezoid shaped brake pads are sintered in a continuous belt furnace at a peak temperature of 1024 °C (1297 K). Heating is at 22 °C/min in an atmosphere of 95 % nitrogen—5 % hydrogen and the peak temperature is held for 45 min. The structure has large pores and almost 25 vol% porosity. Final sintered density is about 4.2 g/cm^3 or almost the same as the starting solid density, so little dimensional change occurs during heating. The final strength is 30 MPa.

The key properties are density, thermal conductivity, and strength. Performance is measured by the mass loss per unit of kinetic energy absorption. Depending on the formulation, wear is from 0.2 to 16 mg per MJ of braking energy. In some regions copper contamination of ground water is leading to bans on disk brakes containing copper.

The newer friction products include combinations such as aluminum with boron carbide, silicon-silicon carbide-graphite, and carbon-carbon. The Al-B$_4$C composite is formed from mixed powders using hot pressing, while the Si-SiC-C composites rely on infiltration of molten silicon into a graphite preform followed by a bake to partially react the two phases. In carbon-carbon composites, graphite fibers are bonded using resin infiltration. That mixture is baked to convert the resin into graphite. Although expensive, graphite composites are widely adopted for aircraft brakes.

Inertial Heavy Alloys

Following the discovery of radioactivity in the late 1890s, needs arose for safe means to ship and handle radioactive materials [34]. The solution came from high tungsten content composites. Usually, the composite is consolidated using a liquid phase at temperatures near 1500 °C (1773 K). Normally, tungsten is brittle at room temperature, but when alloyed with the transition metals such as Ni, Cu, Co, or Fe, there is surprising ductility imparted by the matrix phase [34–38]. These tungsten heavy alloys are abbreviated WHA, W being the chemical symbol for tungsten.

Most tungsten heavy alloys range from 15 to 19 g/cm^3, density levels considerably higher than lead. Due to the high density, the tungsten composites provide excellent containment for radioactive materials. Although lead is soft and easily fabricated, tungsten composites require less volume and avoid toxicity associated with lead, mercury, or uranium. Tungsten heavy alloys are significantly lower cost when compared to high atomic number options, such as platinum, rhenium, osmium, iridium, or gold. Tungsten is the safest and lowest cost option among the high density metals.

The tungsten heavy alloys consist of tungsten grains in an alloy matrix. Figure 11.23 is a microstructure showing single crystal tungsten grains surrounded by the alloy matrix. In this case the composition is W-3.5Ni-1.5Fe, corresponding to about 88 vol% tungsten. Note the tungsten grains bond to form a three-dimensional skeleton. The matrix alloy occupies the spaces between the tungsten grains to form a second interpenetrating network. The matrix alloy is designed to avoid intermetallic formation during cooling from the sintering temperature, ensuring a high ductility. For this reason the alloys favor specific ratios such as Ni:Cu = 6:4 or Ni:Fe = 7:3. An example is W-7Ni-3Fe. When sintered at 1500 °C (1773 K) this composite exceeds a density of 17.4 g/cm^3 with a yield strength of 675 MPa and 29 % fracture elongation. Further property adjustments are possible through post-sintering heat treatments.

Inertial materials have high densities for used in applications such as in radiation absorption, medical X-ray shielding, sporting equipment weights, projectiles, vibration dampening weights, watch winding weights, gyroscope balances, and aircraft wing counterbalances. They are also used in armor piercing projectiles. Long rod shapes are produced by deformation and heat treated to reach high strength levels. Several applications are summarized in Table 11.11.

Fig. 11.23 Microstructure of a tungsten heavy alloy composite consisting of W-3.5Ni-1.5Fe. The tungsten grains form a bonded skeletal structure with matrix alloy (Ni-Fe-W) filling the intergranular spaces. This composition corresponds to 88 vol% tungsten with 12 vol% of the structure consisting of solidified 53Ni-24W-23Fe

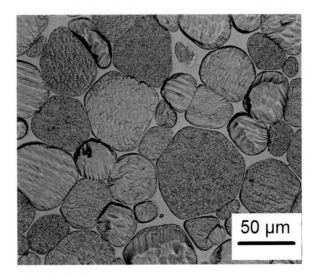

Table 11.11 Heavy alloy composites and uses

Application	Key property	Typical composition
Armor piercing projectiles	Density, strength, toughness, hardness	W-Ni-Co-Fe
Boring bars	Elastic modulus, vibration dampening	W-Ni-Fe
Casing for oil well logging	High density, corrosion resistance	W-Ni-Fe
Cell phone, pager, vibrator	Low cost, high density	W-Ni-Cu
Collimator for radiation tumor treatment	High tungsten content, radiation absorption	W-Ni-Fe
Counterbalance and golf club weights	Density, inertial response, machinability	W-Ni-Cu
Die cast tooling	Resistance to molten metal, high temperature strength	W-Mo-Ni-Fe
Inertial gyroscope guidance devices	Density, inertia, low thermal expansion	W-Ni-Cu
Isotope storage container	Tungsten content, easy machining, radiation opacity	W-Ni-Cu
Lead replacement birdshot	Low cost, easy forming, density	W-Cu-Sn-Fe
Race car crankshaft adjustment	Density, vibration dampening	W-Ni-Cu, W-Ni-Fe
Radiation containment	High tungsten content	W-Ni-Fe
Throwing dart bodies	Density, low cost, aesthetics	W-Ni-Cu
Vibration damping devices	High stiffness, inertia	W-Ni-Fe
Watch self-winding weight	Low cost, density	W-Ni-Cu

Table 11.12 Nominal
compositions for tungsten
heavy alloy composites

W-Cu-Sn-Fe
W-Ni-Fe
W-Ni-Co
W-Ni-Cr
W-Ni-Cu
W-Ni-Fe-Co
W-Ni-Mn-Cu
W-Mo-Ni-Fe

Tungsten contents are from 88 to 98 wt%, but compositions with as little as 50 wt% tungsten are possible. These combinations suppress formation of brittle phases during cooling from the sintering temperature. Alloying enables property adjustments over a wide range, as summarized in Table 11.12 [1, 34, 38, 39]. For example, the addition of molybdenum increases strength retention to high temperatures as useful for hot tooling. The W-Cu-Sn-Fe composition is used for lead-free birdshot; with 50 wt% tungsten the density is 12 g/cm^3. The twofold difference in elastic modulus between tungsten and the matrix provides excellent vibration attenuation for tool holders in precision machining.

The WHA composites are formed from mixed elemental powders nominally from 1 to 5 μm in size. After mixing, the powders are compacted at 200–600 MPa. If the shape is a small and flat, such as for self-winding watch weights, then uniaxial die compaction is employed. Injection molding is applied to complicated three-dimensional shapes, such as fishing sinkers. Long rods are formed using cold isostatic pressing. The pressed compact can be machined prior to sintering. Near full density is attained by sintering at temperatures near 1500 °C (1773 K), usually with a 30–60 min hold. Hydrogen is the typical sintering atmosphere, but vacuum is also used. Prolonged sintering causes microstructure coarsening and reduced strength; pores coalesce and grow over time, resulting in swelling and ductility. After 120 min at temperature, bubbles are evident inside the component as the pores coarsen.

After sintering, the tungsten grain size is 20–50 μm. This represents thousands of particles fusing into one grain during sintering. Dopants are used to reduce the grain size, such as molybdenum, tantalum, or rhenium. Each grain is a single crystal bonded to neighboring grains. Usually as the degree of tungsten-tungsten bonding or contiguity increases the ductility decreases. For W-Ni-Fe compositions, the matrix is approximately 53Ni–23Fe–24W.

The mechanical properties of representative alloys are summarized in Table 11.13 [1, 38, 39]. The elastic modulus ranges from 320 GPa for the low tungsten contents (82 wt% W) to 380 GPa for the high tungsten contents (97 wt% W). Thermal and electrical characteristics change little with composition. Thermal conductivity is about 120 W/(m °C) and the electrical conductivity is about 13 % of the International Annealed Copper Standard (IACS). The thermal expansion coefficient is low due to the high tungsten content, averaging near $5 \cdot 10^{-6}$ 1/°C (or 5 ppm/°C, where ppm implies parts per million). The heat capacity is near 1.5 J/(g °C). These attributes are desirable in measurement gauges where stability is desired independent of ambient temperature.

Table 11.13 Representative properties of heavy alloy composites [1, 38, 39]

Composition	Density, g/cm^3	Hardness, HV	Yield strength, MPa	Tensile strength, MPa	Elongation, %
74W-16Mo-8Ni-2Fe	15.3	365	850	1150	10
82W-8Mo-8Ni-2Fe	16.2	315	690	980	24
85W-5Ta-7Ni-3Fe	16.6	354	740	1025	3
86W-4Mo-7Ni-3Fe	16.6	280	625	980	24
90W-8Ni-2Fe	17.2	290	550	920	36
90W-5Mo-3Ni-2Fe	17.4	280	620	1000	17
90W-7Ni-3Fe	17.1	270	530	920	29
93W-5Ni-2Fe	17.7	280	590	930	30
94W-3Ni-1.5Fe-0.5Co	18.3	325	645	940	25
95W-3.5Ni-1.5Fe	18.1	280	600	920	18
97W-2Ni-1Cu	18.6	280	600	660	3
97W-2Ni-1Fe	18.6	300	610	900	19

As outlined in Fig. 11.24, composition impacts mechanical properties. Tungsten provides strength while the matrix provides toughness. As sketched in Fig. 11.25, four possible fracture paths exist within the microstructure. The actual behavior depends on the relative interface properties; failure occurs by a mixture of matrix rupture, tungsten grain cleavage, tungsten-tungsten grain boundary failure, and tungsten-matrix interface failure. Figure 11.26 is a scanning electron micrograph fracture surface showing mixed fracture paths. Weak interfaces trace to impurity segregation to interfaces, a problem that is cured by rapid cooling. Properties such as density, elastic modulus, and yield strength, are relatively immune to interface characteristics, but elongation, fracture toughness, ultimate tensile strength, and impact toughness vary with interface quality.

The yield strength for annealed tungsten is about 550 MPa with an elastic modulus of 400 GPa. For the 7:3 Ni:Fe alloy, the matrix yield strength is 205 MPa and elastic modulus is 225 GPa. A composite of 85 W-10.5Ni-4.5Fe corresponds to 67 vol% tungsten grains and delivers a yield strength of 411 MPa and elastic modulus of 330 GPa. These properties are as expected based on the constituents. On the other hand, properties related to plastic deformation are less predictable. Tungsten is brittle at room temperature, but the heavy alloy composite exhibits ductility with excellent work hardening. Although tungsten strain softens at high strain rates, meaning that strength declines during deformation, the composite exhibits the opposite behavior with strain hardening. Accordingly, the composite hardens when used as an armor piercing projectile.

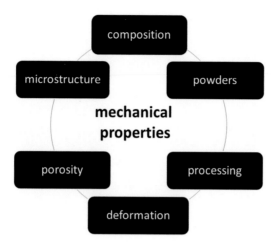

Fig. 11.24 Six factors determine the mechanical properties of tungsten heavy alloys, especially elongation and toughness. These include the (1) composition in terms of W content, matrix type, and ratio of ingredients in the matrix; (2) powders in terms of the particle size and purity; (3) processing cycle including the forming pressure, heating rate, peak temperature, and hold time; (4) post-sintering deformation and heat treatment cycles; (5) defects including residual porosity; and (6) microstructure grain size

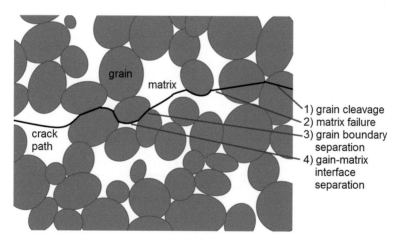

Fig. 11.25 A schematic of the four possible fracture paths in a composite consisting of (1) brittle grain cleavage, (2) ductile matrix failure, (3) grain boundary separation, and (4) interface separation. The latter two paths are sensitive to impurity segregation and are manipulated by heat treatments and cooling rates

Fig. 11.26 Fracture surface imaged by backscatter scanning electron microscopy, where the tungsten is *white* and the lower atomic number matrix is *dark*. The fracture paths noted in Fig. 11.25 are evident here. The predominance of grain-matrix interface failure indicates this material was not optimized for fracture resistance, since impurity segregation on cooling is a typical cause of interface separation

Fig. 11.27 Correlation between yield strength and fracture elongation for tungsten heavy alloy composites. Although the results are scattered, the general trade-off between strength and ductility is evident

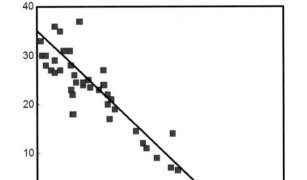

Property trade-offs are possible via composition, deformation, and heat treatment. Figure 11.27 is a plot of yield strength versus fracture elongation. Although scattered, the behavior illustrates how the composite provides either a high strength or high ductility, but not both. For low ductility compositions the ultimate tensile

Fig. 11.28 Fracture elongation as a function of the tungsten grain volume for W-Ni-Fe compositions fabricated with the same processing conditions. The trend line passes through zero ductility at 100 % tungsten

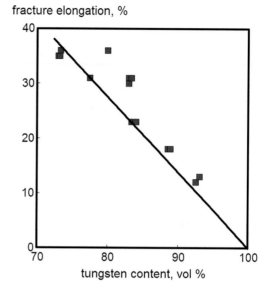

strength is also low. Figure 11.28 plots fracture elongation versus composition for the same materials shown in Fig. 11.27. Underlying these plots are changes in the tungsten grain boundary area, a typical weakness in the microstructure. A few applications are pictured in Fig. 11.29, including weighted throwing darts, vibration weights, birdshot, fragmentation projectiles, and fishing weights.

Iron Neodymium Boron Magnets

Exceptional hard magnetic properties for the $Fe_{14}Nd_2B$ intermetallic compound were identified in the 1980s. The compound provides the largest energy product attainable in a magnet; about 20-fold better than magnetic steel and eightfold better than earlier magnets [40, 41]. The contour plot in Fig. 11.30 indicates the magnetic field strength versus composition. The sweet spot is slightly off the $Fe_{14}Nd_2B$ compound, since a two phase composite microstructure is optimal. The excess rare earth, usually about 5 wt% Nd, Dy, Pr, or Tb, is intentionally oxidized and segregated to grain boundaries to decrease electrical conduction between the $Fe_{14}Nd_2B$ grains. The oxide constitutes about 5–8 vol% of the structure. Residual pores also act to disrupt conduction. These magnets are widely employed in cameras, generators, automotive systems, pumps, motors, business machines, loud speakers, headsets, telephones, power tools, and home appliances.

If the powder has anisotropic magnetic behavior, then a magnetic field is applied during powder pressing. Isotropic powders are magnetized after consolidation.

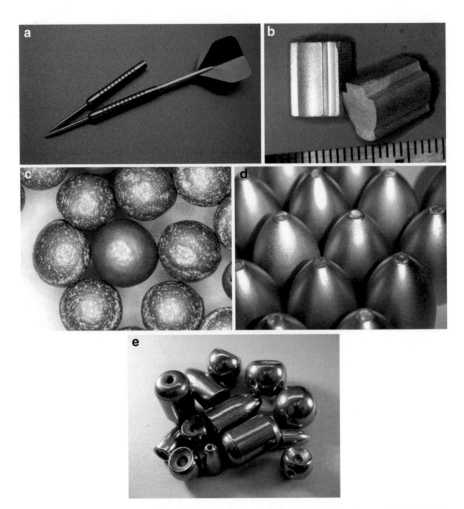

Fig. 11.29 Photographs of tungsten composite components; (**a**) throwing darts with a W-Ni-Cu center shaft, (**b**) vibrator weights from 90W-6Ni-4Cu, (**c**) birdshot from 50W-40Cu-9Sn-1Fe with a 12 g/cm^3 density, (**d**) fragmentation projectiles from W-Ni-Fe, and (**e**) fishing weights from W-Ni-Cu

With isotropic magnets, the particles are spherical and a common option is to mix the powder with a polymer. The powder-polymer mixture is either compression molded or injection molded, leading to a bonded magnet. The polymer phase remains a part of the structure. A final density near 5.5 g/cm^3 is typical for a bonded magnet. In one variant, the powder-polymer mixture is extruded to form rods, tubes, or other simple shapes that are sliced to length. Typical bonding polymers are epoxy, nylon, polyphenylene sulfide, polyester, nitrile rubber, and ethylene copol-

Fig. 11.30 A portion of the Fe-Nd-B ternary phase diagram with contours showing various levels of magnetic energy product in kJ/m³. The location of the $Fe_{14}Nd_2B$ compound is marked. Note the highest energy product is located away from the stoichiometric compound, corresponding to a two phase composite

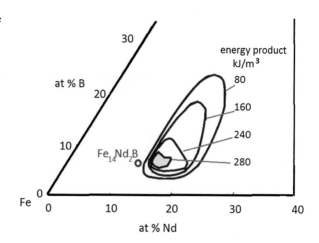

ymers for flexible magnets.

Anisotropic magnets require milling a rapidly cast thin ingot into particles sized below 3 μm using a sequence of first hydrogen embrittlement, milling, and hydrogen removal, followed by nitrogen jet milling. Hydrogen exposure makes the compound brittle and easily pulverized, and under heat and vacuum the hydrogen is removed from the solid. Jet milling shoots the particles at one another using high speed gas jets. The gas flow carries the desired small particles out of the milling chamber.

After reaching the desired size, the powder is compacted under an orienting magnetic field. Liquid phase sintering at 1050–1100 °C (1323–1373 K) for several hours in vacuum delivers near full density, near 7.5 g/cm³. Typically, the final shape is machined from the sintered block, annealed to remove strain, and again magnetized. An alternative is to hot press the powder. The $Fe_{14}Nd_2B$ composition reacts with water, hydrogen, and oxygen, so the magnets are coated to protect against reactions.

The magnets rely on a two phase microstructure as evident in Fig. 11.31. The grain boundaries contain rare earth oxides to isolate the magnetic grains, explaining why peak magnetic properties occur with an excess of rare earth metal. The $Fe_{14}Nd_2B$ compound provides the magnetic behavior while the grain boundary phase minimizes electrical conductivity. In the bonded magnets, the grain boundaries are polymers, but the grain packing is less efficient so properties are lower. For example at a coercive force of 1000 kA/m the sintered magnets have a remnant magnetization of 1.3 T while the bonded magnets are 0.7 T. Of course the bonded magnets are less expensive. Table 11.14 is a comparative tabulation of three important characteristics for bonded and sintered $Fe_{14}Nd_2B$ magnets [41–43].

Fig. 11.31 Microstructure for the $Fe_{14}Nd_2B$ magnets with an Nd-Fe second phase oxide (*white*) and residual porosity (*black*)

30 µm

Table 11.14 Comparative magnetic properties of bonded and sintered $Fe_{14}Nd_2B$ magnets [41–43]

Attribute	Bonded	Sintered
Remanence, B_R (T)	0.69	1.37
Intrinsic coercive force, H_C (kA/m)	720	1035
Energy product, $(BH)_{MAX}$ (kJ/m^3)	360	80

Soft Magnetic Composites

The emergence of electrical systems to displace mechanical systems is a profound shift in engineering design. This is most evident in the automobile. Electronic systems now account for more of the new vehicle cost versus mechanical systems. Hybrid and all-electric vehicles speed this technology shift. The newer vehicles rely on wide band gap semiconductors and magnetic inductors, motors, and sensors. In turn, the systems create demand for improved soft magnetic materials. A soft magnet does not retain magnetization once an applied field is removed; such behavior is evident in transformer cores. In contrast hard magnets retain magnetization after the field is removed.

Edison had the idea to stack sheets of iron-silicon alloy to form a laminated transformer core. The laminates provide two-dimensional magnetic behavior with electrical resistance arising from the silicon oxide insulator films between the laminates. However, considerably more energy efficiency comes from three-dimensional, complex shaped soft magnets. As one example, a small compact motor is formed as the hub of an electric bicycle. During downhill descents the motor delivers regenerative power to a battery. Then during uphill climbs the motor draws that power to provide supplemental energy. For this application, the motor has four stators (each 274 g) forming 24 prongs wrapped in copper and encased in aluminum. This design gives the magnetic output versus applied field curve plotted

Fig. 11.32 Soft magnetic composite response in an electric bicycle motor design showing the magnetic output versus applied field

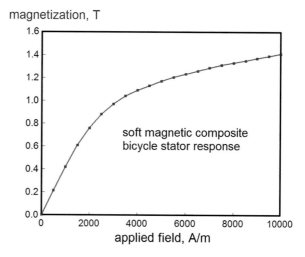

in Fig. 11.32. The motor stators are die compacted using polymer coated iron composite powder.

The demand for soft magnetic materials is expanding at 10 % per year. The mainstay are still two-dimensional designs formed from laminated iron-silicon sheets. However, about 20 % of the demand is for three-dimensional magnets. For improved performance, attention is directed toward soft magnetic composites (SMC) consisting of iron particles encapsulated in about 1 μm coating of nonconductive polymer or ceramic [44–46]. They come in three variants:

1. annealed iron powder coated with thermoplastic glue,
2. annealed iron powder coated with thermosetting compound,
3. annealed iron powder coated with ceramic (dust).

The desire is to minimize eddy current loss during cyclic magnetization and demagnetization. By making the magnet nonconductive, the losses are significantly reduced, especially at frequencies over 1500 Hz when eddy current losses are large. The nonconductor coating makes the composite an insulator, better than stacked iron-silicon sheets. When employed in an air gap motor, the magnetic performance is superior for the SMC.

The SMC composite is pressed to shape, then heated to induce bonding between particles. The keys for success are a pure iron powder for magnetic performance and an insulating coating. A schematic illustration of the microstructure is shown in Fig. 11.33. The individual iron grains are compacted under conditions that retain the insulator layer separating grains. For frequencies over 1500 Hz, iron particles from 150 to 200 μm are favored, but for the high frequencies associated with microwave and radar, in the 3–30 MHz range, smaller particles are favored. The powders are low in impurities with carbon below 0.01 wt%, oxygen below 0.06 wt %, and sulfur below 0.01 wt%.

Fig. 11.33 This schematic illustrates the compacted structure for soft magnetic composites, where the iron particles occupy most of the volume, but the insulator coating keeps the composite nonconductive. A small amount of porosity remains, since the high compaction pressures required to remove the pores also can rupture the insulator film

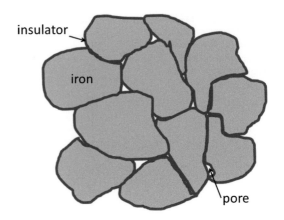

Applications for this composite are in automotive, portable tools, lawn and garden devices such as hedge trimmers, home appliances, and electric bicycles. The shapes are designed for easy die compaction. The greatest success is in high frequency motors. This creates a problem since the composite is not very strong, so the centrifugal force imposed by high rotation rates requires high bond strength.

A complete coating is required on each particle to minimize eddy current losses. Coating technologies include tumbling, spraying, milling, or fluid bed approaches. In the latter, the powder is levitated by an upward gas flow while a coating material dissolved in solvent is sprayed on the agitated particles. After solvent evaporation, the coating remains on the particles. Tooling is designed for the high pressure needed to reach full density, usually reaching a density near 7.30 g/cm^3 at 900–1000 MPa pressure. Heating the powder and die helps by softening the phases, using about 140 °C (413 K). To help polymer viscous flow, the peak compaction pressure is held for a few seconds. Additional heating after compaction improves strength. Options depend on the coating chemistry:

- heating to the 160–200 °C range (433–473 K) for thermoplastic polymers,
- heating to 200–400 °C (473–673 K) to cure thermosetting polymers,
- heating to 600–800 °C (873–1073 K) for ceramic and glass coatings.

The lower temperature heat treatments are in air, but the higher temperatures require protection against oxidation using nitrogen. Slow cooling ensures the iron is free of residual strains. The strength after annealing reaches 100 MPa, giving sufficient strength for machining, drilling, or milling.

Efficiency is a key concern in soft magnetic composites. There are three main loss factors; hysteresis loss from the oscillating magnetic field, eddy current loss, and high frequency loss. Minimized hysteresis loss arises from high purity and large iron grains. The eddy current loss is minimized by the nonconductor phase between grains. Little is done for the high frequency loss. For alternating current devices, the low eddy current loss is dominant.

Fig. 11.34 Plot of the energy loss versus applied frequency to compare the soft magnetic composite with that attained using laminated sheets

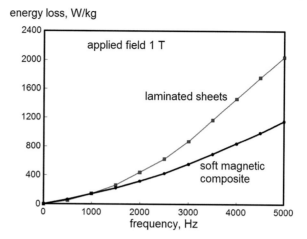

Table 11.15 Representative properties of soft magnetic composites [44–48]

Parameter	Low pressure	High pressure
Compaction pressure, MPa	560	840
Density, g/cm^3	7.30	7.50
Bending strength, MPa	75	125
Resistivity, μΩ-m	280	280
Permeability	455	650
Magnetization B_{MAX} at 4000 A/m (T)	1.2	1.6
Magnetization B_{MAX} at 10,000 A/m (T)	1.5	1.7
Coercive force, A/m	200	200
Core loss at 1 T and 1000 Hz, W/kg	147	92

A typical SMC motor involves multiple poles designed to operate at 20,000 rpm. The core loss in W/kg is plotted in Fig. 11.34 versus the test frequency at 1 T applied magnetization for both the SMC and stacked sheet. At frequencies over 1500 Hz the SMC is more efficient. The SMC magnets require a fast magnetization response where saturation is reached in 4 ms. Table 11.15 summarizes the properties attained in two formulations [44–48]. For complex multiple pole shapes, the SMC provides performance and design flexibility not possible with stacked iron-silicon sheets. Indeed, new direct drive motors are able to provide torque at low speeds, thereby eliminating gearing. However, from a mechanical viewpoint, the SMC strength is low since the iron is unalloyed and the bonds are weak. Software used in motor design is only starting to embrace the advances possible with three-dimensional soft magnetic composite structures.

Thermal Management Materials

Thermal management arises in a variety of situations, ranging from computers to space vehicles. Inherently, these applications rely on a high thermal conductivity material to distribute heat. Early issues arose in electrical switches, especially those used in high current—high voltage circuit breakers.

To satisfy the need for reliable electrical contacts, refractory metals were combined with high conductivity metals [49–52]. Favorites were W-Cu and W-Ag. Similar issues arose in space vehicle reentry where cooling was necessary to successfully reenter the atmosphere. That resulted in transpiration cooling with tungsten-copper composites. In transpiration cooling, the tungsten provides high temperature rigidity while the copper melts and evaporates to consume heat, similar to how evaporation of sweat cools an athlete.

As microelectronics circuits for computers shrank in size, issues arose in heat spreading, since semiconductors are not tolerant to heating. Various ideas arose on heat dissipation, but one limitation was from thermal fatigue arising from on-off cycles. An early solution was extruded aluminum in a high surface area fin array. But heating and cooling cycles create strains between the high thermal expansion aluminum and the low thermal expansion semiconductor. Thermal management devices required a high thermal conductivity and low thermal expansion coefficient similar to silicon. The options expanded to include materials better suited to the thermal loads of semiconductors, including:

aluminum—silicon carbide (Al-SiC)
aluminum nitride—copper (AlN-Cu)
aluminum nitride—yttria (AlN-Y_2O_3)
copper—graphite or copper—diamond (Cu-C)
copper—silicon carbide (Cu-SiC)
graphite—silicon carbide (C-SiC)
molybdenum—copper (Mo-Cu)
silicon carbide—copper (SiC-Cu)
silver—Invar (Ag-(Fe-36Ni))
tungsten—copper (W-Cu).

Tungsten-copper is good for stationary devices due to cost and availability. But for mobile devices, such as in hybrid vehicles, a lower density is needed, such as aluminum nitride with yttria, or aluminum with silicon carbide. For the highest demands, copper with diamond is an option. An example shape used in hybrid vehicles is pictured in Fig. 11.35. Air is circulated over the pin array to remove heat.

Thermal management applications dictate both thermal expansion and thermal conductivity. Thermal expansion is measured in $10^{-6}/°C$ or parts per million per °C (abbreviated as ppm/°C). Silicon has a low thermal expansion coefficient at room temperature, below 5 ppm/°C. A difficulty is encountered because the high-thermal-conductivity materials such as copper or aluminum have high thermal expansion coefficients of 17 and 22 ppm/°C, respectively. The low-expansion materials

Fig. 11.35 This heat sink is used to extract thermal energy from the semiconductor control circuit used on hybrid vehicles [courtesy F. Ahmad]

Table 11.16 Thermal properties of W-Cu composites [51–53]

Copper content, wt %	Copper content, vol%	Density, g/cm^3	Thermal conductivity, W/(m °C)	Thermal expansion, 10^{-6}/°C
100	100	8.96	403	17
65	80	11.0	306	15
30	48	14.3	260	11
20	35	15.6	247	7
15	28	16.4	221	6
10	19	17.3	209	6
0	0	19.3	172	5

are either expensive or brittle (silicon carbide, tungsten, diamond). The W-Cu thermal management composites are around 10–20 wt% copper, corresponding to 19–35 vol% copper. For a thermal expansion coefficient of 7 ppm/°C, the ideal composition is near W-15Cu. Table 11.16 compiles experimental thermal properties of W-Cu composites at different copper contents [51–53]. When mechanical properties are a concern, compositions of 50–90 wt% tungsten are favorites, such as used for electrical discharge machining. From 50 to 90 wt% W the elastic modulus increases from 193 to 340 GPa while strength ranges from 480 to 800 MPa. Fracture toughness changes slightly with composition over this range and generally is near 60 MPa√m. The ductility is below 1 % elongation and Poisson's ratio is about 0.30–0.33.

Fabrication relies on mixing and milling the tungsten and copper particles, largely to homogenize the copper. The typical tungsten particle size is from 0.5 to 2 μm but the copper particle size is larger near 20–40 μm. At low concentrations, the copper particles are not homogeneously dispersed, but milling smears the copper over the tungsten. Other options mix tungsten with small copper oxide particles, then the oxide is reduced in hydrogen to form copper. At high copper contents, infiltration is an option, where first the tungsten is sintered into a porous

Fig. 11.36 Cross-sectional micrograph of the W-20Cu composite structure used in thermal management applications

Fig. 11.37 A picture of copper bleeding from the surface of a W-20Cu composite as the result of excess copper used during infiltration

skeleton into which molten copper is infiltrated. An example of the W-Cu micro-structure is given in Fig. 11.36.

Thermal management devices are simple shapes, such as plates, so die compaction is the appropriate shaping step. Cold isostatic pressing is used for long, thin structures and powder injection molding is used for complicated three-dimensional shapes. In pressing, compaction pressures are 150 MPa, giving 45–55 % green density. Heating at rates near 5 °C/min evaporates any lubricant from the compact. Final sintering is at 1300–1400 °C (1573–1673 K) for 60 min, almost always in hydrogen to remove oxides. Sintering in other atmospheres is possible, but thermal conductivity falls due to retained impurities. This sintering temperature is above the melting point of copper. Densification occurs in the tungsten skeleton, letting the molten copper fill the interparticle voids. Too high a sintering temperature, or too long a sintering time, leads to excessive tungsten skeleton shrinkage, resulting in copper being squeezed out of the compact. Figure 11.37 is a picture of copper exudation to the surface in a W-20Cu compact heated to 1350 °C (1623 K) for 60 min in hydrogen.

Depending on the details, sintered strength ranges from 420 to 1100 MPa, but there is little ductility. Because of the high density, most W-Cu compositions are applied to stationary systems, such as internet servers, microwave relay stations, or radar base stations. Aluminum nitride composites are more expensive, but are used in vehicles.

Zirconia-Toughened Alumina

Alumina (Al_2O_3) is a widely used ceramic with excellent elevated temperature strength, high hardness, and low cost. However, alumina is brittle and susceptible to fracture under high strain rate loading. Various additives are used to improve the hardness, strength, toughness, or wear resistance of alumina. Many of the additives, such as titanium carbide or silicon carbide, fail to improve fracture toughness.

On the other hand, silica (SiO_2) is a low cost additive that forms mullite ($3Al_2O_3 2SiO_2$). Unfortunately, mullite is not effective at temperatures over about 1000 °C (1273 K). One alternative is to add zirconia (ZrO_2), with low levels of yttria (Y_2O_3), to promote both strength and toughness in alumina [54]. Zirconia is a polymorph that transforms from monoclinic to tetragonal near 1170 °C (1443 K). Alloying zirconia with yttria stabilizes the tetragonal phase to lower temperatures. In a composite with alumina, the stress induced transformation of yttria-stabilized zirconia acts to blunt crack propagation. The transformation stresses improve fracture toughness and fracture resistance.

Zirconia is more expensive than alumina, so zirconia-toughened alumina (ZTA) composites limit the zirconia content to 20 vol% or less. The increased crack resistance for ZTA is analogous to ductile metals, where energy is consumed in deforming the material at the crack tip. The net result is high hardness and toughness. One application is in ballistic armor, usually in the form of simple plate shapes. When a bullet strikes ZTA armor, the ceramic fragments to convert kinetic energy into fracture energy. Additional use is in situations that exhibit rapid wear in high temperature corrosive environments, such as petrochemical and chemical equipment, bearings, bushings, cutters, metal cutting tools, and artificial body joints.

The composites are formed from mixed powders using compaction and sintering. Other fabrication routes such as hot pressing and hot isostatic pressing are used. It is common to rely on submicrometer particles to ensure a desirable small final grain size after sintering (about 0.3–0.4 μm). Properties are improved using coated powders, but this is not frequently employed. Sintered densities are usually more than 98 % of theoretical, relying on peak consolidation temperatures

Table 11.17 Properties of zirconia-toughened alumina compared to pure alumina and pure yttria stabilized zirconia [54, 55]

ZrO$_2$composition, vol%	0	10	15	20	100
Density, g/cm^3	3.9	4.1	4.2	4.3	5.9
Elastic modulus, GPa	390	370	350	340	100
Poisson's ratio	0.23	0.24	0.25	0.24	0.28
Bending strength, MPa	350	–	450	800	200
Compressive strength, MPa	2470	2000	2300	2900	800
Hardness, HV	1880	1705	1650	1600	270
Fracture toughness, MPa√m	3	6	6	7	10
Thermal expansion coefficient, 10–6/°C	6.9	8.0	8.1	8.3	8.0
Thermal conductivity, W/(m °C)	29	25	23	22	2

near 1500 °C (1773 K). Higher sintering temperatures induce coarsening with a decline in properties. Full density by hot isostatic pressing is used in high performance applications, such as implant devices.

Properties of zirconia toughened alumina composites are given in Table 11.17 [54, 55]. Measured properties are sensitive to processing history and test sample preparation, accordingly the properties exhibit a high variability. For comparison with the ZTA, pure alumina and toughened zirconia are included in the table (0 and 100 % compositions). Zirconia is denser than alumina (5.9 versus 3.9 g/cm^3) with a lower hardness and elastic modulus. In hot pressed compositions using unstabilized zirconia, the peak fracture toughness (7–9 MPa√m) was at 10 vol% ZrO$_2$. With attention to grain size, processing, composition, and homogeneity values up to 12 MPa√m are reported, four times that of alumina. The composites are not recommended for high temperature or high stress applications and are best suited to temperatures under 600 °C (873 K) and stresses under 350 MPa. To a first approximation, other factors such as strength, elastic modulus, density, and hardness exhibit linear dependent on composition. Thus, composites exhibit toughness, otherwise pure alumina is often superior.

The performance of ZTA as armor results in applications for body vests, helicopter seat liners, vehicle siding, and related devices [56]. As ideas advance on fragmentation energy and reduced penetration, new composite systems will emerge to displace ZTA. Along these lines, efforts to include SiC reinforcements are the current area of effort.

Note, ZTA has two weaknesses. On heating, the composite undergoes a significant loss of toughness, in some compositions dropping by 50 % from room temperature to 600 °C (873 K). Further, cyclic heating induces differential strains between the two phases, leading to significant strength loss in what is termed thermal fatigue. Thus, the composition should avoid applications that experience large temperature variations.

Parting Comments

The two major factors driving interest in particulate composites are improved properties and cost reduction. For example, silver electrical contacts are desirably low in contact resistance, but they weld with high currents. The addition of tungsten to the silver decreases conductivity, but solves the contact welding issue. Additionally, tungsten is lower in cost compared to silver. Thus, W-Ag composites are used for high reliability applications [57]. Likewise, there is little question as to the unique hardness and wear resistance of diamond. However, the synthesis of monolithic diamond structures involves extreme expense due to the temperature-pressure combination required to avoid decomposition into graphite. Diamond-metal composites greatly lower fabrication costs and deliver improved toughness [58]. We see this idea of a composite solution in many applications. For example, heat sinks from Cu-SiC present unique high thermal conductivity and low thermal expansion [59].

This chapter organizes several application examples for particulate composites. At the same time, many examples are missing. Even so, the sense of broad design customization emerges. Generally, the applications are dominated by situations such as summarized here:

1. where toughness is desired in a hard material (cemented carbides)
2. where unique property combinations are needed (thermal management materials with balanced thermal conductivity and thermal expansion)
3. where lower cost is desired, possibly from dilution without loss of properties (mineral filled polymers).

The most demanding situations focus on customized property combinations. The key is to identify a problem and then to synthesize a solution. This is fundamentally the basis for invention.

Study Questions

11.1. In a composite of alumina (Al_2O_3) toughened with up to 10 vol% zirconia (ZrO_2), both the strength and toughness increase. Over this same range, what is the anticipated elastic modulus?

11.2. The density for WC-10Co is 14.6 g/cm^3. What is the volume fraction of each phase assuming there is no intersolubility?

11.3. For the same WC-10Co composition, at 1200 °C (1473 K) the solubility of WC in cobalt is 2.75 at.% W and 2.75 at.% C. Estimate the volume fraction of each phase for this case?

11.4. Particulate composites often find early success in sporting devices, such as darts, horseshoes, birdshot, and golf clubs. Identify a new application in sporting, outline the engineering requirements, and suggest a possible solution. Some examples might be baseball shoe cleats, bicycle peddles, surfboard fins, football helmet pads, and arrowheads.

11.5. It is proposed to form a Ti-TiC composite to serve as the blade on ice hockey skates. The idea is to make the skates rust resistant and perpetually sharp. Comment on the merits and any concerns that come with this application.

11.6. A durable shoe sole is desired for hiking boots. The required attributes include flexibility, wear resistance, easy fabrication, low density, and low cost. One suggestion is to supplement the flexibility of urethane rubber with the added wear resistance of hollow ceramic particles. Outline how a development project might be planned around this problem, including the tests relevant to optimizing the composite for this application.

11.7. In planning the shoe sole development project in Study Question 11.6, make short lists of wear resistant compositions that might be a good starting point for the composite based on (a) low cost, (b) high hardness, and (c) marketing appeal?

11.8. Following a 1923 earthquake in Japan, it was decided to record the names of the dead on a scroll that would resist deterioration for 10,000 years. The 548 pages listing the victims were sealed in fused quartz, wrapped in asbestos cloth, encapsulated in lead, and packed in a silicon carbide container. If you were tasked with a similar challenge today, what would be single composite that could replace the several layers?

11.9. Particle board is widely used in home construction. What are the main constituents?

11.10. Wood particles are about 12 GPa in elastic modulus, with a density of 1.4 g/cm^3 and cost of $0.03 per kg. On the other hand, plastics used to form man-made lumber are about 1.5 GPa, 0.96 g/cm^3, and $2.00 per kg. A typical composite consists of 60 wt% wood particles. What is the expected density, elastic modulus, and material cost?

References

1. W.F. Gale, T.C. Totemeier (eds.), *Smithells Metals Reference Book*, 8th edn. (Elsevier, Amsterdam, 2004)

2. G. Popescu, M. Zsigmond, P. Moldovan, Processing of a composite material like AlSi/SiCp through powder metallurgy. J. Adv. Mater. **35**, 16–19 (2003)

3. H. Shen, C.J. Lissenden, 3D finite element analysis of particle-reinforced aluminum. Mater. Sci. Eng. **A338**, 271–281 (2003)

4. J.E. Spowart, D.B. Miracle, The influence of reinforcement morphology on the tensile response of 6061/SiC/25p discontinuously-reinforced aluminum. Mater. Sci. Eng. **A357**, 111–123 (2003)

5. R.L. Deuis, C. Subramanian, J.M. Yellup, Dry sliding wear of aluminum composites - a review. Compos. Sci. Technol. **57**, 415–435 (1997)

6. T.W. Cline, P.J. Withers, *An Introduction to Metal Matrix Composites* (Cambridge University Press, Cambridge, 1993)

7. R.L. Matthews, R.D. Rawlings, *Composite Materials: Engineering and Science* (CRC Press, Boca Raton, 1999)

8. Anonymous, *Cambridge Engineering Selector* (Granta Design, Cambridge, updated annually)
9. K.J.A. Brookes, Half a century of hardmetals. Met. Powder Rep. **50**(12), 22–28 (1995)
10. K.J.A. Brookes, *Hardmetals and Other Hard Materials*, 3rd edn. (International Carbide Data, Hertsfordshire, 1998)
11. P. Ettmayer, Hardmetals and cermets. Annu. Rev. Mater. Sci. **19**, 145–164 (1989)
12. H.E. Exner, Physical and chemical nature of cemented carbides. Int. Met. Rev. **24**, 149–173 (1979)
13. S. Luyckx, The hardness of tungsten carbide - cobalt hardmetal, in *Handbook of Ceramic Hard Materials*, ed. by R. Riedel, vol. 2 (Wiley-VCH, Weinheim, 2000), pp. 946–964
14. V.K. Sarin, Cemented carbide cutting tools, in *Advances in Powder Technology*, ed. by G.Y. Chin (American Society for Metals, Metals Park, 1982), pp. 253–288
15. W. Hoeland, G. Beall, *Glass-Ceramic Technology* (American Ceramic Society, Westerville, 2002)
16. J.R. Kelly, I. Nishimura, S.D. Campbell, Ceramics in dentistry: historical roots and current perspectives. J. Prosthet. Dent. **75**, 18–32 (1996)
17. J.M. Meyer, W.J. O'Brien, C.U. Yu, Sintering of dental porcelain enamels. J. Dent. Res. **55**, 696–699 (1976)
18. D. Belnap, A. Griffo, Homogeneous and structured PCD WC-Co Materials for drilling. Diam. Relat. Mater. **13**, 1914–1922 (2004)
19. H. Katzman, W.F. Libby, Sintered diamond compacts with a cobalt binder. Science **172**, 1132–1134 (1971)
20. J. Konstanty, *Cobalt as a Matrix in Diamond Impregnated Tools for Stone Sawing Applications*, 2nd edn. (Wydawnictwa AGH, Krakow, 2003)
21. J. Konstanty, *Powder Metallurgy Diamond Tools* (Elsevier, Amsterdam, 2005)
22. P. Slade (ed.), *Electrical Contacts. Principles and Applications*, vol. 2 (CRC Press, Boca Raton, 2013)
23. X. Wang, H. Yang, M. Chen, J. Zou, S. Liang, Fabrication and arc erosion behaviors of Ag-TiB$_2$ contact materials. Powder Technol. **256**, 20–24 (2014)
24. S.Y. Chang, J.H. Lin, S.J. Lin, T.Z. Kattamis, Processing copper and silver matrix composites by electroless plating and hot pressing. Metall. Mater. Trans. **30A**, 1119–1136 (1999)
25. C.D. Desforges, Sintered materials for electrical contacts. Powder Metall. **22**, 139–144 (1979)
26. S.K. Bhattacharya, *Metal Filled Polymers* (CRC Press, Boca Raton, 1986)
27. J.V. Milewski, J.V. Katz (eds.), *Handbook of Fillers and Reinforcements for Plastics* (Van Nostrand Reinhold, New York, 1978)
28. S. Ahmed, F.R. Jones, A review of particulate reinforcement theories for polymer composites. J. Mater. Sci. **25**, 4933–4942 (1990)
29. H.S. Tekce, D. Kumlutas, H. Tavman, Effect of particle shape on the thermal conductivity of copper reinforced polymer composites. J. Reinf. Plast. Compos. **26**, 113–121 (2007)
30. F. Carmona, Conducting filled polymers. Physica A **157**, 461–469 (1989)
31. S.J. Schneider (ed.), *Ceramics and Glasses*. Engineered Materials Handbook, vol. 4 (ASM International, Materials Park, 1991)
32. D. Chan, G.W. Stachowiak, Review of automotive brake friction materials. J. Automob. Eng. **218**, 953–966 (2004)
33. F.A. Lloyd, M.A. DiPino, *Advances in Wet Friction Materials – 75 Years of Progress*, SAE Technical Paper No. 800977, Warrendale, PA (1980)
34. C.J. Smithells, A new alloy of high density. Nature **139**, 490–491 (1937)
35. H.J. Ryu, S.H. Hong, W.H. Baek, Microstructure and mechanical properties of mechanically alloyed and solid-state sintered tungsten heavy alloys. Mater. Sci. Eng. **291A**, 91–96 (2000)
36. R.M. German, A. Bose, S.S. Mani, Sintering time and atmosphere influences on the microstructure and mechanical properties of tungsten heavy alloys. Metall. Trans. **23A**, 211–219 (1992)

37. Z.C. Cordero, E.L. Huskins, M. Park, S. Livers, M. Frary, B.E. Schuster, C.A. Schuh, Powder-route synthesis and mechanical testing of ultrafine grain tungsten alloys. Metall. Mater. Trans. **45A**, 3609–3618 (2014)
38. A. Bose, Alloying and powder injection molding of tungsten heavy alloys: a review, in *Tungsten and Refractory Metals*, ed. by A. Bose, R. Dowding (Metal Powder Industries Federation, Princeton, 1995), pp. 21–33
39. B.H. Rabin, R.M. German, Microstructure effects on tensile properties of tungsten-nickel-iron composites. Metall. Trans. **19A**, 1523–1532 (1988)
40. B.M. Ma, J.W. Herchenroeder, B. Smith, M. Suda, D.N. Brown, Z. Chen, Recent development in bonded NdFeB magnets. J. Magn. Magn. Mater. **239**, 418–423 (2002)
41. J. Ormerod, The physical metallurgy and processing of sintered rare earth permanent magnets. J. Less-Common Met. **111**, 49–69 (1985)
42. J. Ormerod, S. Constantinides, Bonded permanent magnets: current status and future opportunities. J. Appl. Phys. **81**, 4816–4820 (1997)
43. J.S. Cook, P.L. Rossiter, Rare-earth iron boron supermagnets. Crit. Rev. Solid State Mater. Sci. **15**, 509–550 (1989)
44. S. Tiller, Soft magnetic composites in the development of a new compact transversal flux electric motor. Powder Metall. Rev. **2**(3), 75–77 (2013)
45. Y.G. Guo, J.G. Zhu, P.A. Watterson, Development of a PM transverse flux motor with soft magnetic composite core. IEEE Trans. Energy Convers. **21**, 426–434 (2006)
46. Y. Huang, J. Zhu, Y. Guo, Z. Lin, Q. Hu, Design and analysis of a high speed claw pole motor with soft magnetic composite core. IEEE Trans. Magn. **43**, 2492–2494 (2007)
47. I. Hemmati, H.R.M. Hosseini, A. Kianvash, The correlations between processing parameters and magnetic properties of iron-resin soft magnetic composite. J. Magn. Magn. Mater. **305**, 147–151 (2006)
48. H. Shkorollahi, K. Janghorban, Soft magnetic composite materials (SMCs). J. Mater. Process. Technol. **189**, 1–12 (2007)
49. V. Josef, L.K. Tan, Thermal performance of MIM thermal management device. Powder Injection Moulding Int. **1**, 59–62 (2007)
50. W. Nakayama, Thermal management of electronic equipment: a review of technology and research topics. Appl. Mech. Rev. **39**, 1847–1868 (1986)
51. R.M. German, K.F. Hens, J.L. Johnson, Powder metallurgy processing of thermal management materials for microelectronic applications. Int. J. Powder Metall. **30**, 205–215 (1994)
52. C. Zweben, Advances in high performance thermal management materials: a review. J. Adv. Mater. **39**, 3–10 (2007)
53. A. Bothate, R.M. German, W. Li, E.A. Olevsky, W.M. Daoush, S. Moustafa, D. Whychell, Advances in W-Cu: new powder systems, in *Advances in Powder Metallurgy and Particulate Materials* (Metal Powder Industries Federation, Princeton, 2010), pp. 7.6–7.20
54. J. Wang, R. Stevens, Review zirconia-toughened alumina (ZTA) ceramics. J. Mater. Sci. **24**, 3421–3440 (1987)
55. K. Kageyama, Y. Harada, H. Kato, Preparation and mechanical properties of alumina-zirconia composites with agglomerated structures using presintered powder. Mater. Trans. **44**, 1571–1576 (2003)
56. E. Medvedovski, Alumina ceramics for ballistic protection, part 2. Ceram. Bull. **81**(4), 45–50 (2002)
57. F. Findik, H. Uzun, Silver-based refractory contact materials. Mater. Des. **24**, 489–492 (2003)
58. H. Moriguchi, K. Tsuduki, A. Ikegaya, Y. Miyomoto, Y. Morisada, Sintering behavior and properties of diamond/cemented carbides. Int. J. Refract. Met. Hard Mater. **25**, 237–243 (2007)
59. T. Schubert, A. Brendel, K. Schmid, T. Koeck, L. Ciupinski, W. Zielinski, T. Weissgarber, B. Kieback, Interfacial design of Cu/SiC composites prepared by powder metallurgy for heat sink applications. Compos. Part A **38**, 2398–2403 (2007)

Chapter 12
Prospects

Developments in the field of particulate composites provide assurances for the future. Factors contributing to expanded applications include a variety of new materials, powder treatments, fabrication routes, and application needs. This chapter introduces some prospects to help identify areas to watch.

Growth Background

Technological growth comes from a combination of market and engineering efforts. Disproportionate credit goes to technical advances, yet marketing is equally important, leading to the concept of *techno-marketing*. Without proper balance, great technical discoveries struggle in obscurity. On the other hand, marketing with a weak technical base is not fruitful. A popular difficulty is to have an invention looking for a problem. The opposite is best, where a problem leads to an invention. This book emphasizes the technical aspects of particulate composites, but also realizes marketing is needed for commercial success.

The techno-marketing concept helps rationalize project goals. Inputs from technical staff are melded with marketing evaluations. Marketing further evaluates promotion, pricing, and distribution options. Ideally, the cost and performance combination are positioned for rapid adoption. The map in Fig. 12.1 outlines the best target in terms of projected cost and performance. Rapid adoption occurs when cost is low and performance is high. Success is linked to attacking the right applications, at the right price, while delivering the best performance. Some of the world's largest firms excel by rationalizing technical advances with market needs.

Many combinations of phases are possible in forming a composite. There are probably 10,000 common polymer, metal, ceramic, glass, and mineral phases, providing a tremendous number of possible combinations. When rationalized to applications, some of the combinations are unappealing, but that same combination

© Springer International Publishing Switzerland 2016
R.M. German, *Particulate Composites*, DOI 10.1007/978-3-319-29917-4_12

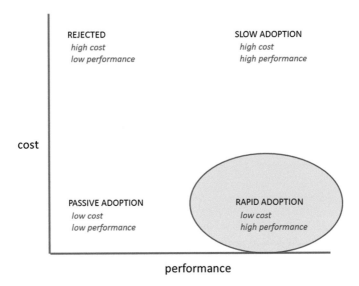

Fig. 12.1 The techno-marketing idea leads to mapping different solutions based on cost to the customer and technical performance. Rapid growth occurs when cost is reduced and performance improves, typically such disruptive advances come from outside an industry

might be valuable to another application. For example sawdust mixed with polyethylene has a relatively low strength of 20–50 MPa. However, the ability to extrude that composite into simple shapes is useful for forming exterior weather resistance products, such as outside flooring. Similar formulations have moved into children's toys, such as the classic alphabet blocks. Instead of machining wood with the generation of waste, the blocks are injection molded from the wood-plastic composite. The blocks look and feel like wood, but are fabricated at a lower cost with little waste and no machining.

In structural materials, reinforced polymers are displacing metals, in part because of a high specific strength as measured by strength divided by density. Polymer composites are approaching 300 MPa/(g/cm^3), higher than heat treated carbon steel at 200 MPa/(g/cm^3). Currently, widespread use of structural polymer composites is inhibited by a cost that is about twice that of the steel on a volume basis. However, the mass reduction and concomitant energy savings is adding to the applications. Depending on the cost of fuel, automotive designers value mass reduction at about \$7–8 per kg. High strength polymer composites are expensive, but some systems are embraced in automobiles [1]. This is following behind aerospace uses where high performance composites are widely adopted [2–4]. As a counter option, metallic components with ceramic additions are a means to lower density while increasing hardness and elastic modulus [5–8]. The interplay of materials for automotive components is an illustration of how a competition exists between various materials targeting the same application.

Growth is expected as more particulate composites are developed as strong, hard, and tough compositions [9–11]. This combination of features provides considerable promise. The addition of constraints, such as low cost or low density, makes optimization a challenging problem. In other prospects, biomedical applications place a high valuation on performance that easily justifies particulate composites [12–15]. But product qualification times are long. Likewise, considerable effort is directed at composites with tailored magnetic and electrical response [16, 17]. Each application field has a different cost sensitivity. Aerospace and biomedical applications are least sensitive, while construction is probably the most sensitive to cost. Between the extremes are sporting, electronic, communication, and automotive fields. Thus, there is differing enthusiasm for a new and expensive formulations. Automotive applications rely on significant cost reductions before adopting technologies. Thus, materials long ago adopted in bicycles are still not adopted by mainstream automobiles. All of this reflects how market adoption paces technical success.

Growth Focus

As mentioned, continued growth is anticipated in hard composites. Hard materials, such as diamond, are already important particulate composites. The production of diamond particles amounts to a \$13 billion industry in 2015, of which 70 % are for use in metal bonded composites [9, 10]. The metallic matrix provides toughness needed for industrial use. Alternative hard phases include WC, SiC, TiC, VC, TaC, and compounds such as BN, TiN, and Al_2O_3. Several metals and ceramics are in use as the bonding phase, including cobalt, steel, stainless steel, zirconia, titanium alloys, bronze, and nickel aluminide. The intent is to deliver exceptional combinations of hardness and toughness, outside the realm possible with either phase alone. As energy and mineral supplies become more difficult to extract there is every assurance of more need for hard composites in mining, drilling, and excavation.

One approach to composite design relies on mimicking biological structures. These bio-inspired structures assess the phases and morphological relations in natural structures. Tough-hard composites are a favorite, including bamboo, turtle shells, and abalone shells. For example, Mother of Pearl on the inside of oyster and abalone shells consists of calcite ($CaCO_3$) bricks and conchiolin biopolymer mortar. A simplified conceptualization is given in Fig. 12.2. The structure is similar to how brick walls are constructed. In one effort to mimic this morphology, alumina powder is plasma sprayed to deposit overlapping plates and the gaps are infiltrated with epoxy [16, 17]. After curing the epoxy, the transverse rupture strength is 225 MPa and fracture toughness of 5.5 MPa$\sqrt{m}$; the perpendicular orientation strength is 50 MPa and toughness is 1 MPa$\sqrt{m}$. Although interesting, the material is not impressive in properties, since standard alumina has an isotropic strength of 350 MPa with fracture toughness of 3.5 MPa$\sqrt{m}$, at a lower cost. Even so, the

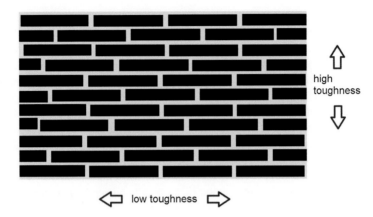

toughness

low toughness

Fig. 12.2 A schematic conceptualization of a bio-inspired composite similar to oyster or abalone shells. The structure relies on a high content of layered hard phase with intermediate tough biopolymer. The structure is anisotropic, with one direction having high toughness, but the perpendicular direction having low toughness

bio-inspired approach identifies new composite morphologies as an inspiration for research [18].

Target applications for the hard-tough composites are industrial components. Engineering these systems starts at the atomic scale using coated powders to control interfaces [19]. A homogeneous microstructure is gained by coating the hard particle with the tough phase. When densified the structure has continuous tough phase and dispersed hard phase, even with 50 vol% hard phase. The composition is adjusted to provide needed property combinations. When fractured, the matrix phase provides tough ligaments, as evident by the crack propagation captured in Fig. 12.3. Further gains arise by forming functionally graded structures. High cost and very hard material is placed strategically at the cutting tip and is backed by a lower cost, tougher substrate. One machine tool tip design is illustrated in Fig. 12.4. Demonstrations along these lines generate 15-fold longer service life. In a production manufacturing operation, extended service life means less down-time for replacement. In one situation the overall cost savings because of fewer interruptions was 45-fold.

New property combinations arises from nanoscale materials [20–22]. To promote durability, attention is directed toward diamond-cemented carbide composites with microstructures reaching into the nanoscale [23]. Nanoscale powders tend to disappoint when consolidated using traditional cycles. Microstructure coarsening results in property combinations similar to what is attained using larger powders [24]. Efforts at rapid consolidation try to avoid coarsening using dynamic compaction [25, 26], plasma processing [27], spark sintering [28–31], reactive synthesis [32], laser melting [33], and reactive hot pressing [34]. Fast consolidation is helping to retain the benefit from the nanoscale particles. Further, composites naturally resist coarsening due to interface pinning. Thus, a homogeneous starting dispersion

Fig. 12.3 Scanning electron micrograph of crack propagation in a composite consisting of hard (*dark*) alumina grains embedded in a tough cemented carbide matrix. Normally the alumina fails in a brittle manner at low strains, but the composite design enables a much higher toughness since crack propagation must pass through the cemented carbide

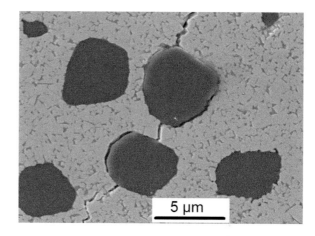

Fig. 12.4 The idea of selective placement of the high cost composite in a metal cutting tip, where the insert is bonded to a rigid substrate

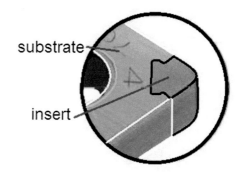

allows traditional consolidation cycles using nanoscale coated powders to avoid coarsening, leading to property gains.

Magnetic components go hand in hand with the growth in electrical systems, including electric automobiles. Transportation systems are moving to more electronics versus the historical focus on mechanical systems. One sign of the shift is the massive growth in magnets fabricated using the $Fe_{14}Nd_2B$ composite. It was discovered in 1982 and by 2014 shipments of these magnets exceeded $1.25 billion, reflecting an average price of $57 per kg. We can anticipate similar growth from the next discovery in magnets. Just as with super-abrasives (diamond-based), there will be some new generations of super-magnets, super-conductors, and super-capacitors, leading to rapid rewards [35].

Composites provide a means to create structures that would not exist otherwise, especially where properties exceed competing materials. Maps help examine key properties—such as hardness versus fracture toughness, as given in Fig. 12.5 for cemented carbides. On this plot is shown a new composite consisting of cubic boron

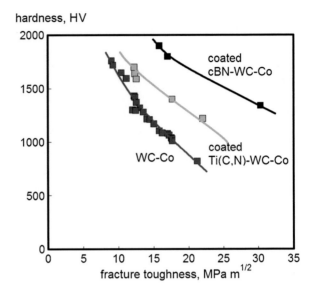

Fig. 12.5 The Vickers hardness plotted against fracture toughness for traditional WC-Co cemented carbides, coated Ti(C,N) with cemented carbide, and coated cubic boron nitride with a cemented carbide. The highest performance is associated with the coated composites

nitride grains coated with a composite of WC-10Co, effectively a two level composite. The property combination provides a means to attack new opportunities. The challenge is to synthesize combinations that push into new zones using traditional processing cycles.

New forming approaches are enabling particulate composite fabrication. The techniques borrow from the existing processes to apply different combinations of pressure, temperature, strain rate, and time. Much growth is seen in binder-assisted shaping ideas, such as extrusion, slurry casting, and injection molding [36]. Improved forming technologies are a fruitful avenue for research, especially net-shape forming [37]. Likewise, high intensity consolidation routes such as spark sintering allow for densification of materials otherwise difficult to process. Although the shapes are simple the properties can be outstanding [38].

Example Opportunities

The production of new powders or the development of new fabrication processes adds to the opportunities for particulate composites. Invention arises at the intersection of materials, processing, and applications. A current few ideas illustrate the context for the future [39]:

- biomaterials, implants, artificial bone, and dental restorations; reasonable high toughness and matched elastic modulus [40],
- biodegradable biomaterials that serve for limited times, then dissolve to avoid the need for surgical removal after healing,

- light-weight and low-cost composites with high strength, high elastic modulus, and high fracture toughness, targeted at transportation systems ranging from aircraft to automobiles [4],
- nuclear energy has a need for neutron capture, heat protection, and wear resistance under intense fields [41],
- high durability structures designed for adverse conditions, such as deep drilling, mining, and seafloor excavation [19],
- ultra-high temperature composites are required for rocket engines, jet turbines, and extreme environments [42],
- high performance electrical and magnetic components, awaiting the next round of discovery in conductors, insulators, thermoelectrics, and electro-magnet devices [43],
- stealth technology relies on composites to absorb radar; opportunities are emerging to provide optical cloaking, a technology of great value on the battlefield [44],
- sound attenuation materials, low-cost, low-density, high attenuation in large structures [45],
- solar materials based on energy absorbing particles embedded in a conductive matrix are the dream of paintable solar cells, a field of immense impact,
- thermal barrier materials to inhibit heat loss, able to operate at very high surface temperatures [46],
- high thermal conductivity structures for heat dissipation in microelectronics [47].

In several instances coated particles extend composition and microstructure options. The idea is to increase phase content without inducing phase connectivity. Bonds between grains influence the composite response, but are generally not important to density, hardness, or elastic modulus. However, connectivity enters into properties such as electrical and thermal conduction, strength, and toughness. As a simple example, Al-15SiC varies in strength by 30 % with differences in phase dispersion, yet at the same time differs nearly sevenfold in ductility [48]. Changes in phase connectivity provide property adjustments for several composites.

Some schemes struggle to gain traction. In spite of repeated efforts to form composites using rapid forming or rapid synthesis techniques, slow processes deliver more consistency. Rapid heating is sensitive to component geometry. Accordingly, struggles arise in scaling from simple test coupons to more complex shapes. Fast fabrication options, such as microwave processing, reduce processing time, but forego product uniformity. Thus, slow processes producing dimensionally correct structures are difficult to displace, and these same slow processes reduce the benefits from nanoscale powders.

A dilemma in structural composites comes from the sometimes disappointing mechanical properties, especially toughness and ductility, in polymer composites [49]. Some systems exhibit significant strength gains over that of the constituents; however, other systems exhibit degraded ductility with little strength or hardness gain. An accurate means to predict strength, ductility, and toughness for composites is needed. Consider a high strength stainless steels where ceramic additions results in marginal strength gain, but significant ductility and toughness loss [50]. The addition of 11 vol% Ni_3Al to a ferritic stainless steel resulted in 15 % yield strength gain with 72 % loss in ductility and eightfold faster corrosion. Predictive models are missing, so empirical trials are the only means available to sort options. Chemical reactions at interfaces are probably a missing factor in the models. Partial dissolution at the interface is often beneficial. This may be a key to improved polymers using fillers that are properly surface treated prior to mixing.

Some of the other opportunities arise in wear resistant materials for applications that range from brakes to machine tools. Global production of 70 million new automobiles creates much opportunity in both fields. New particulate composites containing hard phases are most attractive [51, 52]. New hard composites benchmark against WC-Co compositions, giving 70 % hardness increase and 140 % toughness increase, resulting in much longer service life.

Biomedical components are moving toward materials and structures tailored to the individual patient needs [40, 53]. The opportunity extends to surgical tools. Particulate composites using high strength polymers, such as polyetheretherketone (PEEK), with antibacterial silver particles are an option, giving 100 MPa strength and almost total elimination of surface bacteria. The ability to stop bacterial spread is of great value, making PEEK-Ag composites one example of the future.

Laser processing is an option for creating complicated geometries. Metal structures were formed by the early 1990s targeted at injection molding tools. Advantages were realized by selective placement of cooling channels. By 2005 the ideas extended to biological structures. Laser forming of particulate composites relies on both coated and mixed particles. The powders are deposited in thin layers to grow a three-dimensional structure layer by layer, allowing for property gradients. The gradients are customized to the application needs.

Functional gradients are arising from the intersection of additive manufacturing and particulate composites, leading to selective function with position—magnetic spots in nonmagnetic components, electrically conductive heaters embedded in insulators, or porous tips for drag control on aerodynamic surfaces. One early success is in automotive transmission gears, where hard, fatigue resistant teeth are located on low density hubs.

Summary of Key Points

This chapter is a prospectus for investment. Students may have little money to invest, but they do have time and effort to invest in future careers. For the student, the question is where to study to position for future growth and employment?

Predicting the future is difficult. However, we know where large problems exist and from problems arise opportunities. Much attention is given to energy use and conservation. Enormous energy consumption occurs in the USA where consumption exceeds 10^{20} J/y. About 59 % goes into waste heat. Energy efficiency and

Fig. 12.6 A scanning
electron micrograph of
polycrystalline diamond,
consisting of hard diamond
grains (*black* due to low
atomic number) and
cemented carbide tough
layer (*white* due to high
atomic number) [courtesy
A. Griffo]

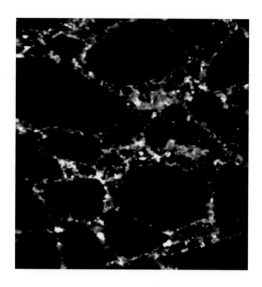

reduced waste are ideal focal points for new technologies. Thermoelectric devices
are a means to generate electricity from waste heat. The challenge is to form the
large, inexpensive structures of n-type and p-type semiconductors into composite
structures. Other discussions identify photocatalytic composites for converting
sunlight into electrical energy. Further, lightweight yet very strong composites
are always desired. The challenge is to synthesize strong polymers laced with low
density metals to form composite structures. One idea is to combine sufficient metal
powders with the polymer so that spark sintering can be applied to the
consolidation.

Multiple level composites are realized already wherein a composite is used as
reinforcement for a composite. One idea involves laminated composites, where the
layers have different ratios of ingredients. Low density armor is one potential
product. The layer of tough phase blunts cracks while the layer of hard phase
consumes energy by fracturing. Within a few grains the structure disseminates
projectile energy without penetration. Indeed, polycrystalline diamond as depicted
in Fig. 12.6, a micrograph showing one desirable mixture of hard and tough phases.

A challenge in particulate composites is to improve interfacial bonding at low
cost. Current efforts rely on acid washing, plasma cleaning, or chemical vapor
deposition. What is needed is a means to activate surfaces for improved bonding.
High temperatures work for several systems, such as Ti-TiB, W-Ni-Fe, WC-Co,
diamond-Co, Al_2O_3-ZrO_3, and TiC-Fe. They are fabricated at temperatures where
the two phases undergo diffusive bonding. On the other hand, composites where
diffusive bonding is missing often lack attractive property combinations.

Coated particles are attractive for microstructure control, but so far these are
either simple (such as electroless copper) or expensive (such as chemical vapor
deposition). As an example, Fig. 12.7 shows an impressive case where cubic boron
nitride particles, in the first frame, are vapor coated with WC shown in the second

Fig. 12.7 An example of chemical vapor coated hard particles, where cubic boron nitride (raw powder is on the *left*) is treated to grow a WC coating (*center*) that forms a shell as evident via cross section on the *left* [courtesy J. Keane]

frame, giving shells on the particles as evident in Fig. 12.7. Due to the slow deposition rate, such an approach must be reserved for high value products. Needed are coating approaches that enable surface activation at low costs, possibly focused on polymer matrix structures. Cost-effective production of small coated powders is a route to outstanding properties.

Several points are mentioned and to conclude these seem to be the current challenges:

- How can we predict if a particular combination of phases will be beneficial?
- Is diffusion across the interface required to induce high interface strength?
- Where are coated particles best applied in composites?
- Why are high strength matrix phases often degraded by a second phase?
- What is a low cost surface activation treatment to improve interface bonding?
- Are there new options for joining particulate composites to other materials?

Study Questions

12.1. Describe how sliding electrical contacts work.

12.2. A high current capacity sliding electrical contact relies on three phases, silver, cadmium oxide, and polytetrafluoroethylene. What are the roles for each phase?

12.3. Describe how a particulate composite fabricated by additive (freeform) fabrication could be used to replace damaged bone. What phases would be most useful?

12.4. Sound attenuation is of great value in automobiles and airplanes. The challenge is to create structures that provide a large sound suppression but at the same time do not sacrifice weight and fracture resistance. Is this a case where the criteria are simply incompatible or is there an opportunity for novel composites?

12.5. Advanced nuclear fusion devices use hundreds of simultaneous laser pulses to implode small tritium capsules to ignite a thermonuclear reaction, releasing thermal neutrons to heat water to drive a steam turbine. The lining on the inner wall of the reaction chamber must withstand demanding conditions. A composite is envisioned as the "first wall" material. What are the composites and properties currently conjectured for this application?

12.6. A perpetual dream is to form a composite of extremely high fracture toughness. What is the highest fracture toughness material available today? Is there a second phase that might be added to improve this material?

12.7. Composite armor consists of laminates of tough and hard materials, similar to automobile windshields (glass-polymer-glass). One option combines high toughness titanium alloy Ti-6Al-4V with hard TiC to replace armor steel, which is 4340 with a yield strength of 1600 MPa, fracture toughness of 50 MPa√m, and hardness of 540 HV. Conceptualize the design of a low density composite offering similar properties?

12.8. Connecting rods linking the piston to the crankshaft in an automobile engine are composites of steel and copper. Efforts try to improve performance using aluminum composites. What are the barriers to aluminum composites for this application?

12.9. A composite consisting of chewing gum and graphite is formed into an electrical sensor for flexible monitoring of humans. What attribute of chewing gum makes for such flexibility?

12.10. Currently valve seats for automobile engines are formed from tool steels infiltrated with copper alloy. For improved durability it is proposed that WC-10Co be infiltrated with the same copper alloy. Argue for and against this change using technical and economic factors, possibly by comparing costs and identifying if valve seat failure is a large problem.

References

1. A. Taub, Automotive materials: technology trends and challenges in the 21st century. Mater. Res. Soc. Bull. **31**, 336–343 (2006)
2. M. Balasubramanian, *Composite Materials and Processing* (CRC Press, Boca Raton, 2014)
3. F. Hussain, M. Hojjati, M. Okamoto, R.E. Gorga, Polymer matrix nanocomposites, processing, manufacturing and application: an overview. J. Compos. Mater. **40**, 1511–1575 (2006)
4. A. Mortensen (ed.), *Concise Encyclopedia of Composite Materials* (Elsevier, Oxford, 2006)
5. H. Ye, X.Y. Liu, H. Hong, Fabrication of metal matrix composites by metal injection molding - a review. J. Mater. Process. Technol. **200**, 12–24 (2008)
6. D. Embury, O. Bouaziz, Steel based composites: driving forces and classifications. Annu. Rev. Mater. Res. **40**, 213–241 (2010)
7. A. Mortensen, J. Llorca, Metal matrix composites. Annu. Rev. Mater. Res. **40**, 243–270 (2010)
8. T. Saito, The automotive application of discontinuously reinforced TiB-Ti composites. J. Met. **56**, 33–36 (2004)
9. J.N. Boland, X.S. Li, Microstructural characterisation and wear behaviour of diamond composite materials. Materials **3**, 1390–1419 (2010)

10. J. Konstanty, Sintered diamond tools: trends, challenges and prospects. Powder Metall. **56**, 184–188 (2013)
11. H. Mei, S. Xiao, Q. Bai, H. Wang, H. Li, L. Cheng, The effect of specimen cross-sectional area on the strength and toughness of two-dimensional C/SiC composites. Ceram. Int. **41**, 2963–2967 (2015)
12. A. Arifin, A.B. Sulong, N. Muhamad, J. Syarif, M.I. Ramli, Material processing of hydroxyapatite and titanium alloy (HA/Ti) composite as implant materials using powder metallurgy: a review. Mater. Des. **55**, 165–175 (2014)
13. E.J. Lee, H.E. Kim, H.W. Kim, Production of hydroxyapatite/bioactive glass biomedical composite by the hot pressing technique. J. Am. Ceram. Soc. **89**, 3593–3596 (2006)
14. S. Ramakrishna, J. Mayer, E. Wintermantel, K.W. Leong, Biomedical applications of polymer-composite materials: a Review. Compos. Sci. Technol. **61**, 1189–1224 (2001)
15. K. Kondoh, Titanium metal matrix composites by powder metallurgy (PM) routes, in *Titanium Powder Metallurgy*, ed. by M.A. Qian, F.H. Froes (Elsevier, Oxford, 2015), pp. 277–297
16. G. Dwivedi, K. Flynn, M. Resnick, S. Sampath, A. Gouldstone, Bioinspired hybrid materials from spray-formed ceramic templates. Adv. Mater. **27**, 3073–3078 (2015)
17. X. Wang, H. Yang, M. Chen, J. Zou, S. Liang, Fabrication and arc erosion behaviors of Ag-TiB$_2$ contact materials. Powder Technol. **256**, 20–24 (2014)
18. M.A. Meyers, J. McKittrick, P.Y. Chen, Structural biological materials: critical mechanics-materials connections. Science **339**, 773–779 (2013)
19. J.M. Keane, D. Evans, A.L. Hancox, Thermal Spray and Laser Applied TCHP Barrier Coatings for Extended Life, in *Proceedings International Conference on Refractory Metals and Hard Materials*, 18th Plansee Seminar, Reutte, Austria, 2013, pp. 2071–2082
20. Z.Z. Fang, H. Wang, Sintering of ultrafine and nanoscale particles, in *Sintering of Advanced Materials*, ed. by Z.Z. Fang (Woodhead Publishing, Oxford, 2010), pp. 434–473
21. K. Lu, Sintering of nanoceramics. Int. Mater. Rev. **53**, 21–38 (2008)
22. W. Li, K. Lu, J.Y. Walz, Effects of solids loading on sintering and properties of freeze-cast kaolinite-silica porous composites. J. Am. Ceram. Soc. **96**, 1763–1771 (2013)
23. Z.J. Lin, J.Z. Zhang, B.S. Li, L.P. Wang, H.K. Mao, R.J. Hemley, Y. Zhao, Superhard diamond tungsten carbide nanocomposites. Appl. Phys. Lett. **98**, 121914 (2011)
24. M. Nygren, Z. Shen, On the preparation of bio-, nano- and structural ceramics and composites by spark plasma sintering. Solid State Sci. **5**, 125–131 (2003)
25. W. Xu, X. Wu, X. Wei, E.W. Liu, K. Xia, Nanostructured multiphase titanium-based particulate composites consolidated by severe plastic deformation. Int. J. Powder Metall. **50**(1), 49–56 (2014)
26. G. Sethi, N. Myers, R.M. German, An overview of dynamic compaction in powder metallurgy. Int. Mater. Rev. **53**, 219–234 (2008)
27. K. Upadhya, Sintering kinetics of ceramics and composites in the plasma environment. J. Met. **39**(12), 11–13 (1987)
28. Z.A. Munir, D.V. Quach, M. Ohyanagi, Electric current activation of sintering: a review of the pulsed electric current sintering process. J. Am. Ceram. Soc. **94**, 1–19 (2011)
29. Z.A. Munir, U. Anselmi-Tamburini, M. Ohyanagi, The effect of electric field and pressure on the synthesis and consolidation of materials: a review of the spark plasma sintering method. J. Mater. Sci. **41**, 763–777 (2006)
30. D. Zheng, X. Li, Y. Tang, T. Cao, WC-Si$_3$N$_4$ composites prepared by two-step spark plasma sintering. Int. J. Refract. Met. Hard Mater. **50**, 133–139 (2015)
31. J.F. Garay, Current activated, pressure assisted densification of materials. Annu. Rev. Mater. Res. **40**, 445–468 (2010)
32. K. Morsi, The diversity of combustion synthesis processing: a review. J. Mater. Sci. **47**, 68–92 (2012)
33. H. Attar, M. Bonisch, M. Calin, L.C. Zhang, S. Scudino, J. Eckert, Selective laser melting of in situ titanium - titanium boride composites: processing, microstructure, and mechanical properties. Acta Mater. **76**, 13–22 (2014)

34. J.M. Lonergan, W.G. Fahrenholtz, G.E. Hilmas, Sintering mechanisms and kinetics for reaction hot pressed ZrB_2. J. Am. Ceram. Soc. **98**, 2344–2351 (2015)

35. S. Deville, A.J. Stevenson, Mapping ceramics research and its evolution. J. Am. Ceram. Soc. **98**, 2324–2332 (2015)

36. R.M. German, Alternatives to powder injection moulding: variants on almost the same theme. Powder Injection Moulding Int. **4**(2), 31–40 (2010)

37. A. Bose, Overview of several non-conventional rapid hot consolidation techniques, in *Reviews in Particulate Materials*, ed. by A. Bose, R.M. German, A. Lawley, vol. 3 (Metal Powder Industries Federation, Princeton, 1995), pp. 133–170

38. S. Grasso, C. Hu, G. Maizza, Y. Sakka, Spark plasma sintering of diamond binderless WC composite. J. Am. Ceram. Soc. **95**, 2423–2428 (2011)

39. J.L. Johnson, Opportunities for PM processing of metal matrix composites. Int. J. Powder Metall. **47**(2), 19–28 (2011)

40. Q. Chen, G.A. Thouas, Metallic implant biomaterials. Mater. Sci. Eng. **R87**, 1–57 (2015)

41. I. Smid, M. Akiba, G. Vieider, L. Plochl, Development of tungsten armor and bonding to copper for plasma-interactive components. J. Nucl. Mater. **258**, 160–172 (1998)

42. H.J. Brown-Shaklee, W.F. Fahrenholtz, G.E. Hilmas, Densification behavior and thermal properties of hafnium diboride with the addition of boron carbides. J. Am. Ceram. Soc. **95**, 2035–2043 (2012)

43. S. Shiga, M. Umemoto, Effect of the composite of good electrical conductor with thermoelectric semiconductor on effective maximum power. J. Jpn. Soc. Powder Powder Metall. **45**, 1318–1322 (1995)

44. P.J. Patel, G.A. Gilde, P.G. Dehmer, J.W. McCauley, Transparent ceramics for armor and EM window applications, in *Inorganic Optical Materials II*, eds. by A. J. Marker, E. G. Arthurs. Proceedings SPIE, vol. 4102 (2000), pp. 1–14

45. Y. Okudaira, Y. Kurihara, H. Ando, M. Satoh, K. Miyanami, Sound absorption characteristics of powder beds comprised of binary powder mixtures. J. Jpn. Soc. Powder Powder Metall. **42**, 239–244 (1995)

46. M. Eriksson, M. Radwan, Z. Shen, Spark plasma sintering of WC, cemented carbide, and functional graded materials. Int. J. Refract. Met. Hard Mater. **36**, 31–37 (2013)

47. R.M. German, K.F. Hens, J.L. Johnson, Powder metallurgy processing of thermal management materials for microelectronic applications. Int. J. Powder Metall. **30**, 205–215 (1994)

48. Z. Wang, R.J. Zhang, Microscopic characteristics of fatigue crack propagation in aluminum alloy based particulate reinforced metal matrix composites. Acta Metall. Mater. **42**, 1433–1445 (1994)

49. D.D.L. Chung, *Composite Materials Science and Applications*, 2nd edn. (Springer, London, 2010)

50. A. Upadhyaya, S. Balaji, Sintered intermetallic reinforced 434L ferric stainless steel composite. Metall. Mater. Trans. **40A**, 673–683 (2009)

51. Z. Qiao, J. Rathel, L.M. Berger, M. Herrmann, Investigation of binderless WC-TiC-Cr_3C_2 hard materials prepared by spark plasma sintering (SPS). Int. J. Refract. Met. Hard Mater. **31**, 7–14 (2013)

52. H. Qu, S. Zhu, P. Di, C. Ouyang, Q. Li, Microstructure and mechanical properties of WC-40 vol% Al_2O_3 composites hot pressed with MgO and CeO_2 additives. Ceram. Int. **39**, 1931–1942 (2013)

53. J.R. Davis (ed.), *Handbook of Materials for Medical Devices* (ASM International, Materials Park, 2003)

Index

© Springer International Publishing Switzerland 2016
R.M. German, *Particulate Composites*, DOI 10.1007/978-3-319-29917-4

Printed in the United States
By Bookmasters